THE HISTORY OF MINING

THE EVENTS, TECHNOLOGY AND PEOPLE INVOLVED IN THE INDUSTRY THAT FORGED THE MODERN WORLD

BY MICHAEL COULSON

Hh

HARRIMAN HOUSE LTD
3A Penns Road
Petersfield
Hampshire
GU32 2EW
GREAT BRITAIN

Tel: +44 (0)1730 233870
Fax: +44 (0)1730 233880
Email: enquiries@harriman-house.com
Website: www.harriman-house.com

First published in Great Britain in 2012

ISBN: 978-1897597903

British Library Cataloguing in Publication Data
A CIP catalogue record for this book can be obtained from the British Library.

Printed and bound in the UK by Lightning Source

•••••

This book is dedicated to my grandmother, Alice, and my mother, Vera, whose enthusiasm as private investors for the mining share market was the spur that led ultimately to the writing of this history.

CONTENTS

LIFE PORTRAITS

ABOUT THE AUTHOR

Born in 1945 Michael Coulson has been associated with the mining sector for over 30 years, although his university background is in economics where he holds a BSc from the University of London. He first worked as a graduate trainee on the legendary mining desk at James Capel in 1970, for many years the leading mining stockbroker in the City. After that he became a mining salesman at Sterling & Co and also developed the firm's research coverage of the sector. In 1973 he joined Fielding Newson-Smith (later to become NatWest Markets) as a gold mining analyst where he began a long association with the South African gold mining industry. Two years later he became senior mining analyst at L Messel (latterly Lehman Bros) where he started to produce an annual gold review which he published every year until 1991. In 1979 he moved to Panmure Gordon and in 1982 he left and joined Phillips & Drew (UBS) with the task of establishing the firm in the mining market.

After a successful four years there, where two years running he was voted No 2 gold analyst in the Extel Analysts Survey, he moved to Kitcat & Aitken where he set up a highly regarded integrated mining desk. In 1990 K&A's Canadian owners closed the firm and he was briefly with County NatWest. The following year he set up a small mining team at Durlacher, but in 1992 was back in the mainstream at Credit Lyonnais Laing where he was a salesman/analyst on the firm's specialist mining team and established an expertise in African shares. He was then approached by South African bank, Nedcor, to join a start-up broking operation the bank was establishing in London. This operation was closed in 1997 and the following year he joined Paribas to head its Global Mining Team.

He left Paribas in 2000 following the completion of the merger with BNP. Since then he has been doing independent research, mainly on a commissioned basis, primarily for small UK brokers lacking mining expertise. He also wrote *An Insider's Guide to the Mining Sector* for Harriman House, which has run to two editions and has been translated into Chinese. He is currently a non-executive director of City Natural Resources High Yield Trust.

Michael lives in Wandsworth with his wife and has three daughters and two granddaughters. He has a lifelong passion for cricket and football and as befits a veteran miner, albeit of the armchair variety, also enjoys a glass or two of wine as he contemplates strategies and projects for the future.

FOREWORD

If men could learn from history, the English poet and philosopher Coleridge exclaimed, what lessons it might teach us. There is certainly much in Michael Coulson's magisterial history of mining that will be of inestimable value to anyone with an interest in this industry that touches all our lives. Equally importantly, by charting mining's crucial role in the rise of civilisation and economic growth, it stands as a salutary corrective to those who would portray miners as rapacious plunderers of the earth and exploiters of its people.

Not that the industry's conduct has always or universally been impeccable, of course: over the centuries it has hosted perhaps more than its fair share of rogues and charlatans, drawn by the prospect of instant riches. As Michael explains in this book, however, mining has been responsive to societal change, and I think there is now a widespread acceptance within the industry that heedless self-interest is not only morally wrong but bad for business. This is particularly true in those developing regions where so much of the world's remaining resources are stored.

I first met Michael some 25 years ago when I was still working in the platinum industry in South Africa. Since then I have come to know him as an especially astute analyst, incisive in investigation, energetic in action and forthright in holding managements to account when he deemed this necessary. As the two books he has published to date attest, he is also a gifted scholar, and this combination of qualities makes him one of a rare breed which the industry sadly no longer produces.

The History of Mining appears at a time when the world is a turbulent place, rocked by conflicts, the resurgence of a nuclear threat, the near-collapse of the global banking system and the staggering debt burdens on its once key economies. Meanwhile, the rise of powerful new emerging-market economies has driven a decade-long bull run in commodities. Amid this confusion, we look to the past for lessons and signs of what the future might hold. After all, if you don't know where you've come from, how will you know where you should be going? Michael's timely book answers the first part of that conundrum and points the way for the road ahead.

Mark Bristow, Chief Executive, Randgold Resources
London, September 2012

ACKNOWLEDGEMENTS

Writing a history of mining is tantamount to trying to judge the length of a piece of string. What does one include; what does one leave out; has one said enough about this or too much about that? And what sources does one use? Does one concentrate on familiar, conventional sources or try to root out more original material? In the end the internet as a source has provided a number of gems, which I hope both enliven and enrich the narrative, and it is only right, at the outset, that I should acknowledge the help I received from this powerful tool.

It is at this stage that the author thanks all those who have given direct assistance in the writing of the book, and my thanks to those who provided invaluable help are below. From the mining industry I am particularly grateful to Mary Magafakis at BHP Billiton and to BHP Billiton itself for their help and for permission to use images from BHP's extensive photo archive, and to Claire Carr on behalf of Randgold Resources for her assistance, and particularly to Randgold's CEO, Mark Bristow, for agreeing to write a most generous foreword.

I must also thank my wife, Hilary, for her sterling work in preparing the photos used in the book and also for the use of a couple of her own photos; my daughter, Lisa, made a valuable contribution here as well. I also want to thank Stephen Eckett at Harriman House for his encouragement in getting the project off the ground and to his colleague Craig Pearce who laboured on the text and suggested many improvements which have contributed to making the narrative both tighter and more readable. Also thanks are due to Nick Read at Harriman for his work on the photos. Unfortunately I was unable to call on the assistance of my old friend and colleague Charles Kernot due to his illness; he provided invaluable help with my last book, *An Insider's Guide to the Mining Sector*. Sadly he passed away before this book was published.

Although every effort has been made to trace and get hold of all copyright holders prior to publication, contact was not possible in every case. If notified the publisher will be pleased to rectify any omissions at the earliest opportunity.

Michael Coulson, London, 2012

PREFACE

This book will trace the history of mining from those early moments when man first started using tools and weapons to the present day where metals continue to underpin economic activity in the so-called post industrial age. Before we conclude our journey we will also take a look into the future in an effort to chart the direction this great industry might take in years to come.

As we move through the history we will also take a look at some of the great mining entrepreneurs who, particularly over the last century and a half, have given the mining industry much of its colour and controversy. The book will also try and assess the cause and effect of technological advances and the discovery of new metals and the revival of old ones. This is an important area of the industry's development.

Many books have been written about mining; the majority have focused on a particular metal, geographical area, mining event or mining personality. This book is broad in its scope and as a result decisions, not always easy, have been made as to what aspects of the industry's history are emphasised and which are mentioned in passing or even ignored. Not all readers will agree with the choices made but the subject is a large one and editorial selectivity had to be extensively employed.

Two particular decisions had to be made. The first related to the handling of the material and whether the book should simply go from A to Z, jumping all over the place as years and centuries went by, or whether it should deal with metals, geographical areas and mining events (the 19th century gold rushes for example) as discrete sections.

The second concerned the weighting of the ancient against the modern, or at least the fairly modern. Some may recall that British history books 50 or more years ago tended to compress the dramatic events of the post-Boer War period into a modest few pages with by far the greater part of the book dedicated to the preceding centuries. The reason usually given for this imbalance was that recent events were far too fresh in minds for proper historical context and analysis. This book has a different balance as the period from the outset of the Industrial Revolution to the current day represents a period of huge growth and development for the mining industry, far greater and more significant than anything that went before.

So the structure of the book mirrors the way that mankind has developed. The first section covers several millennia from the Stone Age to the middle of the 11th century AD when the Normans invaded Britain. It is short in comparison to later sections because reliable original records are relatively thin on the ground and progress in mining methods, treatment technology and the uses for metals remained unchanged for centuries or changed only very slowly during this time. Much of what we know about ancient mining comes from the work of archaeologists and although their observations on mining and smelting are important and illuminating, at times because their accounts are often very similar it can be a case of 'read one, you've read them all'. I have taken the view that readers wanting to know how mining has developed through the ages will be grateful that too much repetition is avoided and the narrative is kept moving.

After the ancient world we move through the Middle Ages to the Industrial Revolution and then to the modern age. At each stage the level of detail included increases, as should be expected bearing in mind both industrial and social advances over the period and the expanding role mining has played in the process of these industrial and social developments. The widening range of metals available to fuel technological advances and the huge increase in metal consumption, in line with an exploding growth in population worldwide, have also influenced the size of each succeeding section as more topics and issues are explored.

Over the last hundred years or so there has been an enormous expansion in mining activity, and information on the industry and the book's structure and emphasis reflects this as well. Readers will also notice that at times for reasons of context and logic I have continued the narrative in certain areas beyond, or indeed in some cases started it before, the historical era that the subject under discussion falls within.

As regards the layout and structure of the book, I should say at the outset that though my intention in writing a history of the mining industry was to cover it from the dawn of time to the present day, it must be understood that reader accessibility has also been a key objective. So as well as the historical editing mentioned above there are also areas covered on which some readers may have desired more information than I have provided, particularly in terms of technical detail; it ought to be born in mind that the classic book on mining in Germany in the Middle Ages, *De Re Metallica* by Agricola, is on its own longer than this book.

On the other hand I hope readers will approve of the potted biographies of miners and other industry figures (virtually all men I'm afraid to say) that pepper the book. Since mining is very much about people and their dreams, it seemed right to tell readers a little bit about them, perhaps whet their appetite for some personal research, without a wish to clutter the text.

I have also tried throughout the book to erect signposts along the way to draw attention to the very close links between changes or developments in metal uses and the advance of technology as it progresses hand in hand with civilisation. I believe that the history of the mining industry is inextricably linked with economic growth and the continuing improvement in standards of living in many parts of the world. This might contrast with at least some people's view of mining, which is that it is a dirty, dangerous activity which poisons and despoils the environment and kills people, sometimes directly, sometimes indirectly.

Whilst such sentiments cannot be dismissed out of hand, there is another, perhaps somewhat callous view, that even desirable progress has its victims, a price that society must pay for that progress and the generally better life for all that usually accompanies it. We must, however, be careful here as mining and the environment have become big political issues, and a cavalier approach to running a mine, as might have been the situation a century ago, is no longer considered acceptable behaviour.

The writing of *The History of Mining* has been very much a labour of love, but one that has been equally a journey of discovery. In researching what is one of the oldest activities of mankind I uncovered large quantities of information that were new to me, very much underlining the adage that life is just one long learning process. It also made me realise that even after more than 40 years of covering one sector as an analyst, my knowledge of the history of the mining sector was, perhaps, more than adequate but in spite of that by no means comprehensive. The research effort was much helped by the fact that a large body of writing on the industry exists, although I should add a lot of it is out of print or exceptionally expensive to acquire, and in many cases highly technical and therefore rather inaccessible to the lay reader.

Finally, to those who think that there is too much of this or too little of that in the pages ahead I apologise in advance. However, I hope that those who read this book will find it both interesting and illuminating, a good read and a useful reference work. I have attempted to use the experience gained over many years to put together a substantial history of this great industry, to make comments on how the industry has evolved and to draw conclusions as to its importance in economic development over the centuries, and its role in the future. It is for you to judge how well I have succeeded.

A NOTE ON CURRENCIES AND METALS PRICES

Most metals are priced in US dollars, which is also often the currency of share prices and corporate earnings reports even if the reporting company is sometimes not a US company. The reader, however, should be aware that other currencies, not all of them still used today, creep into the text. This is deliberate and is often as a result of the way historic information has come to me. The dollars referred to in the text are usually US, but keep an eye out for the Australian and Canadian versions. Sterling is often used in keeping with the UK's position, both historic and current, at the centre of international mining, and at times the old pre-decimalisation system is also refered to. But the euro does not intrude very much in these pages.

As to weights and measurements the book reflects the real world, not the bureaucrats' version, so imperial/American and metric are both used and often in the same breath. Thus a gold mine's reserves are measured in grams per metric tonne (tonne) but the gold in those reserves is valued on the basis of troy ounces. Similar examples of mixing measurements can be found with base metals such as copper, nickel and zinc where metric tonnes and lbs exist easily together. As regards tonne/ton, the former is the metric measurement and the latter is a short ton in North America and a long ton usually in colonial South Africa and Australia.

M.C.

INTRODUCTION

> *"Those who decry mining are ignorant of history. If they knew anything about metals, they would know that all business, all industry and all human progress depend on mines."*
>
> *William Sulzer, Governor of New York 1938*

Film buffs will surely remember the opening minutes of Stanley Kubrik's 1968 classic *2001: A Space Odyssey*. Prehistoric man picks up an object from the ground and starts to use it to break up a skeleton in front of him. In that moment man first discovers the tools and weapons that will enable him to move from a simple and precarious existence to one where he shapes his future through his own technological development. The opening frames of the film then give way to the picture of a futuristic space ship on its journey into the unknown, underlining the technological development of centuries that has enabled man to leave his planet and venture into outer space.

Back on Earth, technology continues to revolutionise the way we live. All of these developments and advances would have been impossible without the use of the raw materials which lie beneath our planet's surface in the form of metals, minerals and energy.

However, in the last two decades of the 20th century interest in raw materials faded as economists looked to new industries such as mobile communications and the internet to be the *sunrise* activities delivering growth and prosperity way into the future. Raw materials would take a backseat and become sunset industries, generating below average growth with supply more than adequate to satisfy demand for the foreseeable future. Turning this development on its head again, in the last ten years there has been a sensational surge in demand for and interest in raw materials. The reason for this has been explosive economic growth from China, primarily, and also India. These are developments that should have been foreseen but which were not fully recognised until the new millennium was well underway.

These last few years have therefore seen a major revival in interest in mining both as an industry and as an investment sector, and though information sources have now begun to expand, publications in the early years of the 21st century were rather thin on the ground. The literature on the sector has also been getting rather long in the tooth, with only the odd specialist work published during mining's dog years.

In addition, the financial sector had materially reduced its mining coverage in the later years of the 20th century and many practitioners steeped in the industry had disappeared from the scene, although in recent years the trend has reversed as the financial and media sectors have begun to rebuild their interest. With this revival the time seemed right to consider a comprehensive history of the mining industry from its earliest beginnings to provide context for some of the industry's exciting developments, and also to provide background for those coming new to the subject of mining.

Those who have read my earlier book on mining, *An Insider's Guide to the Mining Sector*, will be aware of the importance that I place on history when following events in the mining industry, particularly where these events are unfolding, and where predicting their path may have a critical impact on investment returns or corporate decisions. So another idea behind this book is creating a work that will provide an historical backdrop allowing readers to observe the mining industry's long-term trends, draw conclusions, and apply those to present and future developments.

An example of this would be the revival of gold since 2002, having been largely dormant in price terms since 1980. Those who lost patience and interest over this period might have gained heart by studying the flat period that followed the end of World War II until the start of the 1970s, when western gold production outside South Africa virtually disappeared as the price was officially locked at $35/oz, a price established in 1934 during the Great Depression.

A similar lesson can be learnt from the substantial revival in capital-raising for junior exploration stocks in London, which has been a feature of the last ten years or so. In the period before that, exchange controls, regulatory market hurdles and low levels of activity in grassroots exploration had left London as a wasteland for all but the funding of big ticket mining developments. Those who knew their history would have been aware of the central role played by London in mining exploration in Australia, South Africa and Latin America in the 19th century, and when restrictions were loosened in the aftermath of the Thatcher Conservative government in the UK in the 1980s they could have anticipated a revival in junior activity in line with historical tradition.

But although I think a history of the mining industry is useful in its own right I have also aimed in the book's concluding sections to look forward to what future historians might have to chew on in terms of the direction that the industry might take over the next few decades. On the geopolitical front we have begun to see fast growing countries like China doing an increasing number of deals to secure long-term access to raw materials in areas such as Africa, Australia and Latin America. Will these deals really secure the Middle Kingdom's long-term access to metals and minerals, energy and possibly even agricultural products, or will they eventually come to be seen as exploitive arrangements akin to past British deals in Africa and American deals in Latin America?

We also now find ourselves in an age of rising environmental concern and this is an area where the exploitation of raw materials is centre stage. Mineral deposits are becoming increasingly low grade, requiring very large quantities of ore to be mined to produce worthwhile amounts of metals. This is bound to have an impact on the environment; perhaps ocean mining can play its part in the future in opening up new sources of raw materials which are less environmentally damaging. These are all areas for analysis and comment and they are likely to make a large mark on the mining industry in the future.

Mining has always created political issues for countries rich in resources anxious to maximise returns, sometimes for the good of the people, too often just for the good of the rulers. In terms of how mining companies square up to political pressure, some have a much better record than others. In the past, especially in colonial times, the mining industry had the straightforward task of delivering metals and minerals to the colonial master sufficient for it to achieve its geopolitical aims. If there was room for the providing colony to benefit on a trickle-down basis so much the better, but that was a subsidiary issue.

Today the mining industry has to be increasingly careful that it pays close attention to host governments and make sure that their treasuries receive a satisfactory share of the spoils. But also governments have to be careful not to squeeze mining companies so hard that they pack their bags and go to more encouraging climes to do business. In recent years Australia, the Congo and Zambia have all found that mining companies will fight back if they feel they are being pushed too far. On the other side, mining is often carried out by privately-owned, often foreign-based, companies who, after substantial upfront investment, want to repatriate profits for their shareholders, whilst governments tend to want to restrict this flow in order to maximise the state's use of the profits. This leads to a continual tension between mining company and host government which has to be managed for the benefit of all parties. If the right balance can be reached then all benefit and the industry can prosper, enabling it to meet the demands of both customers and stakeholders.

The important thing then is to create the right conditions for the mining industry to prosper and fulfil its role as the provider of metals and minerals critical to the world's continued social and economic development. I firmly believe that mining is a desirable and essential activity and will have a central strategic role in producing the raw materials that may save the Earth and its occupants in the centuries to come. In the circumstances I am hopeful this book will help newcomer and expert alike to understand the history of this great industry and therefore play a useful part in providing the context that often improves decision making.

To end with another movie analogy, some may remember *Slumdog Millionaire* director Danny Boyle's startling but less well-known film *Sunshine.* In the film, a lifesaving heat shield that protects a huge spaceship on a mission to re-fire the dying Sun has taken much of the world's molybdenum to provide the shield's protection. This allows the ship to get near enough to the Sun to launch a nuclear device, as large as Manhattan, which had absorbed all of the world's fission material. The explosion of this device is intended to reverse the star's decline. So to save the world the ship's shield must have *moly* to stop it melting as it nears the sun and *uranium* for the bomb needed to re-ignite the Sun. Science fiction maybe, but underlying it is a truth about the critical role that metals, and therefore mining, play in the life and destiny of our planet, as they have done throughout history.

M.C.

THE ANCIENT WORLD (FROM THE BEGINNING TO 1066)

1. THE STONE AGE

IN THE BEGINNING

Precision in identifying the various stages of the Stone Age is difficult to achieve but much evidence has accumulated over the years, often as a result of advanced archaeological work, that suggests man developed basic skills at a very early stage in his existence. Amongst those skills was the ability to use what materials were available to fashion tools and then weapons, the latter probably first to kill animals for food and then, as tribes and communities came into contact with each other, to make weapons for defence and attack. It is now accepted that even early man was capable of structured thought, albeit of a fairly basic kind, and this enabled him to identify raw materials that he needed which had to be extracted from the earth. Thus mining was probably one of the earliest of man's *industrial* activities.

The Stone Age is quite difficult to date, certainly on the basis of the development of societies. The society that explorers found in Australia in the 17th century was still essentially Stone Age in that its primary tools and weapons were made from stone. However, the Bronze Age, when man began to mine and work metal for the first time, arrived in Asia, Europe and north Africa at the latest around 2500 BC, and probably the advance from the Stone Age started rather earlier than that, around 4000 BC. The earliest post Stone Age societies were to be found in the Middle East and SE Asia and were earlier again, around 6000 BC. But as we will see mining per se started long before man became aware of the industrial potential of metal.

TOOLS AND WEAPONS

The first tools and weapons that prehistoric man used were made of stone. Over the centuries archaeologists have discovered a large number of these ancient and rudimentary objects, and have had a great deal of fun in working out what function the different shaped tools fulfilled. They have also had to speculate as to how these stones were acquired – *were they picked up off the ground or were some of them perhaps mined?*

It seems probable that many of the tools were shaped from stones found lying on the ground but it is also likely that rock formations would have attracted the attention of ancient man when he was looking for a particular shaped stone for use as a tool or weapon. It is also a subject of some discussion as to whether ancient man actually acquired stone in a manner that we would recognise today as mining, but there have been a number of sites uncovered by archaeologists that suggest very early organised mining.

MINING

One of the most famous ancient mines is Lion Cave in Swaziland in Africa. The site in the Ngwenya Mountains became a modern day iron ore mine, but before that in the late 1960s evidence of a serious ancient mining operation was uncovered. Further work by a team of geologists discovered a tunnel which went about 50 feet into the side of the hill. Inside it was found that whoever had created the tunnel had clearly done so to access *specularite*, a form of iron ore that smears and would have been used as a cosmetic for personal adornment. When it came to dating the cave, the final calculation suggested that it could have been dug around 40,000 BC, which would make it perhaps the oldest underground mine yet discovered.

Widespread evidence has been found of the usage of red ochre (haematite) in this period; in the modern world it is an important iron ore but in the ancient world it was primarily used as a pigment for cosmetics and for cave paintings. It is also believed that it was used in primitive religious rituals as a powder, its red colour perhaps symbolising blood. Ochre has been found at ancient sites as far distant from each other as Tanzania, the Czech Republic, France and Spain, with evidence in some locations of a basic crushing process to make ochre powder.

Quarrying activities go back even further, to as early as 60,000 years ago. In Egypt's Nile Valley a number of flint diggings have been found, an important building material that was used in making sharp tools in the Stone Age and before. Flint workings in the form of ground scraping have also been uncovered in England at Beer in Devon, where outcrops of the valuable stone were found. The Egyptian workings at Nazlet Khater 4 in the western Nile Valley of Upper Egypt between Asyut and Sawhaj were particularly interesting. Work done in 1982 showed that ancient miners exploited the site at least 33,000 years ago. The miners were seeking flint and had sufficient geological knowledge to know that there was a flint seam overlain by silt, sand and gravel. The flint seam could be seen outcropping on the surface before plunging under the silt and sand, and the miners had dug a 30-foot by 7-foot trench and seven vertical shafts to reach the flint.

Although it is beyond the scope of this book to stray into the area of anthropology, it is nonetheless interesting to note that some of the dating of this basic mining activity goes back beyond the time when the forerunner of *modern* man appeared, generally thought to be between 50,000 and 100,000 years ago. It is thought that some of the mining activity was carried out by Neanderthal man, or fossil man. Such people were generally thought to be *pre-human* and thus incapable of any sort of sophisticated activity such as mining, as it was suspected that they would have been unable to instigate the planning necessary to develop a mine successfully. Some anthropologists thus believe that the uncovering of ancient mining activity has thrown new light on the issue of when recognisable humans appeared on Earth.

2. THE BRONZE AGE

As with the Stone Age, the Bronze Age is difficult to pin down in terms of dates, particularly as there is strong evidence that stone weapons and tools were still being used when man had begun to smelt and fashion copper into weapons, creating an extended passage of time from the Stone to the Bronze Age. It is also the case that gold was well known in the Bronze Age, so whilst we tend to define the period as one where there was a major technological breakthrough enabling copper to be alloyed with tin to create bronze for weapons and metallic items, the mining and usage of other metals, including lead and zinc, went on in tandem.

It is also the case that copper objects were found long before the onset of the Bronze Age; the oldest found, a copper pendant from Shanidar in Iraq, is believed to date from around 6500 BC. Such copper objects were unlikely to have been made with mined metal but would have been worked using *native* copper picked up off the ground.

The Bronze Age is usually split into periods which depend on the area of the world being discussed. The early Bronze Age in the Near East, then far more advanced culturally and technologically than Europe, began around 3500 BC and the final stage, the late Bronze Age, ended around 1200 BC when the Iron Age began to emerge. In Europe the Bronze Age continued in some places up until 700 BC. The early Bronze Age also saw the invention of writing in Mesopotamia (now Iraq), which helped provide the means for chronicling the development of society and commercial life.

Gold is mentioned, along with silver, in the Code of Menes which was promulgated by King Menes of Memphis – effectively the first Egyptian Pharaoh – in around 3100 BC. It is probable that there was knowledge of gold for millennia before because the sight of gold nuggets gleaming in alluvial settings must have caught many an ancient eye. However, the ability to work the material, other than in its *native* form using crude beating methods, would not come until the technological advances of the Bronze Age.

MINING METHODS

The crucible of the technology of metallurgy, providing the ability for man to work metal, was Anatolia (now Turkey) where tin deposits allowed the alloying of this key metal with copper in order to make bronze. This technology spread to the area which today encompasses the eastern Mediterranean/Middle East, with copper being mined in Cyprus, the Negev desert of Israel, Egypt, Persia (Iran) and the surrounding Gulf, all providing material for the metalworkers.

Interestingly the appearance of bronze did not lead to a material breakthrough in mining methods; that moment had to await the Iron Age. Bronze was considered too precious and too soft a metal to be used for mining tools, and mining continued to be done using basic stone tools. The depths mined remained modest, usually no more than 30 feet, although as the Bronze Age proceeded and demand for all metals increased, concomitantly the volumes of bronze needed increased, requiring a more organised structure with larger, probably slave, workforces.

The mining methods used were related to the tools available so stone and animal bones such as antlers were predominant. Stone hammers, made from as hard a material as possible and usually round in shape, were used to break up softer rocks in which mineralisation had been identified. Bones were used to lever loose rocks and to wedge into cracked rock in order to break it into smaller, more easily worked pieces.

As might be expected, these stone hammers were vulnerable to breakage or chipping and probably didn't have a long life; they were also difficult to make as the stone heads had to be attached firmly to the handles. Even so, such hammers were used extensively, and in Greenland over 10,000 stone hammers were found in the 19th century close to fragments of a meteorite which hit the earth in the Stone Age. These hammers had been used to try and break pieces off the meteorite, which was essentially made of iron.

EXTRACTING METAL

If the rock that Bronze Age miners were attempting to break up was particularly hard a technique called *fire setting* was used. We will come across this technique – which consisted of lighting large fires close to the rock face where the metal could be seen – frequently later in the book. The fires were tended for many hours and as the rock got hotter and hotter it began to crack. The process was then completed by throwing cold water onto the hot rock face, intensifying the cracking and allowing miners to access and lever out fragments containing metal from the rock face.

A copper mine at Rudna Glava in the mountains of north-east Serbia used fire setting to mine malachite (copper carbonate ore) in very hard rock at what were then considerable depths of around 70 feet. Earlier, and at greater depths of perhaps 300 feet, fire setting was used in Egyptian gold mines to break up the hard, gold-bearing rock face. It is unlikely that there was much in the way of ventilation in such early underground mines so working conditions must have been particularly tough.

We have seen that in mining for flint the ancients had accumulated a body of geological knowledge that helped them identify promising environments for finding flint. They were also likely to use visible markers such as water colour to spot where certain metals might lie close to the surface, the water having become stained by the ore as it passed over it. Copper was known in the Stone Age but then would have been found in small accumulations in the form of nuggets and raw (native) copper on the ground, or washed out of rivers on to the bank. In due course larger quantities in the form of copper veins were found. During this extended period man would have experimented with the copper, finding that it was soft enough to shape by beating it with stone hammers.

The onset of the Bronze Age came about as a result of man developing smelting techniques to work the copper into objects of use. The key to copper smelting was to create fires with sufficient heat to turn the copper ore into molten metal and molten slag. When this had been achieved the metal and slag could be separated. The earliest *smelters* consisted of an open hearth made out of fire-resistant bricks. Charcoal was then loaded on to the hearth and lit, and after that copper ore was placed on top of the charcoal. The use of charcoal required a large quantity of wood – it has been calculated that 5 cubic metres of wood would produce 300 kg of charcoal, which in its turn would produce just 1 kg (2.3lbs) of copper from the ore.

The next step in the development of smelting technology was the use of *crucibles*. The charcoal and copper ore were put in a crucible which was made out of ceramic material. The crucible pot was then placed in a crude furnace and the temperature raised by the use of bellows. The ore would then reduce to metal and slag as in the open-hearth process. When the copper had been freed it could then be alloyed with tin, usually within a range of 3% to 8% added tin, to produce bronze. Another property of bronze was that it was much easier to cast than copper on its own – copper tends to contract as it cools becoming fractionally porous and adding tin counters this tendency allowing easy separation of the bronze item from its mould.

As the casting of bronze developed other metals were tried as additives, such as lead and zinc, but the resultant metal was not really bronze, and it had faults in terms of reduced strength and a tendency to brittleness if the added metal content rose above 3%.

Over the years ancient bronze pieces have been unearthed in a wide variety of places from the Middle and Near East, to Britain and South America. Analysis of these objects shows that different percentages of tin were used in the bronze-making process depending on what kind of object was being made. For example, a sword requiring a hard, sharp edge would use no more than 3% tin but other items such as plate might contain more. It is worth remembering that although the start of bronze making constituted a major technological step forward in the use of metals, metallurgical precision in terms of blending took centuries to perfect.

Even as the Bronze Age progressed other metals were catching the eye of blacksmiths and forgers. One of the key new metals was iron, which we come to next. Of course the coming of the Iron Age did not see the use of bronze disappear but iron ultimately increased the range of products that man could develop and so opened up another avenue for social and economic advance.

3. THE IRON AGE

With the coming of the Iron Age, the final period of pre-history, mining took a step forward, certainly as far as efficiency was concerned. One of the problems with earlier mining was the fact that stone implements were the main tools for digging and breaking rock, and they were not particularly robust and were thus prone to become unusable quickly. Bronze was too valuable and too soft to be a realistic substitute in the making of such heavy-duty tools, but the coming of iron introduced an altogether tougher and more durable metal, ideal for tool making. The Iron Age brought to four the number of important metallic compounds that could be used to make weapons and other durable tools – namely copper, brass, bronze and iron. These proved of material help in the advancement of civilisation, particularly in the hands of the Romans.

The Iron Age itself, like previous ages, emerged in staggered sequence – in the 12th century BC in Greece and the Near and Middle East; a century later in India; and between the 8th and 6th centuries BC in the rest of Europe, with central and southern Europe ahead of the north. But iron was known and used in small quantities many centuries before that, with the earliest use being in the 4th millennium BC by the Egyptians. We have also seen that meteorites rich in iron were the source that ancient Egyptians tapped for the limited iron objects that they fashioned – spear tips for example. Much of this use of iron was ornamental and in no sense was iron widely used to make primary products such as tools and weapons. It was also more expensive than gold in those times because of its rarity, and also was viewed by some as a metal sent from the gods (it came out of the sky after all) and therefore sacred.

Once conventional sources of iron ore were found, which proved far easier to work than the earlier meteorite sources, the possibilities for making iron objects multiplied. The earliest method of turning ore into iron metal was to combine it with carbon and then melt the treated ore in a basic furnace until it became a spongy mass – sponge iron. The iron was then beaten with hammers and folded until all the carbon was released through oxidisation. The resulting product was wrought iron with very little in the way of impurities. Further heating of the wrought iron in charcoal, followed by water-cooling, produced a harder metal due to the process which added a steel surface to the wrought iron. Whilst this process did not completely achieve the hard finished product that was to follow through the making of cast iron, the ability to achieve the sponge iron stage did mean that, with relatively low temperatures in the early furnaces, an iron product could still be made.

The first objects made from mined iron ore came from Anatolia in the 12th century BC and this spread throughout the Middle East, and it was from here that it is thought iron ore mining and iron making spread to China where wrought iron objects made in the 8th century BC have been found in the north west of the country near Xinjiang. Sources of iron ore were reasonably available once the technology for making finished iron had been mastered. In the 5th century BC, during the Zhou Dynasty, kiln technology was developed which enabled the iron ore to be heated to much higher temperatures than formerly, which led to the manufacture of cast iron. The technology had spread to India by the 2nd century BC.

4. CHINA

In the latter years of the 20th century and the opening decade of the 21st, one of the most dramatic developments affecting the global economy has been the rise of China to economic superpower status. However, China's position as an economic power goes back millennia and along with this goes an equally long history of mining. China's reputation as a centre for inventions and thus technological advance is well deserved, with paper, printing, gunpowder and the compass as four important examples. Paper was invented early in the 1st century AD in the Han Dynasty, and the other three items appeared in the middle of the 11th century, just before the cut-off date used in this book when we move from the Ancient World to the Middle Ages. These four were, though, inventions where metals had only a minor role.

It is interesting to note that in those ancient times China was responsible for the development of a considerable number of technologies that were metal related and this underpinned its position then as one of the most developed, if somewhat opaque, countries in the world. Amongst those inventions that benefited from the advances made possible by the working of metals were crossbows and arrows from the 5th century BC, and a rather sophisticated bronze steamer from the 10th century BC. Nearer the millennium, around 200 BC, a number of technical advances were made by the Chinese in the area of iron smelting, leading to the co-fusion process for making rudimentary steel in the 6th century AD.

Archaeological investigations have provided some broad parameters regarding the development of metal use in ancient China. One or two objects have been found that suggest the usage of metals in China in the 3rd millennium BC, but the evidence is not conclusive. Copper artefacts found in the north east of China in Shandong and Inner Mongolia provinces date from the Longshan Dynasty in the 2nd millennium BC. Bronze metallurgy existed in the Shang Dynasty in the latter part of the 2nd millennium BC, and it has been suggested that the Chinese may have in part skipped the copper age and gone straight to bronze due to their early appreciation of the results of alloying tin with copper.

Written records of historical Chinese mining are small in number, partly as a result of the country's social hierarchy system which meant that mining was not considered an activity for the educated. Mines tended to be owned by rich landowners and although as the 1st millennium AD progressed mining was one of China's most important industries behind agriculture and textiles, the landowning class did not get involved in it directly. So prospective ground was developed and the mine workings were leased to interested parties, with the landowner taking a rent or profit share from the operator but staying well clear of the business of mining.

The upshot of this disinterested attitude was that as little capital as possible was employed in the mines, which consequently were dangerous with poor working conditions. Mines were also mostly small scale and labour was both plentiful and cheap, so there was little incentive for mine owners and operators to spend much capital in upgrading working methods. Interestingly, the workforce often contained women and children.

Though China invented gunpowder there are no signs that it was deployed in ancient mining – for centuries rock continued to be broken using fire setting and basic leverage tools. Exploration was also fairly rudimentary, with traditional cultural techniques such as divining and fengshui being used, along with more scientific methods such as looking for surface indications of mineralisation, promising outcrop features and even associated plant growth.

MINING AT TONGLUSHAN

One of the best preserved of the ancient Chinese mining sites is Tonglushan, a copper mining area only a few miles from Huangshi in Hubei Province, dating back at least 3000 years to the 1st millennium BC. This area is still a substantial producer of a wide range of metals today but its ancient expertise was in the mining and smelting of copper ores.

The methods used by the Chinese in mining beneath Tonglu Mountain, where the rich copper ore lay, appear to have been quite sophisticated. Substantial numbers of shafts were sunk to access the ore, and drives and tunnels were constructed to transport the ore to the numerous smelting furnaces attached to the mining operation. Some of the tunnels were also used for bringing water to the workings, as well as removing water table inflows caused by some shafts having been sunk below the water table's level. The mines were worked between 1000 BC, the time of the Western Zhou Dynasty, and the early part of the 1st century BC, the time of the Han Dynasty. It is estimated that Tonglu Mountain could have produced between 80,000 and 120,000 tonnes of copper over this long period.

In 2007 extensive sampling of lake sediments from Liangzhi Lake near Tonglushan threw more light on the history of mining in this province. A study, written following this exercise, suggests that before 3000 BC, the pre-Bronze Age, metal shows in the sediments were very low, indicating natural slow leaching from the background mineralisation but no mining. Entering the Bronze Age metal shows in the sediments increased significantly, suggesting the start of organised mining of copper, nickel and lead, a sure indication that metallic items including bronze were being manufactured by the ancient Chinese.

The period from around 500 BC to 200 BC was one of conflict within China, and the use of copper and lead increased significantly indicating an upsurge in weapons manufacturing. During the Han Dynasty there was strong growth in agriculture, which led to an increase in the use of iron tools, and this continued through until 200 AD. Also during this period, ceramics and iron began to replace bronze as the raw materials for domestic vessels and tools.

MINING AT OTHER SITES

Another significant copper mining and smelting site was found at Tongling in Jiangxi Province, which carbon dating places in the middle period of the Shang Dynasty around the 14th century BC. Jiangxi abuts Hubei and also Anhui Province, where further 1st millennium BC copper activities have been uncovered at the Muyushan site. This whole area in the east of China, south of Beijing, clearly has a long history of mining and this continues to this day.

In the period of the Shang Dynasty there was a significant and advanced civilisation in the north of China not far from the last Shang capital of Anyang. Here archaeological work has uncovered the remains of a bronze foundry and furnace; records of the supplying mines were also found, some close to Anyang and none more than 200 miles away. All this points to a relatively sophisticated civilisation, not dissimilar from those of Asia Minor, and one with considerable skills in mining and the working of metals.

Copper and bronze artefacts dating back to the 1st millennium BC have also been found in Xinjiang Province in the far north west of China on the border with Kazakhastan. Much of the evidence of mining, smelting and manufacturing, as we have and will continue to find, comes from the artefacts and metal slag uncovered during archaeological digs. Some of the finds in Xinjiang, due to its proximity to Kazakhstan, have led to speculation that there may have been interaction in terms of trade between China and the Near East. As a consequence of this we cannot be sure of the provenance of the technology that led to the establishment of mining and metallurgy in regions like Xinjiang, which lie on ancient trade routes.

There is evidence of iron mines in the north of China and a number of 1st millennium BC iron ore smelting sites have been uncovered. The process used in those days would almost certainly have led to the production of wrought iron and later, as Chinese technology advanced and higher smelting temperatures were attained, cast iron. The source of the iron ore would have been a combination of meteoritic iron and surface ores. The essential requirement was that the iron ore deposit would have been relatively high grade, as there tended to be a considerable loss of dross metal in the early smelting process. In the 1st century BC ironworks unearthed in Henan Province also pointed to coal mining activity in the region, with coal being used in the works but probably not in the smelting process, as coal only supplanted charcoal in the process in 7th century AD.

The progress made in iron working in China over the centuries allowed local engineers to develop sophisticated techniques in a variety of areas, one example being the building of bridges. To begin with the first bridges were more like pontoons than bridges as we understand them, and this meant that crossings were of rivers rather than aerial gaps. It is, however, thought that around 600 AD Chinese engineers may have built the first cast iron chain suspension bridge in Yunnan province, although some historians think that the first iron chain suspension bridge could well have been constructed even earlier.

The table summarises some of the more significant Chinese metal developments in the ancient period.

SIGNIFICANT HISTORICAL CHINESE METAL DEVELOPMENTS

Development	Date
Cooking steamer clay and bronze	1000/5000 BC
Dagger axe	2000/3000 BC
Cast iron	5th century BC
Crossbow metal arrows and stock	5th century BC
Chromium use (arrow tips)	200 BC
Cupola furnace iron ore	200 BC
Finery forge wrought iron	200 BC
Pig iron blast furnace	100/200 BC
Borehole drilling (to 2,000 feet)	200 BC/200 AD
Co-fusion steel process	6th century AD

5. EARLY MINING IN INDIA

One of the interesting aspects of mining in India is the relatively early use of zinc, which meant that brass for coins, sculptures, images and other objects was being used quite widely before other ancient societies such as China and Greece made brass. Indeed, some Indian brass objects found have been dated as early as the 3rd century BC. One of the best-preserved zinc mining sites in India is Zawar in Rajasthan in Western India. There, shafts were found down to a depth of 300 feet and wooden residues indicate the use of ladders, supports and drainage conduits in the operations. The workings are thought to date back to the latter half of the 1st millennium BC and were linked to an advanced system of zinc smelting involving retorts and furnaces which produced a zinc vapour, which was then collected in vessels and condensed, leaving behind the contained zinc metal.

India's historic mining and metallurgical expertise in copper, tin and thus bronze goes back even further to at least the 4th millennium BC. The ancient workings of Khetri and Rajpura-Dariba in the north west of the country demonstrate that sophisticated copper mining and smelting operations existed in ancient India. It is believed though that the tin used to make bronze did not come from the region itself but possibly from the centre of the country, or perhaps was imported from the rich tin areas of Malaya or Indonesia. It is thought that Indian expertise in copper may well have been introduced from Persia, and in support of this theory many copper and bronze articles have been found in the east of modern day Iran, in the north west of India and some in the south west of Afghanistan. Indeed, archaeological evidence indicates that there was a metal working site at Mundigak in Afghanistan and that techniques developed there, or perhaps developed further west, travelled down to the Indus Valley during the 3rd millennium BC.

The earliest bronze artefacts in India were found in the north west of the country in the Indus River Valley; these were largely weapons like knives and spears but also tools like axes and domestic items such as mirrors. In this we can see that technological progress was slow as the Bronze Age spread round the known world, so for thousands of years bronze technology was aimed primarily at manufacturing equipment for war. The artefacts are the clearest sign of a copper mining, alloying and smelting industry in this part of India, with remains of the ancient operations rather limited after so many centuries. It is therefore difficult to ascertain the likely levels of production of copper, but it would be a reasonable guess that annual output would have been in the hundreds of tonnes rather than anything more. Certainly in recent times a small-scale basic copper mining industry has operated in the Indus Valley area producing modest amounts of copper and this may well mirror the scale of the ancient mining. So the presence of small-scale copper mining today provides a link with much earlier operations in the Indus Valley.

Gold mining in the state of Karnataka, south west India, was carried out around the end of the 1st century BC where evidence of fire setting techniques from that era has been found at the operating Hutti gold mine, as well as evidence of charcoal for smelting and also gold residues. At about the same time – the end of the 1st century BC – a gold mine was established at Uti, near to Hutti. Around the 3rd century AD a number of small pits were dug to extract surface gold in Karnataka at Kolar and by the 9th century AD a larger operation had been developed there. Also in Karnataka state, the now-closed Ingaldhal copper mine in the Chitradurga district was first worked around the time of the establishment of the Hutti gold mine.

6. MINING IN ANCIENT EGYPT

The very essence of modern Egypt to outsiders and visitors is its fabulous ancient cities and monuments, and the treasures that were found in the country's most famous ancient structures, the Pyramids. These great antiquities have fascinated people for centuries and have led to much controversy as many of the priceless artefacts uncovered by archaeological expeditions, often foreign, have found their way outside Egypt. There are a number of different points of view about the history of this traffic in antiquities which, thought provoking though they are, lie well outside the scope of this particular history. However, it is beyond doubt that these great treasures would not have come into existence if the ancient Egyptians had not developed mining skills.

As we noted earlier, evidence suggests mining was first carried out in Egypt at least 40,000 years ago and possibly before that. In the Stone Age the targets for ancient miners were primarily materials such as flint for tool and weapon making, and other basic stone. Although it was the Bronze Age that heralded the rapid spread of the use of metals, the mining of copper, gold and other minerals in Egypt probably occurred long before that, as there is evidence of copper mines at Bir Nasib in the southern Sinai and turquoise mines at nearby Serabit el-Khadim.

The latter mines, developed first around 3500 BC as the Bronze Age was dawning, consisted of large hollowed-out galleries where the miners carved out turquoise for shipping to Egypt, to be made into jewellery and pigments for paint. The copper mines of the Sinai were established considerably earlier than the turquoise mines and there is archaeological evidence of copper smelting in the 6th millennium BC, probably using native copper, and copper mining 3,000 years later in the time of the Old Pharaohs. The mines lasted for centuries and disgorged large quantities of minerals. Indeed in the 19th century the British, who occupied Egypt for several decades, attempted to re-open the mines of Serabit.

STONE QUARRIES

In Ancient Egypt there were also many stone quarries littered around the eastern and western deserts and also down in the south at Aswan. The latter were critical in the construction of the temples and tombs in the Valley of the Kings near Luxor, one of the world's great tourist attractions and historical sites, and also the temples at Abu Simbel to the south near the Sudan border. The granite quarried at Aswan was exported by the Romans, when they occupied Egypt, to many Mediterranean sites, and indeed granite mining around Aswan continues to this day. The great Pyramid of Giza just outside Cairo used granite transported 500 miles from Aswan, demonstrating that not only were the ancient Egyptians able to mine and cut stone but also had the means to transport the mined materials large distances. The use of slaves for mining and hauling would have helped make this economically viable.

Other varieties of stone were also extensively mined by the ancient Egyptians. Although granite was a first class basic building block many of the temples required more ornate and beautiful stone for facing and other uses. Mines at Tura and Ma'sara, south of Cairo, provided high quality white limestone used for facing the tombs of the grandest occupants. At Tura the limestone was found at depth and could not be mined from the surface, so the ancient miners tunnelled underground and then cut the limestone out in large blocks, the limestone left behind acting as the support pillars for the mine. This demonstrates significant sophistication in mining, in keeping with the technical achievements in construction of the ancient Egyptians. Other decorative stone quarries and mines can be found the length of the Nile and during the period of the Pharaohs they were active in providing raw materials for the continuing building programme.

METALS IN THE DESERT

The need to transport building stone considerable distances meant that many of the stone quarries and mines developed were those close to the Nile; the river being the main transport route between the north and the south of ancient Egypt. The copper and gold mines were often to be found further inland and the tough working and living conditions meant that slaves formed the main part of the labour force.

One of the most extensive mining sites was at Wadi Hammamat between the Nile and the Red Sea in the Luxor area. There, over the years, gold, granite, slate and eventually iron ore were mined. Whilst mining itself was both tough and dangerous there were other dangers that increased with the remoteness of the mining site – these dangers related to the marauding Bedouin tribesmen who were in the habit of ambushing supply camel trains going in to mines and metal delivery trains leaving the mines.

One of the biggest copper mining centres was in the Timna Valley near Eilat on the eastern edge of the Sinai Peninsular. Mines had been known in the area since the Late Neolithic period (4000 BC) and production was particularly buoyant around 1300 BC when the Egyptians took control. The mines that the Egyptians developed at Timna were complex and technically advanced, which is little surprise due to the construction achievements of theirs that we noted earlier.

The first Late Neolithic copper mining consisted of crudely beating out an opening in rock, where the copper seams were observed to run, with hammers, and then excavating galleries to dig out the copper. The Egyptians mined in an altogether more organised manner. They used metal implements rather than stone hammers and dug regular round shafts and cut out steps so that miners could access the workings. The shafts went down sometimes as deep as 100 feet, depending on where the copper-bearing sandstone was to be found. Narrow galleries or drives followed the ore and where substantial quantities of ore were located the drive was widened into a cave in order to work the face. There was also some basic ventilation in the mine. Once mined the ore was dragged through the drives and then hauled to the surface. The Egyptians also organised the treatment process so that the furnaces ran 24 hours a day, which economised on the huge consumption of wood/charcoal under the earlier *stop-and-start* system when fires were left to burn out overnight before being started again the next day.

MINING FOR PRECIOUS GOLD

If the great kings of ancient Egypt needed spectacular tombs to mark their passing, and thus large quantities of stone, they also needed precious metals for adornment both in life and in death. The main gold mines were to be found in the south of the country in Nubia and in the Eastern Desert along the Nile. The mining methods used were not greatly different from those used in the copper mines, with shafts being sunk and galleries and drives being excavated to access the gold ore veins.

When the rock reached the surface it was first heated to make it brittle enough to break and then pulverised into a dusty substance which could then be agitated with water over a receptacle, the heavy gold sinking to the bottom. The gold dust, which could also be made finer by grinding with a corn millstone, was sometimes washed through a sieve; it is thought that sheepskin was also sometimes used to capture the gold as it was washed. Alluvial gold operations, as well as hard rock, were also known. Egyptian gold production levels at that time are clearly difficult to calculate but it has been suggested that it was around 1.5 tonnes a year.

The Greek historian Diodorus Siculus provided an extensive description of these Egyptian gold mining techniques at the gold mines of Bir Umm Fawakhir in the Central Eastern Desert region near Luxor in the 2nd century BC. It is likely that the mining methods were little changed from those employed in previous centuries and indeed millennia, perhaps only different in that by the time of Diodorus's observations tools had become more effective.

Diodorus pictures the work force as made up almost entirely of slaves who were basically worked until they died, with little or no allowance being made for age, physical condition or health. These dire working conditions and cruel use of slaves suggests that a royal control structure was probably essential in the ancient Egyptian gold mining industry to give legality to such activities. It is also the case that gold, right from the start, had a monetary role as well as a decorative role, which would have been an even more crucial reason for royal interest.

7. THE EASTERN MEDITERRANEAN AND THE NEAR/MIDDLE EAST

Apart from Egypt, a giant in ancient times, other important and prosperous states were located at the eastern end of the Mediterranean and some of them like Crete and Mesopotamia accumulated considerable quantities of gold. This gold was largely under the ownership of the royal houses, as it was in Egypt, underlining the idea that wealth in these early times was significantly in the hands of kings and princes.

Crete and Mesopotamia were insignificant gold producers so their gold would have come from neighbouring states such as Egypt, indicating an active trading environment in that part of the world. Crete, which was a substantial sea power, would have obtained gold from the northern Aegean and from the Balkans and the Danube states, where gold was mined extensively. This gold would have been shipped through Aegean ports.

There are also theories that the Egyptians launched expeditions in the 11th century BC to eastern Africa and brought back gold from mines in the area that is now Mozambique and Zimbabwe. The Phoenicians in the 6th century BC are believed to have circumnavigated Africa, starting in the Red Sea, and brought back gold from both the eastern and western sides of the continent. It is also more than likely that large quantities of gold reached the Middle East from the neighbouring Arabian Peninsular where ancient gold workings of considerable size have been identified at Mahd adh Dhahab in Saudi Arabia near Mecca. The current mine there is a high-grade one by today's standards (plus 20gms per tonne), and it is believed the ancient mine operated during the reign of King Solomon.

JORDAN

The advanced state of Egyptian and Mesopotamian society and the sophisticated nature of their buildings and the mines and quarries that provided the raw materials suggests that the Mediterranean/Middle East region, until the rise of the Roman Republic and Empire, was the centre of the civilised world, in contrast to today. We should therefore not be surprised that other parts of the region, and in particular Jordan, have a very long history of mining, going back to the 5th millennium BC.

The copper mines of Feinan, 40 miles south of the Dead Sea, underline the importance of Jordan as a mining region at that time. Amongst the ancient mining sites uncovered are Wadi al Abiad, Wadi Ratiye and Qalb Ratiye. Evidence of more than 100 mines has been found in this broad area, where ancient miners cut up to 30 feet drives into hillsides following the visible copper veins. The adits and drives that have been uncovered by archaeologists appear to have been kept low and were no more than 8 feet wide, with pillars to support the roof. This enabled miners to gain access to the veins, mining them in a gallery setting, maximising the amount of metal taken out and minimising the amount of digging required.

This room and pillar method of mining was accompanied by *backfilling* in a number of cases, which is an unusually advanced concept for the time and required considerable physical effort on the part of the miners. This was effort without any obvious economic return, except to avoid the galleries' roofs collapsing – perhaps an early example of mine reclamation. Flint picks and stone hammers found around the site, very much as elsewhere, provided the tools needed for mining in these early days.

As well as gallery mining there were a number of small-scale copper diggings at Madsus, another regional mining site. Here copper ore had been washed down over time from higher levels to terraces below, where the miners dug small pits to retrieve the copper-enriched material. These diggings were close to the ancient settlement of Wadi Fidan 4 where evidence has been found of residues from copper smelting, suggesting an integrated settlement that mined, treated and made copper objects such as weapons.

The investigation of Wadi Fidan suggests that the smelting process was the familiar one whereby a crucible containing copper ore was covered by charcoal and fired to start the smelting of the high grade ore. The slag that has been found at Wadi Fidan suggests that copper metal in this ancient process did not, as would become the case in later eras, become fully molten.

It is also worth noting that at Feinan the mines were not in close proximity to the smelting settlements and ore would have been transported some miles to be treated. This probably occurred in order to get economies of scale at the smelting end; in due course mine output would have expanded and it would then have been worthwhile having smelting capability close to the actual mines.

These mines were clearly operational for centuries and were worked again by the Romans when they applied their advanced mining technology in the 1st century BC.

GREECE

One of the most famous mines of the Iron Age was the great silver and lead operation at Laurion, a few miles south of Athens, developed in the 5th century BC. Greece was a major area for silver mining and its coinage and regional wealth was based on the metal. In this it was different from Egypt and indeed also from rival Persia, both of whose wealth was built on gold. Gold was also the financial key to Rome's wealth and power later. Laurion was originally worked in the 2nd millennium BC but really came into its own under the Greeks, whose main sources of silver in northern Greece had been cut off by the Persians in 512 BC.

The Laurion mines were first worked as shallow pits going down no more than a few feet. In due course the Greeks started sinking shafts and as many as 2000 have been identified at the ancient site. The shafts were sunk in pairs and with multiple crosscuts, which also aided ventilation. As the shaft system deepened it was joined together by drifts and the shafts themselves went down as much as 300 feet; it has been calculated that progress could have been at a rate of 5 feet a month. However, it is important to remember that both shaft sinking and mining was spread over several centuries so the relatively slow shaft-sinking rate was not really a handicap in economic terms.

At 300 feet the ancient miners, many of them slaves, would have reached the water table inhibiting further shaft deepening. The shafts themselves were well-constructed, often perpendicular and around 6 feet by 3 feet in diameter. The mining method was fairly rudimentary but since this was now the Iron Age the mining tools were often picks and chisels of wrought iron, which were more effective than the old stone implements used previously. Underground the drifts followed the veins of ore and galleries supported by un-mined rock pillars provided the miners with enough room to prise the ore from the rock.

Once the ore had been mined it was carried to the surface where it was crushed and then washed on grooved tables allowing the metal grains to be caught in the grooves and the waste crushed rock to be washed away. Since the mine was in a dry region, water was precious and the Greeks stored their water for washing the ore in tanks built into the hillside. They also recycled used water back into the tanks after the treatment sludge had been removed in settling basins. The remaining concentrate was treated in small furnaces to free the silver and then fired again in clay crucibles, which allowed the lead to oxidise forming a disposable slag and leaving behind pure silver. The Greeks may have been cruel taskmasters when it came to the conditions under which their slave miners worked, but these mining techniques nonetheless displayed technically advanced knowledge.

The Laurion silver mines eventually closed in the 2nd century AD, partly due to exhaustion and partly due to new, richer silver mines being found in Macedonia and Thrace to the north. It is interesting that over the seven or so centuries silver was mined at Laurion political and military struggles between the Greek city states of Athens and Sparta often had silver at their centre, as one of the main targets of the conflicts. Laurion was also central to the ability of Athens to see off Xerxes and the Persian army when they invaded in 480 BC, for it was the wealth from Laurion that enabled Athens to build a navy that effectively cut off the Persians after they had sacked Athens and it was also the wealth produced by Laurion that enabled Athens to be rebuilt.

ASIA MINOR

The area historically known as Asia Minor includes Anatolia (modern Turkey) and in terms of ancient trading relations it also touches on Syria, Mesopotamia (modern Iraq), Persia (modern Iran) and Armenia. Assyria (the ancient empire that covered parts of the region, expanding and contracting over the centuries) was one of the ancient world's most advanced civilisations and was an important factor in the emergence of the Bronze Age in the 3rd millennium BC. At its height in the 1st century BC, the Assyrian Empire was also a highly developed society which had embraced technological change and had pioneered a number of advances including a basic electric battery which may have been used in the process of treating silver. Some historians even call Asia Minor the *Cradle of Civilisation*.

It is therefore not surprising that this region was the centre of a vigorous trading system where Anatolia was the great mining and metal working nation and exported finished metal to its near neighbours. For example, the ancient tin mine at Kestel and the associated smelting site at Goltepe in central Turkey reach back as far as the 3rd millennium BC.

Copper artefacts have been found all over the broad Asia Minor region; the oldest is thought to be a pendant from the 9th millennium BC found at Shanidar in the Zagros Mountains of north eastern Iraq. Archaeologists have also found copper objects at Cayonu Tepesi in Turkey that have been dated to the 7th millennium BC. Further finds of copper artefacts from the 6th and the 5th millennia BC in Iran and Turkey have also been uncovered. Much of this early copper would have been native, released from rock by weathering over thousands of years. As we have observed in other ancient mining areas, the very earliest use of metals was based on a supply of pure material simply picked up off the ground, a phenomenon that we will come across later in the tale of diamonds in Namibia's deserts in the 19th century AD.

A relatively sophisticated trade in copper and copper objects developed over the millennia in the extended area of Asia Minor and the Middle East, and it is quite likely that the growth in this trade put pressure on the supply that was being obtained from native copper findings. Since we are talking about the cradle of technology as well as of civilisation, the response to possible copper supply problems would have been to examine how the pure copper metal had got where it had. Here we are just guessing, but it is likely that very early fossickers saw copper streaks in rock and worked out that these were similar to the native copper that they had been picking up off the ground.

From here it would have been a matter of developing the processes to extract the copper from the rock. The secret would have been the ability to generate sufficient heat in a rock furnace to melt the copper out of the ore and then separate the two – metal and slag – leaving a relatively pure copper residue. It is believed this smelting technique was developed around the 4th millennium BC.

One of the most interesting metal mining, smelting and manufacturing sites found in Asia Minor is the Medzamor complex immediately to the east of Turkey in Armenia, which is thought to date back as much as 5000 years. Medzamor was uncovered in the 1960s by Soviet scientists and archaeologists and some western opinion at the time was a little sniffy about its significance. Dozens of smelting furnaces were found and evidence of multiple working of metals covering both copper and other base metals as well as gold. All sorts of metal products were uncovered – including vases, bracelets, rings, spearheads and knives. This suggests a settlement specialising in industrial activities emanating from mining, contrasting with the traditional farming settlements more associated with simple ancient lifestyles.

PERSIA, THE EMPIRE AND IRAN

In the 7th century BC Persia emerged as a major power in the Middle and Near East and in due course its influence extended both east towards India and also further into the Mediterranean, where it often clashed with Greece. Gold was an important factor in Persia's expansion as it provided the Empire with a legitimate monetary system which supported its political ambitions. Persia's gold came from both wealth acquired as a result of conquest and from mines which came under its control. This wealth in gold was always useful, as Persia's major rival at the time, Greece, was rich in the less valuable silver but had limited access to gold. Thus Persia found it was able to buy influence in the eastern Mediterranean without the need for military successes, which was fortunate, as its few sallies forth against the Greeks had singularly failed to achieve anything.

Persia's gold came from mines in both Asia and the Middle East, its empire stretching from India and the modern *-stans* of central Asia, through Iran to the

Arabian peninsular. Much of the Empire's gold was mined as alluvial gold washed down from the mountains of Anatolia (Turkey) in the west and of Bactria (Uzbekistan) in the east. Apart from using gold as currency Persia also kept gold in ingot form as a sort of foreign currency reserve.

But Persia was a many faceted power. Whilst its interest in gold was of great importance it was also an advanced civilisation with concomitant needs and skills. In the area of metals it embraced the Iron Age earlier than most and quickly built up iron making and working. Hasanlu, a major city in north eastern Iran and now a great archaeological site, was in the 8th century BC a centre for metal working and a pioneer in making iron. Hasanlu was also a crossroads city attracting hosts of invaders throughout the centuries. Importantly for the advance of its technical skills, it also attracted travellers who both brought news of technical developments in mining and metals from foreign parts and took back home news of Hasanlu's own progress in these industries.

In the 9th century BC Hasanlu was destroyed by invaders and abandoned, disappearing under rubble and, over the centuries, sand. In the 1950s the city was the subject of archaeological investigation and numerous metal items were found buried, including gold, silver, bronze and iron. A furnace for making iron was also uncovered and numerous well-preserved iron objects including knives, nails, buttons and pots were found. One of the objects, a knife blade, indicated that the ironworkers of Hasanlu had also made progress in making steel, which is harder than iron and thus more suitable for weapons in particular.

8. THE ROMAN REPUBLIC AND EMPIRE

It is in the nature of great empires to expand at least partly because of their need to acquire natural resources to fuel this expansion; one might say that expansion begets expansion, and such wealth as results from this becomes critical to the health of the imperial power. The Roman Republic (510 BC to 44 BC), which saw Rome begin to expand from its primarily central Italian base, and the Roman Empire (44 BC to 476 AD), which continued the expansion, were no exceptions to this. An important economic part of Roman conquests was the acquisition of sources of essential raw materials, in particular metals and minerals. The main areas providing the Empire with key metals were Spain, Britain, France (Gaul), central Europe along the Danube, Greece and Asia Minor. The metals produced included gold, copper, iron, tin and lead.

The gold mines of Spain were particularly important in providing finance for the Roman Empire as it continued the expansion begun by the Republic. In 31 BC Augustus became Emperor and after military success in subduing Egypt he turned to Spain. Here Rome was already strong in the south and was already exploiting the gold and base metal deposits in the region which had been developed by the Carthaginians before they were ejected by the Romans from southern Spain at the end of the 2nd century BC.

The Greek geographer, Strabo, in his magnum opus, *Geographica*, describes in detail the mines of the region that is now Andalucia. The mines in the region produced gold, silver, copper, lead and iron ore, all of which were of course of material value to the Romans. Strabo is particularly interesting on the issue of treatment of the ores, where on occasion furnaces became overheated leading to a loss of metal due to vaporisation. He also describes the necessity for chimneys over silver smelters due to the poisonous nature of the furnace fumes, probably caused by the lead often associated with silver ore. The area was also a source of very high grade ores with pure gold nuggets commonly found on the surface. Copper ore mined was also of a high grade, as much as 25%, but such figures are obviously difficult to authenticate. It is thought that Strabo never visited Spain, his own travels being confined to the eastern end of the Mediterranean, so he relied on third-party sources for his information.

In the northwest in Galicia, according to Pliny the Elder, the Roman philosopher, naturalist and military commander (see below), gold mining output reached an aggregate figure of over 200,000 ozs. A diversity of mining methods were used, including the traditional alluvial mining techniques along the Spanish run of the Tagus and Douro rivers, and in Galicia larger operations incorporated both underground shaft and gallery mining methods. In the case of the Galician underground operations Pliny records that shafts were sunk through the soft surface earth which was supported by wooden props. The shafts were aimed to intersect gold channels or reefs which were hacked out and the ore was taken to the surface where it went through the familiar process of crushing, washing and burning in a furnace to produce a powder. This powder was then crushed further and heated again in the furnace, which Pliny describes as made of a white material like potter's clay, whereupon gold was finally released. Where large rocks – often of flint – were brought down in the tunnelling process they were broken up with large 150 pound iron rams and wedges, and carried to the surface in baskets.

The most spectacular mining method used was where galleries were driven into mountains or hillsides using a chain mining system; miners worked from deep inside the mountain passing out the mined rock. Left behind was a pillar-supported gallery and the pillars were then part cut to weaken them. The mine was evacuated when a spotter on a nearby hill, who watched for the outside signs that the gallery was on the point of collapse, signalled the workers to leave. When the mountain finally collapsed huge quantities of loose, broken material could then be treated in an attempt to extract any gold trapped in the material. This was a somewhat hit-or-miss operation as gold could not be guaranteed at the end of the process – the Romans would have had some idea of the prospectivity of the mountain or hill for gold, but their geology was rudimentary.

The secret of the treatment process used was being able to bring large quantities of water to the site and then store it in a reservoir so that a strong but controlled stream of water could be brought to bear on the broken material. A complex patchwork of channels and wooden sluices were also constructed to carry water and slag away, and trap any released gold. The system and scale of the operation puts one in mind of the gold workings in California in the mid-19th century where hydraulic mining was pioneered. Certainly historians think tens of millions of tonnes of ore were mined by the Romans in the region.

The technologies the Romans used in their Spanish mines were also sophisticated, as they had to be when shafts were sometimes sunk several hundred feet in pursuit of rich ore. Because of the sometimes large quantities of water at depth, de-watering was often an insurmountable problem, but a screw pump was uncovered at the ancient Centenillo mine near Linares in Andalucia. The pump was designed by Archimedes based on a system he had observed being used on the Nile to raise water. The machine consisted of a screw within a wooden cylinder, which was turned by hand and foot, with the feet working peddles which turned the screw. The turning of the screw enabled water to rise slowly up the cylinder and when it reached the top the water was then directed away from the workings.

The gold mined for the Romans was used in two ways. The primary aim was to provide money in the form of coinage to lubricate the Roman economy and importantly to pay for Rome's imperial ambitions as well as the empire itself. There is much evidence that the second main use of gold was as gold ornaments for the upper layers of society – gold plate and jewellery has been unearthed over the centuries by archaeologists.

Indeed, it is postulated that before the reign of Constantine there was a period when gold supplies being used for coin were under pressure because a rising percentage of gold was being diverted into fabrication for personal use. There is also a view that after Spain the only new sources of gold for the Empire came from mines in Dacia in the Carpathian region of the North Balkans, conquered by Trajan in the 2nd century AD. Over the next two centuries there appears to have been little in the way of additional gold found in the Empire and gold prices, having previously fallen, then strengthened a little. This encouraged the inflationary practice of debasing the coinage by clipping coins in circulation enabling additional coins to be minted without the imperial treasury having to purchase new (and more expensive gold) for the coins.

PLINY THE ELDER (23-79 AD)

Although the great Roman soldier and observer Pliny the Elder was not a mining man, his writings on mining operations that he visited across the Roman Empire, particularly Spain where he was a procurator responsible for mine administration, have created an invaluable record for posterity. The book in which they appear, *Historia Naturalis*, is in fact a comprehensive study of geography and nature in the Roman Empire, and the sections on metals and mining are just a small part of the work. However, they are one of the few sources of directly observed comments on mining techniques in the ancient world that have survived to this day.

Born Gaius Plinius Secundus in 23 AD in Como, northern Italy, Pliny's family held equestrian status indicating that they had money and were deemed aristocrats or knights. Pliny received formal education in Rome where his tutor, Pomponius, who was one of Rome's leading poets, provided Pliny with connections important to his future career. At the age of 21 he went to the Low Countries to serve as a military tribune, an administrative post, but soon became a military officer. For the next few years Pliny served in the Lower Rhine area where he saw active service under his old tutor Pomponius.

At this stage, around 50 AD, Pliny began to write books, which included a biography of Pomponius and a history of the Germanic Wars. He returned to Rome where he might have been expected to continue his career in public service. It was, however, the age of Nero and Pliny's position as a serious historian did not fit into Nero's increasingly debauched society. He wrote further books which did not attract positive comment and as Nero's rule ended with the Emperor's suicide, a vicious civil war broke out. Around that time Pliny's brother-in-law died and Pliny took on the guardianship of his nephew Pliny the Younger.

Vespasian, whose son Titus was a great friend of Pliny's, had taken up arms to press his claim to become emperor as the post-Nero situation became increasingly chaotic. Vespasian entered Rome as emperor in 69 AD and from that point on Pliny's career took a major turn for the better as he assumed a number of important public posts as a procurator, which took him all over the Empire. One of the regions he visited was Spain, where his responsibilities included keeping an eye on the gold mines which had assumed a key role in the Empire's finances. It was here that he made copious notes on the techniques of mining and treating gold at locations such as Las Medulas in Leon, observations which eventually were incorporated into the mining section of his *Historia Naturalis.* He was not remotely a gold bull, however, as he believed that it merely played on man's natural greed. He also made observations about silver mining, base metal (including lead) mining and even diamonds. The *Historia* also contained sections covering the use of metals in what we would now call manufacturing.

Pliny's career continued with his appointment as prefect for the Roman navy in the western Mediterranean and it was during this phase of his life that *Historia Naturalis* was completed and published in 77 AD. In August 79 AD Pliny was holidaying with his sister and Pliny the Younger in Misenum when Mount Vesuvius erupted. Pliny ordered a fleet of ships, which he commanded, to set sail to rescue those who had escaped the disaster and were gathered on the shore outside Pompeii. He, unfortunately, had to spend the night ashore and in the morning was found dead overcome by the sulphurous fumes from Vesuvius.

The only book of Pliny's to have survived is *Historia Naturalis.* Whilst archaeologists have done much work on ancient mining sites around the world and in so doing have provided much useful information on mining construction and techniques, Pliny's observations are the real thing. They reveal the relatively sophisticated mining methods of the Romans, including the use of hydraulic (water pressure) style mining 1700 years before it was introduced into 19th century mining. As such Pliny's work holds an invaluable place in the study of the mining industry's historic development.

ROME'S SPANISH MINES

As well as gold, the Romans operated silver and lead mines around the same time in central Spain at Plasenzuela, Extremadura, for a period of around a hundred years. These mines do not appear to have been visited by Pliny so we must rely once more on the observations of archaeologists and the discoveries of the miners who operated the mines again in the second half of the 19th century. The dating of the Plasenzuela operations to Roman times relates to the toxicity of some of the residues on the site, which contain arsenic and come from smelting of the mine output, a sure sign of Roman involvement. Roman roof slate fragments were also found in the slag heaps. Investigation of some of the slag also suggested that the Romans had experimented with a process to try to recover some of the lead metal lost in the smelting process.

Much evidence has also been found of ancient mining in southwest Spain in Almeria in the Sierra Almagrera, which probably started in the 3rd millennium BC, and over the centuries carried out by a variety of visitors including the Phoenicians, the Romans and the Moors. The mines here produced lead and silver and though they were inactive for centuries during the Middle Ages they came back to life in the 19th century. Today mining is very small scale, carried out by individual entrepreneurial miners.

Another ancient mining site in Almeria was Rodalquila, where gold has been the target since Roman times. The Romans were also responsible for the development of the Ojos Negros iron mines in Teruel in the north of Spain, and of the giant mercury mines of Almeden Puertollano in Ciudad Real. Near Barcelona the mines of Gava were mined from around the 4th millennium BC for a green mineral called variscite that was probably used for making basic but colourful jewellery. Also from this period in Murcia in southeast Spain mining of iron, lead and zinc began and continued up until the 20th century.

Whilst Spain was both the Republic and the Empire's most important provider of metals and minerals, Italy itself was a substantial source of raw materials, particularly of industrial minerals such as salt, stone and clay. Records are thin on Rome's Italian mineral sources but according to Polybius rich gold deposits were found and worked at Aquileia at the top of the Adriatic, near what were to become the lagoons of Venice. The Republic also acquired the gold mines of the Alpine tribe, the Salassi, when they conquered the small territory which lies on the Italian side of the Alps in 138 BC. Sardinia, acquired by the Republic from Carthage in 238 BC, was also a substantial producer of silver, lead and iron ore.

ORGANISATION OF ROMAN MINES

The Romans were an advanced and so very well organised society, a critical attribute for the efficient operation of an empire that lasted many centuries. They firmly believed in structure and mining was an area where they established, both at home and in the conquered territories, a legal structure to enable raw materials to be extracted and for the extraction process to yield tax and royalty revenues for the treasuries of the Empire. The richness of the Spanish gold and silver mines unfortunately led to bureaucratic corruption, although the flow of revenues to Rome meant that a blind eye was turned to official pocket lining.

The Romans developed two main administrative structures for mining. One consisted of a leasing system where Roman revenue officials auctioned mining leases to the highest bidder. Having leased the mines to operators, this bidder was then expected to provide Rome with an agreed flow of revenue from the mines. This revenue was obtained from the mine operators, very often the main lease owner making large amounts from the difference between what Rome expected in payments from him and what he charged the lease operator in royalties.

The other system was similar in terms of lease payments to Rome but the control of the mines was vested in an administrator appointed by Rome who either leased the properties to mine operators or himself took on the task of running the mines. There was plenty of room within this system for personal gain and the chicanery started at the top, with the Caesars in the late 1st century AD particularly rapacious as they took control of many of the Empire's gold and silver mines. In due course, as the Empire began to collapse the mines deteriorated, corrupt practices abounded and eventually the mines became uneconomic.

With the collapse of the Roman Empire, mining, along with everything else, slipped into the Dark Ages and mining techniques reverted to crude, manual methods that were dangerous and inefficient. It was many centuries before the industry regained the heights of operational efficiency reached under the Romans.

9. GREAT BRITAIN

The stirring words of the great English hymn 'Jerusalem' speculate that Jesus Christ may have visited England and more precisely Cornwall before he began his ministry. This idea came about as a result of a theory that the Phoenicians from the eastern Mediterranean used to trade finished goods such as cloth for Cornish minerals, particularly tin. The myth was that Joseph of Arimathea came to Cornwall with his nephew, Jesus, on such a trip.

In fact there is no evidence of trade between Cornwall and the eastern Mediterranean, although it is likely that there was trade in minerals and other goods with the western Mediterranean and particularly with Greek traders who operated from Marseilles. In support of this thesis has been the discovery in both Cornwall and Brittany of silver coins used around Marseilles, and believed to have been minted in the 3rd century BC. A number of ancient writers such as Pytheus and Timaeus also mention Cornish tin.

CORNISH TIN

The earliest indications of a tin mining industry in Cornwall date from the early Bronze Age (around 2000 BC) – tin slag from that period has been found near St Austell. The discovery that tin, when added to copper, produced bronze, a hard but easily workable metal that could be used for weapons, was a major technological breakthrough and Cornwall's substantial tin resources undoubtedly made Britain a magnet for traders from the Mediterranean long before the Roman invasion.

Cornish tin was initially mined using alluvial methods as the tin was often found in streambeds having been washed down from surface outcrops. Plenty of evidence has also been found of continuing tin mining in Cornwall in the middle Bronze Age, with the unearthing of mining tools and bronze objects in old tin environments. Bearing in mind the development of copper mining in Cornwall in the Middle Ages it is likely that some of the surface copper lodes would have been worked during the Bronze Ages and some bronze objects found indicate that likelihood.

As well as alluvial tin mining it is likely that early miners would have mined from tin lodes found on the surface and exposed for quarry-style extraction. There is also strong evidence of a tin smelting capability with discoveries of smelting vessels, some from the Iron Age, indicating a steady development of technology over the centuries. Whilst the greater proportion of ancient Cornish tin output came from alluvial sources, there is clear evidence that in pursuit of richer lodes some underground working was done. Miners at the Wheal Virgin mine in the 19th century came across a wooden shaft 30-feet deep dating from the Bronze Age, suggesting that ancient miners had found a potentially very rich tin lode and followed it underground. There they would probably have driven a gallery along the strike but this would only have been done because of the richness of the ore, there being plentiful surface tin to mine.

Although there were records of mining of copper and iron ore in Wales during Roman times it was Cornwall, and to a lesser extent Devon, where Britain's mining industry was primarily located. Despite that, Roman records regarding their occupation of Britain do not have much to say on the subject of tin mining, probably due to the fact that the Romans had established mines in Spain and these provided them with the bulk of the tin they needed. Cornish mines, however, continued to operate and by the 3rd century AD production expanded as rising prosperity led to increasing use of pewter for household plate and drinking vessels. Tin mining and smelting continued in the West Country through the Dark Ages, from the withdrawal of the Romans around 400 to the Norman invasion in 1066.

MINING IN WALES

One of the major mining sites in Britain beyond the South West was the Parys Mountain area on the Isle of Anglesey in North Wales, where copper rich base metal mining took place spasmodically from 3500 BC until the early 20th century.

Evidence of this was unearthed when underground mining exposed collapsed ancient bell pits, which were surface areas containing minerals that had been mined by crude scraping.

Other copper mining sites in Wales included Great Orme near Llandudno where, as at Parys, charcoal used for fire setting was found, which has enabled carbon dating. There is also evidence at Great Orme of a network of underground tunnels and workings which indicate that it was an extensive mine by the standards of the prehistoric era; there are suggestions of around 200 tonnes of copper having been extracted then. Since these Welsh mines are thought to have operated during the same period as the copper mines of the advanced world (then the Near East and eastern Mediterranean) it is likely that some of the mining techniques were imported, probably by traders or visitors from the European continent and beyond. Another ancient Welsh mining site was the Copa Hill copper mine near Cwmystwyth in west Wales, where extensive radiocarbon dating suggests significant mining took place between the 3rd millennium BC and the 12th century BC.

The Roman invasion of Britain also led to the development of a gold mine at Dolaucothi in the Welsh region of Carmarthenshire. Mining in this part of Wales is believed to go back to the Bronze Age but the Roman operation was on a different scale and used the hydraulic techniques of the Spanish gold mining industry recorded by Pliny. The existence of gold mining at Dolaucothi was uncovered in the 18th and 19th centuries when gold objects and then gold ore were discovered at the site. Later, other discoveries were made at Dolaucothi suggesting that the Romans had constructed a very sophisticated mine.

At Dolaucothi underground tunnels that chased the gold ore veins were discovered, as was part of a water wheel that would have been used to de-water the deeper levels of the mine which went down as much as 80 feet. Scorched timber was also found, indicating that fire setting to crack the rock hosting the gold seams was practiced. A number of pits, which were probably workings of surface veins, were also found.

The most interesting discoveries were of a number of tanks found in the proximity of the network of watercourses (aqueducts and leats) built by the Romans to bring water to the site. These tanks stored water which could then be poured over the loose and crushed rock at various rates of intensity to wash the gold out of the material; at that point there would likely have been a number of large *trays* that would have captured the gold as the water poured away. Some carbon dating work done on residue from the site indicates that the Romans may have worked Dolaucothi until the time that they left Britain in the 4th century AD.

COAL MINING

Coal mining is also thought to have been carried out during the Roman occupation of Britain as there is evidence that Roman troops guarding Hadrian's Wall used coal for heating in the cold of England's north east, digging the coal from surface seams. There is also archaeological evidence that coal was used millennia before then in Wales to fire funeral pyres. It is also possible that coal's heating properties were appreciated as a result of observation of spontaneous combustion of surface coal seams under certain conditions.

It appears though that wood remained the preferred method of heating, cooking and metal smelting in Britain for many centuries, probably as a result of its natural proximity to towns and settlements. This is something we will pick up again later.

BASE METALS

Whilst recorded evidence for ancient mining is difficult to obtain, particularly in the less developed world, which of course included Great Britain, archaeologists have uncovered clear proof of base metal mining activity in Wales and England, as well as in Ireland, going back to the Copper and early Bronze Age (4000 to 2500 BC). In Ireland two copper mines have been identified, at Ross Island and Mount Gabriel in the southwest. The origins of copper mining in Ireland were believed to relate to the Bell Beaker people who travelled throughout Europe and are thought to have developed important skills in mining and the treatment and working of metals.

However, it was another wave of visitors, the Hallstatt people from central Europe, who brought the knowledge and skills necessary to take Britain into the Iron Age. The Hallstatts are believed to have used their commercial position as salt miners and pig farmers to export to other continental markets and in due course were in a position to absorb foreign ironworkers and finance the development of their metal working skills. The Hallstatts were particularly interested in the manufacture of iron for weapons and were knowledgeable about bellows furnaces and carburising, which would have enabled them to make semi-steel.

The Hallstatts eventually arrived in Britain around the 5th century BC, bringing their iron-making skills with them. Britain had abundant supplies of iron ore and in due course a widespread iron-making industry grew up. One of many iron-working sites uncovered by archaeologists was Kestor near Chagford in Devon. Here an ancient settlement has been excavated and within some of the identified dwelling structures were small furnaces and signs of a forge for working the iron.

The quantities of iron used suggest that sources of the raw material came from surface accumulations such as bog iron. Apart from weapons, particularly swords and iron blanks for blades, archaeologists also found standard size iron bars which were used as currency. By the time of the 1st century BC British blacksmiths were also hammering iron into relatively thin circular shapes for use as hoops to strengthen wooden barrels and, critically, iron tyres for chariots and other heavy duty vehicles.

10. CENTRAL/EASTERN EUROPE

THE EASTERN ALPS

Evidence of Bronze Age copper mining and smelting has been found throughout the Alps, stretching from eastern Switzerland across Austria. A particularly large site was found in the 19th century at Muhlbach, in the Mitterberg region in Austria. The chalcopyrite copper deposit was discovered by chance, but in time (the mid-1990s) archaeologists unearthed a Late Bronze Age (1300 to 700 BC) smelting complex which almost certainly was an important user of the Muhlbach copper ore. There is evidence of an extensive set of furnaces and slag heaps, and the site is close to water that would have been used in the washing process of treatment.

The ancient mines in the Mitterberg district have also yielded some of their secrets to archaeologists and surveyors who opened up one of the ancient mines in the late 1990s. The shafts went down almost 600 feet, very deep for such an early period, and galleries and drives ran in a number of directions in pursuit of the copper ore.

GERMANY

What we now know as Germany was historically a region, and at times an empire, consisting of a number of separate kingdoms, including Saxony, Bavaria and Bohemia. With the coming of the Dark Ages at the end of the 5th century AD, as we mentioned before, mining and much else besides fell into decline as the old Roman Empire crumbled. Power in Europe was now based in a socially and economically backward central region, stretching through France, Germany and into Europe's east; conflict was endemic and the main demand for metals was for iron to make weapons.

In 800 AD Charlemagne became Emperor of the Holy Roman Empire and was responsible for the striking of a new silver currency, which itself helped to start the revival of mining. Charlemagne used Saxon slave labour to mine silver from known deposits in the Erzgebirge Mountains of Saxony. In pursuit of precious metals for the Empire's treasury to finance almost constant wars on the continent he also encouraged the re-opening of the gold and silver mines of Schemnitz and Kremnitz, first worked in the 6th and 7th centuries AD, in what is now Slovakia.

Perhaps the most famous mines developed in Germany, the life of which also spanned the ancient period, the Middle Ages, the Industrial Revolution and the modern age, were those at Rammelsberg in Lower Saxony in the Harz Mountains. The mines had been worked intermittently for over a thousand years but fell silent in the Dark Ages, re-opening in the 10th century. They produced copper, zinc, lead and silver over the centuries, the focus on particular metals in the ore changing as monetary requirements favoured silver at the end of the Dark Ages and then base metals as the Industrial Revolution loomed. It is thought that as much as 27 million tonnes of ore was mined over this long period before the last owners, Preussag, closed the mine in the late 1980s. Overall the mines were estimated to have produced over 3.75 million tonnes of zinc, 1.6 million tonnes of lead, almost 550,000 tonnes of copper, and more than 115 million ozs of silver and 825,000 ozs of gold.

THE BALKANS

We have already seen how the Cretans and the Romans obtained gold from the Balkans but long before then in the 5th millennium BC there is strong evidence of copper mining in Bulgaria, Rumania and Serbia. What is less obvious is the question of whether these developments came about as a result of Near East influence or whether the discovery of copper led to the establishment of an indigenous mining and metallurgical industry in the region. What we do know is that advanced civilisations like Crete imported copper from the Balkans, but in the ancient treatment sites uncovered there is some evidence of a local way of doing things in the shape of the copper casting moulds.

The main ancient Balkan copper mining sites were Ai Bunar in Bulgaria and Rudna Glava and Rudnik in Serbia. Their age, around the end of the 5th millennium, has been estimated using carbon dating techniques. These techniques were also important in raising and progressing the idea that Balkan copper mining and metallurgy were more advanced then elsewhere in the Mediterranean and Middle/Near East at that time. Certainly archaeological work over the last few years has uncovered a large number of copper artefacts, which are older and still more sophisticated than those found elsewhere in the region.

Without written records, the sort of civilisation in the Balkans several thousand years ago must be drawn from a mixture of evidence from archaeological diggings and speculation. The evidence from graves containing a large number of jewellery items, uncovered near the Bulgarian town of Varna, can be used. Many of the items were of gold, which suggested a graveyard for the wealthy of the region. It also suggested the likelihood of metal trading, as some of the copper artefacts in the graves were not consistent with local ore types. So in this relatively unconsidered, in today's mining terms, part of eastern Europe we note copper mining, treatment, manufacturing, and metal trading with a social hierarchy thrown in for good measure, and all now 7000 years ago.

OTHER MINING REGIONS

From the 7th to the 11th centuries the silver and lead mines of Melle in central, western France were an important source of wealth for the French crown. Charlemagne, France's first *modern* king, was particularly in need of currency as he waged almost total conflict during his reign (768-814) in his efforts to unite Europe. Before the Carolingian period lead had been the main target of the preceding Merovingian dynasty.

The mines at Melle were extensive with galleries running several miles which had been carved out by the miners over the centuries after accessing the ore via vertical shafts; fire setting and then pick were the traditional mining methods used. From the 11th century silver was mined in the Fournel valley in France's southern Alps near the village of L'Argentiere, named after the silver mines; again the need for silver coinage was the key motivation. The mines operated intermittently over the centuries, finally closing in the 19th century; an extensive network of tunnels where miners followed and extracted the ore remains today and can be visited. Evidence also remains of the surface treatment facilities.

11. NORTH AMERICA

The possibility that the indigenous Indians of North America could have been involved in mining activity is something that early European settlers would have found unlikely given the basic nomadic structure of the lives of these people. However, the discovery of ancient mine workings in the area of Lake Superior in Michigan in the 19th century suggested otherwise, to begin with anyway.

S.O. Knapp, who worked for the Minnesota Mining Company, came across two ancient copper mining sites in Michigan when surveying for the firm. One consisted of an ancient gallery which had been excavated underground and in the gallery were a number of stone hammers and an exposed copper vein. Close by Knapp found evidence of another copper mine where a shaft had been sunk to around 30 feet. In this shaft Knapp came across a large smooth lump of native copper, weighing perhaps 6 tons, which had clearly been worked by miners in the very distant past.

On Isle Royale there are many examples of old mines where pits had been sunk almost 60 feet down to expose the copper vein; these pits extended over a two-mile length. The period of these workings is thought to relate to the 3rd millennium BC and as much as 500,000 tonnes of ore may have been mined, spread over at least 1000 years, although these figures can only be best guesses. Similar ancient copper workings were also unearthed in nearby Wisconsin. In Utah, coal miners at the Lion mine in Wattis found an ancient tunnel system which followed the underground coal seams for several hundred feet. Within this tunnel system there was also a coal ore collection area from where the coal would have been carried to the surface.

Interestingly, in both the cases of Michigan and Utah the indigenous Indians could provide no information from their cultural history of mine working by their ancestors, so Knapp and others had to look elsewhere. It was therefore speculated that these mines were worked by an ancient but technologically advanced people who may have come from as far away as Alaska or indeed Asia and in time continued their journey south across America to Mexico and beyond. No sign has been found of smelting near the ancient Michigan finds, suggesting that whatever was produced was transported – maybe hundreds of miles – to be treated elsewhere and made into copper and possibly bronze items.

In the state of Nevada further evidence has been found of indigenous mining activity in the form of arrowheads and spearheads made from hard rock deposits such as quartz. Spearheads have been found in the Carson Sink in the northwestern part of the state and in the Washoe Valley near Lake Tahoe dating from as far back as the 8th millennium BC. In the 3rd century AD for a couple of hundred years the Anasazi Indians mined turquoise and salt in what is now Clark County, and early prospectors in the 18th and 19th centuries recorded having received assistance from local Indians in locating promising mineral areas.

12. SOUTH AMERICA

Before the coming of the Spanish in the Middle Ages the areas that are now known as Latin, Central and South America were populated by a variety of civilisations that, because of the colour of their skin, some speculate may have made their way from Asia (as described in the previous section) many millennia before over the land bridge that existed across the Bering Strait, and then spread out east and south. Others may have taken a sea route from Asia going southeast on the prevailing Pacific currents leading them eventually to the shores of central and southern America.

Today South America is well known for hosting a wide array of metals and minerals, both industrial like copper and precious like gold, but its historic tradition was mainly associated with precious metals. Ancient civilisations in South America such as the Chavin and Mochican, were capable of building sophisticated stone structures for living and worshiping in. They also had organised farms and had acquired the skill to work precious metals – gold and silver – into the most intricate and beautiful objects. Interestingly, despite these skills with precious metals, it appears that the Iron Age did not come to South America until the arrival of the Spanish conquistadors with their steel weapons in the 16th century AD.

Although there are many gold mines in production in South America today there is little evidence of ancient gold mines as one might find, for example, in Egypt or Spain. However, the importance of gold metalworking and the quality of the objects found suggests that gold was mined, perhaps extensively, by these ancient civilisations, and certainly centuries later the Conquistadors found more than enough gold objects to satisfy their lust for the metal. It is also the case that the Spaniards themselves did not find gold mines; the gold booty they took back to Europe from their conquests was simply the ancient artefacts that they acquired from the indigenous people.

So if the source of the gold used by South American goldsmiths did not come from open pit and underground mines where did it come from? The most likely answer is that the gold was alluvial and had been naturally scoured out of rich deposits high up in the Andes over millennia, and washed down to lower levels where perhaps large nuggets of pure gold caught the eye of local inhabitants. It is not a large jump to where ancient craftsmen, intrigued by the beauty of the gold, started to fashion items of jewellery and plate from a metal that they would have found easy to work due to its softness. It is also likely that the basic furnaces, which they would have used to make molten gold for the moulds of the objects they were to work on, have simply disappeared over time.

There is also a theory that the Incas, who were fine stonemasons, may have developed an extraordinarily advanced technology to cut stones to very high specifications. It is suggested that huge gold sheets that the conquistadors found were used to concentrate sunlight rays to create a source of solar energy powerful enough to cut stone. Unfortunately, though the Incas may well have developed this industrial-type technology they did not have the technology to record the process – i.e. paper and ink.

CHILE

There is an old mining saying that the most likely place to find a mine is where one existed in the past. Therefore, it is perhaps not surprising that evidence was uncovered of mining around 500 BC at the giant Chuquicamata copper mine in northern Chile. Of particular interest were remains uncovered when the current mine was being explored and then developed in the late 19th and early 20th centuries. A number of well-preserved hammers were found, as were the remains of a body of an ancient miner and a basket of copper ore.

Radiocarbon dating suggested that these finds were from around the 3rd century AD, or perhaps a little earlier than that. Unfortunately, although some ideas were formed as to where mining took place, at Chuquicamata no actual workings were uncovered, perhaps indicating that ancient mining was confined to surface shows of copper. Although Chile was later to become a sizeable gold producer in the 18th and early 19th centuries, there is evidence of some gold mining in ancient times, and a little more during the period of the Inca occupation of parts of northern Chile.

PERU

In 2004/05 a major ancient iron mine was excavated by archaeologists from Purdue University in the US, in an area of Peru a couple of hundred miles south of Lima, where centuries ago the Nasca civilisation lived between the 1st and 7th centuries AD. The old mine was in the form of a 700 cubic metre chamber close to a current ochre mine. The chamber would have been dug out using hand tools to break and then scour out the ore. It has been calculated that the ancient mine produced around 3,700 tonnes of haematite over 1,400 years of operation. This is of course a tiny amount by modern mining standards but the Nasca people were artistic and the haematite when crushed and treated would probably have been used for painting, not for producing metal on an industrial scale.

13. AFRICA

The techniques required to treat copper and iron were introduced into southern Africa as much as 2000 years ago, although exact dating is difficult as climatic conditions have eroded quite a lot of the evidence. Although the people of southern Africa did not formally document their social and commercial activities as was done increasingly in the Roman world from the early centuries AD onwards, plenty of evidence has been found of mining activity, particularly in what is now South Africa.

The main sites were uncovered as a result of geological work carried out ahead of the establishment of new mines – two major South African copper mines, Palabora and Messina in the northeast corner of the country, were built on old mine sites. Palabora is thought to have been originally worked around 800 AD when azurite and malachite were the targets, and a network of chambers with linking shafts and adits was uncovered and investigated before Rio Tinto developed a new copper mine in the early 1960s, a mine that continues to operate today. Messina on the Zimbabwe border is also thought to have been originally worked around the same time as Palabora, and in the southern African region other smaller workings in the form of open pits have also been uncovered, although the dating is often less certain than for Palabora.

Earlier copper workings were found near Agades in Niger where it has been estimated that there were mines in the 1st millennium BC and perhaps before. The Kansanshi area of Zambia, whose modern operations we cover later, was also a mining site around the same time as Palabora. With no written records it is difficult to advance investigations much further than observing the evidence of mining activity in sub-Saharan Africa derived from old workings, basic smelting sites and metal objects, and carbon dating techniques used to assess the age of the evidence.

Of particular interest is the identity of the Africans who were involved in this ancient mining. The curiosity comes from a colonial belief that sophisticated commercial activity would have been beyond the abilities of indigenous Africans so there must have been migration from the north, the Arab world in particular, perhaps in the form of technical advice from Arab traders or perhaps in the form of a long disappeared civilisation.

This line of speculation is similar to that relating to ancient mining activity in North America where, as we have already noted, Indian lore does not seem to include information on historic mining in the Lake Michigan area where ancient workings were found. The conclusion, however, must be that at some time technically advanced African tribes occupied areas of southern Africa and mined copper – it is important to remember that Africa has experienced significant tribal movement throughout history (although critics of the continent's colonial period sometimes seem to suggest that the Europeans' arrival disturbed political structures that had existed unchanged for millennia).

Other ancient southern African mining sites have been found in Botswana and Zimbabwe, where copper artefacts including jewellery and beads as well as iron tools have been carbon dated, with some estimated to be from the end of the 1st millennium AD. There is, too, evidence of ancient mining in the techniques for treating low grade iron ores, observed around the turn of the 19th century in West Africa, today a key destination for customers like the Chinese searching for long-term iron ore supplies. The Hausas of Northern Nigeria used galena as a cosmetic powder, which simplified the search for minerals in the region in the colonial period. The development of this traditional adornment is thought to go back many centuries.

There is also evidence of gold mining in the West African region as early as the 5th century, and mining carried out by indigenous miners using techniques not dissimilar from those used elsewhere including fire setting to crack gold bearing ore. The mines themselves were largely relatively shallow pits, alluvial operations and also small-scale underground operations, perhaps going down 30 feet.

14. THE STRUCTURE OF OWNERSHIP AND OPERATION

Today the mining industry's operating and ownership structure is largely settled. There are some government-owned mines but mining worldwide is dominated by private ownership where capital is provided by shareholders and banks, and a board of directors sits on top of an administrative and operating structure.

National and regional governments set tax and royalty rates and also provide a legal base for the industry's operations through legislation. Governments may also make state land available for mining development. This more formal structure had been evolving since the Middle Ages as we will see later on. In the ancient world the situation was far less straightforward and thus less stable, although as we have shown the Roman Empire did have an organised structure for its mining industry.

In the ancient world there were two forms of mine operation – one related to the larger-scale mines which required the labour of slaves and where ownership was either in the hands of the Crown or aristocratic landowners. Where the mining operation was small scale the miners would usually pay a lease rent to the owner for mining the land, although it was not impossible for small miners to mine outcrops without recourse to ownership issues.

For areas like the mining of gold and silver, with their monetary aspects, the state often claimed all rights to any metal mined, and although rights to mine were given to miners the output would often have to be sold to the state at a price determined by the state. On the other hand, in Britain in the Middle Ages there are records of extensive purchases of lead in the commercial market by the Crown for the building of royal residences, indicating that the Crown, certainly in England, did not want to be involved in the mining process and was happy to fulfil its industrial metal requirements in the open market.

The profitability or otherwise of mines in the ancient world would have been affected by broadly the same issues as rule today – the cost of the inputs against the value of the metals extracted. Whilst some ancient mines operated for hundreds of years, there is evidence that on many occasions production was interrupted, often for years, indicating fluctuating financial returns.

The use of slaves in large mining operations would have obviously reduced operating costs considerably. For instance, the quarrying of stone to build the pyramids in Egypt would have been very labour intensive, as would the construction of the pyramids themselves, so a slave labour force would have boosted the economics of the venture considerably. The relative shallowness of most mines should have been helpful also and it is likely that the development of deeper mines was constrained by issues of engineering, water tables and geological factors, in addition to development costs.

As far as the economic importance of mining in the ancient world is concerned, strong data is difficult to find but it is possible to make some intelligent guesses. Agriculture and forestry were almost certainly the primary activities of most ancient economies, with manufacturing being confined to the essentials of life, including weapons and household items. Mining would have provided the raw materials for this manufacturing and also stone for building, but it is in the area of precious metals that its main impact would have been felt. Here the state had an overwhelming interest in obtaining gold and silver so that, in particular, wars of conquest could be financed.

The economic effect of acquiring large quantities of gold and silver is something we will come to later when we look at the Spanish conquest of parts of South America in the later Middle Ages. Suffice it to say at this stage that the huge inflation that accompanied the arrival of South American gold in Spain suggests that mining would have been a substantial industry in terms of its position within the economy. In the ancient world we could draw similar conclusions about the position of metals and minerals, including precious metals, in the economy of the expanding Roman Empire.

15. CONCLUSION

In the 21st century things move very fast; technological advances can be rendered obsolete even before they have achieved widespread use and science seems to have replaced religion as the fountain of truth. In other words our world is completely different from the one that our ancient ancestors lived in.

It is therefore not surprising how slowly things changed all those millennia ago and that the world of mining mirrored that slow pace of change. A number of mining techniques persisted for centuries, passing from Stone to Bronze to Iron Age. The practice of fire setting remained a widely used method of breaking mineralised rock, particularly underground, across these periods. Stone hammers, which were used to break and crush smaller pieces of rock, survived the Stone Age, and treatment of crushed ores was consistently done in open, and later enclosed, furnaces and in some cases crucibles, the fuel to fire these furnaces being charcoal. Some of the treatment techniques even survived into the modern era – hydraulic mining known to the Romans was rediscovered in the 19th century AD and the use of stored water to wash ore was another early development that survives even today.

Mining methods also continued basically unchanged throughout the ancient period. Pits were scoured out and shafts were painstakingly dug using basic tools of stone and then iron. Underground tunnels were driven, galleries dug out and wooden open drains to control water inflows were laid down.

The products that metals were used to make also hardly changed in seven millennia. Basically man needed weapons, protective metal products such as armour a little later, cooking and kitchen utensils (although clay/ceramic items were also used), tools, jewellery and, in the case of gold and silver, money.

As an example of this slow pace of development, we can look at the working of wood which, along with stone, was the most important construction material from the Stone Age to the Middle Ages. For 4000 years until the Bronze Age wood was cut using jagged flint tools; in the next 2000 or so years a hand-sawing blade made of metal emerged. With the coming of iron, saws began to cut stone as well as wood. As the age of the Industrial Revolution dawned the technology of woodcutting – exploiting water and then electrical power – began to change and improve. The speed of this technological change also accelerated as new metal compounds were discovered improving operating efficiency and productivity. Saws were also developed that cut metal as well as wood.

It is also the case in a world where life was often 'nasty, brutish and short' – and one might add also extremely unhealthy – that population growth was glacially slow, so that annual output from individual mines was correspondingly modest, allowing some metal deposits to last for centuries. Indeed, over the period of the Ancient World (roughly 6000 BC to 1000 AD), world population is thought by the US Census Bureau to have risen from 10 million to 300 million, roughly the same increase numerically that the World experienced in the three years from 2005 to 2008.

Mining conditions remained dangerous throughout this period – there was little incentive to improve things as it was common to use an expendable slave workforce in mines and quarries; indeed mining conditions probably only began to improve significantly in the 20th century AD.

A VIRTUALLY UNCHANGED INDUSTRY

This picture of an industry operating in a virtually unchanged manner for so long, supplying metal for the manufacture of products which also remained unchanged over centuries, mirrors the limited technological advances made during *prehistory*. Of course the rise of the Assyrian Empire and the period of Roman supremacy, which followed it, did lead to considerable technological developments but these were in large part lost when the Roman Empire crumbled and Europe in particular entered the Dark Ages. Apart from the Romans, the most advanced region or nation, particularly in the later stages of antiquity around the 10th century AD, was China, but its great technical inventions such as the compass and paper were not metal critical.

Having said that, the achievements of the Romans must have led some thinkers then to ponder whether they could take Roman advances further. The work of the Greek mathematician Hero of Alexandria in the 1st century AD – with his rotating cylinder or aeolipile powered by steam from a furnace connected to the cylinder – was the first steam engine and pre-dated Stephenson by almost two millennia. If the Romans had built on that who knows where technology and therefore metal demand might have got to, and a whole lot else indeed.

Certainly in the next period we will look at, the Middle Ages, the Florentine genius Leonardo da Vinci, showing the same sort of foresight, sketched out numerous futuristic developments including the aeroplane.

THE MIDDLE AGES TO THE INDUSTRIAL REVOLUTION (1066-1800)

1. INTRODUCTION

The ancient world presents a patchwork of contrasts. Over the centuries great technological strides were made by the Romans and the Chinese, only for massive geo-political events in Europe in the Medieval period – such as the rise of the Visigoths and the Vikings – to throw the whole process into reverse. So the developments in hygiene, water, heating and sewage that the Romans had tackled so successfully were lost and the consequences in terms of public health breakdown returned time and again over the centuries to threaten civilisation.

In such circumstances of uncertainty it is not surprising that technological innovation and thus economic advance were so modest. This uncertainty was materially fuelled by continuing political upheaval and in consequence almost permanent conflict. Thus for centuries the demand for metals was driven by the demand for weapons. As the Middle Ages advanced there were some signs that this baleful trend might give way to something more constructive, although as we know to our cost today war and conflict is never very far away, even in times of economic growth and relative peace.

It is accepted that historians disagree in their interpretation of events. In the case of defining the Middle Ages period, we have, as you will have noted, chosen our start date as the Norman invasion of Britain in 1066. As we stated in the previous section on the ancient world, the period from the fall of the Roman Empire to 1066 can be described as the Medieval period, although for many historians the period goes on for a further three centuries. Following that, Europe in particular started something of a revival based on economic growth associated with an accelerating increase in population.

The Middle Ages was also a time of resumed technological advance, which was particularly helpful to the mining industry where deeper mines and growing production required new ways of working to be developed. Initially miners in Europe in the early Middle Ages had enjoyed something of a golden period – in the 12th and 13th centuries miners in Italy, through the Treaty of Trent, and in Germany and England, had their rights to minerals they found, subject to royalties, enshrined in law. But the onset of deeper and bigger mines meant this phase of mining industry development by small scale owner-miners was relatively short-lived.

DEVELOPMENTS IN MINE FINANCE

At the same time as the mines got bigger, in an age where slave labour was no longer as widely used as in the past, these old miner-financed operations were in decline as more sophisticated forms of finance were developed. In particular in Germany (Saxony and Bohemia) mine finance using capital provided by non-participants emerged; the concept of buying equity in a mine, thus spreading the financing risk and drawing dividends from surpluses earned, caught on. The initial idea was that the shareholders would be close to the mine and meet frequently to be briefed as to how things were going. In due course the meetings got less frequent and the shareholders became more remote from the mine, a pattern for what happens today in publicly listed shareholder owned companies, and perhaps an inevitable development.

This new structure also meant that the old link between the miner and direct rewards for his efforts was broken. This historic arrangement had often had the backing of the mineral rights owner, aristocrat or landed gentry, and the miner in essence shared the rewards of his efforts with the owner. Once that structure had been changed the miner more often than not became an employee of the companies who increasingly owned and managed the mines.

SOCIETY AND TECHNOLOGY

The Middle Ages also contained events of great political and cultural importance – the Reformation and the Renaissance are two major examples of events that helped put Europe back on the map in terms of power and influence. But there were also great disasters to overcome. The Black Death in the mid-14th century unsurprisingly slowed progress right down and had a profound impact on economic activity for decades. Despite that, the march of technology was bringing Europe back to where it had been when the Roman Empire had been at its zenith. In mining, the use of water power for pumping and treatment had been refined to increase efficiency and on the treatment front the blast furnace allowed substantially increased throughput in the smelting process.

A body of literature and research on mining and metallurgy was also being built up during this period; perhaps the best-known writer was Georgius Agricola, who we highlight below. Other observers included two Arab alchemists whose westernised names were Geber and Avicenna and who lived and worked in the 9th and 11th centuries respectively; the 13th century philosopher Albertus Magnus who wrote *De Mineralibus*; and the 16th century Venetian alchemist Aurelio Augurelli.

Much of the thinking in those days was based on the theory of alchemy and held that metals, having been expelled from the core of the earth, were trapped in veins and fissures and then went through a process of cooling followed by steady transmutation into higher metal forms over the centuries – so base metals would eventually become gold and silver. These theories lasted until the 17th century when slowly but surely thinking began to change as the age of modern chemistry dawned.

2. MINING IN CENTRAL EUROPE

The German state as we know it today was largely formed in an administrative sense in the 19th century by Otto von Bismark, the Prussian Prime Minister. Before that Germany had consisted of a number of loosely aligned states, independent but at times cooperative. The travels of Georgius Agricola over many years, as he compiled his analysis of German mining in *De Re Metallica*, took him to a number of these states, although his particular emphasis seems to have been on his own Saxony.

Agricola was firmly of the belief that metals and mining lay at the heart of civilisation, stating in *Metallica* that 'man could not do without the mining industry, nor did Divine Providence will that he should'. He does, however, give houseroom in *Metallica* to the opposite view of mining opponents, whose arguments of despoiling and pollution, concerns about the safety of miners and thoughts about the *evil* products of metals such as weapons of war have a very modern ring. Interestingly many of the mines surveyed by Agricola were located in fairly remote mountainous areas and this led him to believe that their environmental impact was very limited; such a view would not be acceptable today.

Agricola divided *Metallica* into 12 separate books which covered, amongst other subjects, miners and mining officials, rock strata, mining and surveying, mining equipment, assaying, treatment of both ore and metals (particularly precious metals), management and finance and the manufacture from minerals of basic compounds such as salt and glass. Agricola does not always identify the mines he refers to, perhaps partly because he did that in earlier works such as *De Veteribus et Novis Metallis*, but he was from Saxony and he does name a number of towns that benefited from the development of mining. These were primarily Saxony towns and included Freiberg, Annaberg and Marienberg.

Many of the mines that Agricola refers to had been in existence for many hundreds of years – the Freiberg silver mines, the Goslar lead mines and the

Schemnitz silver mines. As we have noted elsewhere, one of the features of ancient mines can be their extraordinary long lives and that is a function of the low levels, by today's standards, of production in the pre-industrial era. In this regard the disasters in the 14th century, particularly the Black Death, also had a major and long lasting impact on European metal production and demand. However, with the coming of the industrial revolution mine output took off almost vertically, consequently exhausting historic mines very quickly.

GEORGIUS AGRICOLA (1494-1555)

Georgius Agricola is one of the most influential figures in the history of mining. His book *De Re Metallica* was the core textbook on mining and metallurgy for over two centuries. It remains today a remarkable body of work, both historically significant and an unrivalled description of mining in the Middle Ages and before. In 1912 it was translated into English by American mining engineer Herbert H. Hoover and his wife Lou; Hoover subsequently became President of the USA.

Born Georg Bauer in Glauchau, Saxony, in 1494, Agricola went to Leipzig University and after graduating in 1518 he taught Greek and Latin in the city of Zwickau. He went back to teach at Leipzig but intrigued by the Renaissance he travelled to Italy where he became interested in science and obtained a degree in medicine in 1526, and where he also became known by the Latin translation of his German name. In the late 1520s he returned to Germany to Joachimstal where he took up the position of town doctor. It was at this time that Agricola became interested in mining, having invested with some success in the Gotsgaab silver mine which provided him with a lifetime income. He made an extensive study of mining techniques and metallurgy in the region around Joachimstal, which was a major mining centre.

In the 1530s Agricola moved to Chemnitz to take up another medical appointment and it was around this time that his first book on mineralogy, *De Natura Fossilium*, was published. He continued his study of mining and metallurgical techniques in tandem with his medical practice and began work on *De Re Metallica*. In 1543 he married Anna, a widow, whom it is thought may have been his second wife. Agricola had at least three children by Anna and reference is made in Joachimstal records to other earlier children.

During this period he also found time to serve as burgomaster in Chemnitz, a considerable tribute to his standing as he was a staunch Roman Catholic in a Protestant country. He died in 1555 having completed *Metallica* a few years before and the book was posthumously published a year later. It was his last and greatest work; in total he wrote more than 20 books and pamphlets covering mining, metallurgy, medicine and religion. Anna and at least three of his children survived him and were still living in the 1580s.

De Re Metallica is comprehensive in its scope, covering hundreds of mining operations that Agricola had observed or studied. The book also includes scores of highly detailed woodcuts illustrating aspects of mining and metallurgy, and some of the first observations on geological strata and how rocks occurred in layers that could be traced for many miles. It is an historic and formidable work that stood the test of time, two centuries in fact, and it remains an invaluable work for studying the mining techniques of the Middle Ages today.

SILVER

Whilst both England and Spain had historically important mining industries, central Europe was also a key mining area, encompassing the states of Germany and other countries such as Hungary and Italy. In the early 13th century near the city of Schemnitz, 60 miles northwest of Budapest, silver deposits were developed and production may have reached 5 tonnes a year. The mines continued to be substantial producers until the end of the century and other silver mines were established in the region, but Hungarian production declined as the century waned, setting off a period of silver shortages and rising prices.

The strong silver price also stimulated production in other parts of Europe – for instance silver mining near Coombe Martin in Devon in England was revived. Silver was important because of its critical role as coinage and fluctuating supplies did create considerable problems in terms of financial liquidity, which could lead to substantial fluctuations in price levels if supply expanded or contracted rapidly. It was for this reason that gold became an increasingly sought after metal, and production in the 14th and 15th centuries rose in order to fulfil this role, with mines in Transylvania and Slovakia, then known as Upper Hungary, being developed, with output totalling around 4 tonnes per year.

Perhaps the most important silver discovery in the 13th century was made near Kuttenberg, now Kutna Hora in the Czech Republic. The find generated a lot of interest and something of a rush developed with some chronicles suggesting that at least 10,000 miners had been attracted to the field; other figures were even higher. At its peak in the late 1290s Kuttenberg produced around 200,000 ozs of silver annually but production soon tailed off, although silver continued to be mined for a further 100 years. Attention then switched to a new discovery in the region at Pribram.

When output at Pribam faded too, attention turned in the first half of the 14th century to silver deposits much further south on the island of Sardinia where deliveries to the mint at Villa di Chiesa, which ran at just over 200,000 ozs, helped to make trading in the region more liquid. At the same time silver was also coming from areas to the east in Persia and Afghanistan, suggesting not only favourable regional (in the widest sense) geology but also well-trodden trading routes, and continental shortages. This latter situation (as we mentioned above) led to increasing interest in gold mining and the use of gold in coinage, as well as new sources of silver.

Another problem for European silver miners was the steady fall in silver grade mined over this period, with 100 ozs per tonne of ore common in the 12th century but 200 years later grades of 10 to 15 ozs per tonne were not untypical. Of course grade reduction is something very familiar to today's miners, and also familiar today is the circle where falling grade leads to falling output, which in turn starts pushing the price back up, leading to a re-examination of the viability of old mining areas. This started to happen in the 14th century in the Balkans where in one case the silver mine at Srebnica in Bosnia that had produced around 50,000 ozs per year in mid-century produced a paltry 5,000 ozs in the late 14th century. As the silver price waxed and waned over the decades silver mining tended to shift away from Hungary to Balkan countries like Bosnia and Serbia.

In the second half of the 15th century Balkan silver and gold production was disrupted by the invasion and occupation of the region by Turkey. In the following century Turkey's grip on Balkan commerce weakened, enabling silver mines to sell into buoyant markets, with Bosnian and Serbian production soaring to almost 500,000 ozs in the 1520s. Gold output from central European, Balkan and African mines exporting to Europe was also on the rise and by the mid-15th century had reached 200,000 ozs. However, there was a price for this growth as some of the treatment processes were rather hair-raising, particularly when mercury was used to attract gold and then evaporated off causing air pollution that was dangerous to the health of unwary process workers.

LEAD, TIN AND COPPER

Silver and gold were not the only targets, and deposits of lead, tin and copper were also mined in central Europe during the 13th century, with more copper being found in the Harz Mountains of Germany where silver mining had re-started in the 10th century, tin in Bohemia and lead in Poland. Output, however, began to decline later in the century, and in the 14th and 15th centuries European consumers had to turn to Turkey and the Middle East. The Balkans, Austria and Slovakia also developed copper mines as demand rose in European markets, and these areas became the main suppliers as the 15th century progressed.

One of the new uses for copper at the time was as copper plate for engravings as the printing industry evolved. At the same time there was a revival of lead and silver output from old mines in England located in the Tamar Valley in South West Devon. Here new pits were dug and old shafts de-watered to re-start mining; for a few years this thrust England back into the picture as an important source of metals. The Devon output was primarily silver, but lead mines further north in Derbyshire provided a flow of that sought-after metal to central Europe.

DEVELOPMENT OF MINING TECHNIQUES

Mining techniques were also changing as miners began to dig deeper in pursuit of plunging orebodies with attractive grades at depth. Such orebodies were often below the water table and required de-watering – something, as we have seen before, that was historically done by hand. The coming of waterwheels, then steam power and pumping, meant that this labour intensive and increasingly expensive task could be mechanised and made more efficient and effective. Basic gunpowder also began to be used in the mines in the Erzgebirge Mountains of Hungary in the 17th century to blast the ore from the mineral-carrying seams, replacing fire setting, the traditional method of breaking ore that had been used for centuries.

3. SCANDINAVIA

For much of the Middle Ages Scandinavia was divided into two blocs, the political union of Denmark and Norway, and the kingdom of Sweden, which fully incorporated Finland. During this period Norway became a significant mining and quarrying region with the 16th century witnessing a surge in developments, in part due to the revenue needs of the Crown.

NORWAY

One of the major mining regions was Telemark in the south of Norway and it was here that the Gullnes copper mine was opened in 1524. In 1537 control of the mine passed to the Crown, which imported a number of German miners from Saxony to improve the mine's efficiency. The locals of Hjartdal, who coveted the rich silver lodes that existed in the area, were unhappy with the German presence and the privileges granted them by the Crown, and eventually ejected the miners, who fled back to Saxony. This presumption on the part of the locals led to the Crown prosecuting and hanging the ringleaders.

Although the putting down of the revolt in Telemark was hardly likely to endear the Norwegians to mining as an industry, especially as the King in Copenhagen was behind the introduction of German mining laws and methods, the expansion of mining continued for over two further centuries. By the middle of the 16th century iron ore was being mined in Hadeland, north of Oslo, and ironworks were built in Skien, Telemark, and at Hakadal near the iron ore mines. Copper mines were also established at Ytteroy, Trondheim Fjord, in central Norway and mining of phosphate continued into the 20th century. Two other major copper mines, established in the mid-17th century in the central part of Norway, were the Kvikne and Rorus mines – the former operated for 200 years before being flooded. Other copper mines were opened at Trondelag, also in the central part of the country, and smelters were built near Oslo, the first one at Baerum.

The biggest mine was the Lokken Verk copper mine, developed in 1652 it continued in production until its closure in 1987, by which time it had become primarily a pyrites operation. The mine and its profits formed the historic base for today's giant Orkla Group. Although for many decades copper was the main metal mined in Norway, in the early 17th century silver was discovered in Kongsberg in the south of the country; silver mining continued there until the mines closed in 1957, having produced an estimated 48 million ozs over the period. As the Industrial Revolution loomed in the 18th century, further copper mines at Selbu in Trondelag and at Folldal in the east of Norway were developed, and cobalt was also discovered. The economic impact of Norway's mining industry was widespread; as well as providing raw materials to Denmark and Norway, mine products were also exported to Germany and to the Netherlands, with Amsterdam a major customer of Norwegian stone.

SWEDEN AND FINLAND

There was also a vigorous mining industry in Sweden and Finland in the years before the Industrial Revolution. Iron ore has been mined in the Bergslagen district in the centre of Sweden since the 1st century BC, but it was in the 16th century that larger-scale iron ore mines were developed and at the end of the 17th century iron ore was discovered in the north of Sweden in the Kiruna region by Samuel Mort. These deposits continue to be mined today, 300 years later.

Some base metal mines, amongst these the Garpenberg and Zinkgruvan copper, lead and zinc mines, which are also still in production today, were established around the same time. The former had been originally worked in the 15th century and was restructured in the following century with the arrival of German miners. The Falun copper mine in Bergslagen has probably been mined since the early 11th century and a share issue took place in 1288 making Falun the oldest joint stock company in recorded history. In the 17th century, when Gustavus Adolphus was attempting to unify and conquer Europe, it is thought the Falun mine was responsible for almost 70% of known world copper output. It continued to operate until closure in December 1992 after around 1000 years of life. The 14th century also saw the start of silver mining at Hornkullens near Filipstad in central Sweden, an activity that continued until the 19th century.

For most of its history Finland was an agricultural country, sparsely populated but with basic skills in iron working due to the incidence of surface iron ore. In 1540 the first Finnish iron ore operation, the Ojama mine, was established in Lohja in the south of the country following Finland's integration into Sweden. Finland had substantial iron ore reserves and also huge waterpower potential and massive forests to provide it with charcoal for the iron treatment furnaces, and this led to the Swedish Crown and private interests establishing iron works in Finland to convert its ore resources into iron.

By the middle of the 17th century, mining of Finnish iron ore had become unprofitable and Finnish ironworks began to import Swedish ore. After many decades as a great European power, Sweden faded, and at the end of the 17th century famine hit Scandinavia. Finland was particularly badly affected and also had to contend with aggressive Russian actions during the Great Northern War in the first two decades of the 18th century as Baltic states sought successfully to put an end to Swedish hegemony in the region. When peace returned Finland was unable to sustain its iron industry and Swedish capital took control. However, as the 18th century ended copper was discovered in Orijarvi to the east of Helsinki, promising a new era for Finnish mining.

4. FRANCE

With the ending of the Dark Ages, France, no longer under the thrall of Rome, began to recover and as a result a number of iron-based industries emerged, including weapons, armour and wrought work. By the 12th century coal had been discovered and began to be mined in the region east and west of Liege, where coal mining continued well into the 20th century.

One of France's oldest mines was the Salsigne gold mine near Carcassonne in the southwest. The mine was first exploited by the Romans in the 3rd century, but as an iron mine. The furnaces were fired by charcoal and wind tunnels were used to power them, another example of Roman technology anticipating the modern world. The Romans, however, did not realise that there was gold in the slag that was discarded in the iron-making process, perhaps as much as 1 million ozs, and now lost forever under the town which was eventually built on the mine dumps. During the Middle Ages Salsigne was mined intermittently for its iron ore, but it was only after a period as an arsenic or arsenopyrite mine at the end of the 19th century that gold was finally discovered, and during the 20th century 100 tonnes of gold and 300 tonnes of silver were produced at Salsigne.

By the time of the Middle Ages mining had begun to expand in Europe, with the negative influence of the Dark Ages on social and economic activity beginning to wane. The old silver and copper mines of Aveyron in south central France were re-opened in the 10th century and were worked for some centuries following. The main target for this revival of mining activity in France was silver, which was increasingly in demand for coin as commerce began to revive. Other areas that saw a surge in mining were Lorraine in the northeast, the Auvergne in the centre of the country, and the Pyrenees, where silver and gold were the main metals sought.

In the 14th century the lead and silver mines in the Languedoc in the southeast at La Caunette were re-opened and to the northwest in the Cevennes alluvial gold was mined in the region's river courses. Mines in the Pyrenees had been worked intermittently for centuries for lead and silver by both the Romans and the Moors, and in the early 17th century under Henry IV of Navarre mining resumed at these ancient sites also. It is believed that the mines provided huge profits for both the King and his brother who oversaw the mining operations.

Mining has never been a natural activity for the French, who are far more comfortable with the agrarian way of life, and in the 13th and 14th centuries the mines found themselves the target of rising rental claims from the landowning nobles who were often at odds with one another, inserting another layer of uncertainty for the mines. The mines also had to battle against rising operating and labour costs, the latter being particularly affected by an end to the system of forced labour.

THE PARIS BASIN

Perhaps in hindsight one of the strangest mining areas in Europe was the Paris Basin, where a considerable number of mines were developed in the Middle Ages in the area of what is now substantially the city of Paris. The minerals won from these mines were not the glamorous ones of gold, silver and copper but were industrial minerals such as gypsum and building stone for Paris's expansion. Indeed, it is the gypsum deposits of Montmartre that gave the name to the well known product *plaster of Paris*, and mining of that gypsum was first recorded in the late 14th century. In the 13th and 14th centuries stone quarries were developed well away from the old city centre and they were often accessed by digging a large shaft down and then mining the stone horizontally. This continued until the 15th century when the city began to expand towards these previously mined and now forgotten areas.

As the decades went by major buildings such as the Observatory and the Val de Grace church in the south of Paris were built on top of the old quarries and mines, as were a number of roads going south towards Fontainebleau and Versailles. In time cave-ins occurred, sometimes with considerable loss of life, and in the end the City had to survey the old mine workings, map them and secure them. The underground mines of Paris are sometimes referred to as the Catacombs, and though most of them are legally off limits today not everyone pays attention; unofficial visitors to them are known as *cataphiles*.

5. GREAT BRITAIN

Until the 20th century Great Britain was a major producer of metals. Initially, in the Middle Ages, it was an exporter, but it turned consumer as the Industrial Revolution broke and then finally it became a major importer in its role as the world's leading industrial power. In the Middle Ages England was on the edge of Europe, a feisty independent country with an historical relationship with France going back to the Norman Conquest, in the aftermath of which Britain inherited power over huge swathes of France which it gradually lost.

In due course continental powers like Spain, Holland and France came into conflict with England, which due to its powerful navy was becoming increasingly outward looking, a match militarily for anyone wanting to make trouble, and confident enough to start expanding its global reach through land acquisition and eventually colonisation. This latter development was a key part of the economic imperative to find natural resources to fuel the Industrial Revolution.

The importance of silver in the Middle Ages in its role as specie or coin led to a surge in silver mining, as we have observed earlier. England as an historic source of the metal became a serious producer based on the revival of its past operations in the southwest and new mines developed further north. Following the Norman Conquest in 1066 England's silver mining industry was to be found in Somerset's Mendip Hills, the Derbyshire Peak District, on the Welsh borders and in the North Pennines, which was the location of the most productive mines.

The ownership structure, which had been based for centuries on who owned the land where the silver was found, changed during the 13th century when the Crown claimed control of the silver, and therefore of the mines being developed in Devon at the end of that century. In this still feudal era there were also surprising examples of self-employed miners who leased mines, which they worked for personal returns. In the earlier part of the Middle Ages the mines were in the main small hand dug pits, but as we have seen elsewhere the veins often plunged and so shafts had to be sunk to follow the mineralisation. These shafts were sunk to the level of the water table where the miners had to de-water by hand. In due course rising silver prices on the continent financed the development of mechanised de-watering systems in the late 15th century, often using wheel and bucket lifting methods. There was also an environmental cost to the re-opening of the silver mines of south Devon, namely the stripping of woodland for fuel to be used in the treatment process and to provide mine timbers.

Silver and lead mining continued off and on in Devon well into the 18th century, as did mining in the North Pennines, and new mines were also established in mid-Wales. By the end of the 18th century mining had begun to extend below the water table in Devon as technological advances led to the development of steam-powered pumps capable of greatly exceeding the pumping rate of traditional bucket de-watering methods.

As far as treatment processes were concerned, new smelting techniques gave rise to changing fuel types, from traditional charcoal, on to peat where available, and then later in the 19th century coal, helped by advances in smelting technology, became the prime fuel. This also led to the development of treatment complexes close to the expanding coalfields.

6. SPAIN

Whilst the Spanish were involved in acquiring gold from their South American conquests, at home mining of gold continued along with the mining of other minerals, largely from historic mining districts, some of which had been active for several thousand years. Permission to mine coal in the Seville region at Villeneuva del Rio was given by the Crown in 1621. Initially individual miners worked the coal seams but in 1740 the Royal Artillery Petty Officers of Seville took over the mines as rising industrialisation and therefore rising demand led to a need for more integrated and efficient mining practices. After further changes of control in the 19th century, the mines were nationalised in 1969.

The iron ore mines of Viscaya in the north at Las Encartaciones were worked throughout the Middle Ages, as were the iron ore deposits of the Valle del Sabero in Leon in the north. The gold mines at Rodalquila in Almeria and the iron mines of Ojos Negros in Aragon were other ancient operations which were mined right through from the medieval period to the modern era, as were the Almaden mercury deposits in Castile-la Mancha, the salt mines of Iman in Guadalajara, and the coal mines of the Valle de Sabero in Leon. Many of these areas crop up again later when the Industrial Revolution led to a surge in demand for minerals throughout Europe.

Particularly unusual was the Almadèn mercury mine, which is believed to have operated for an unbroken two millennia, although there is a view that it may have been exploited as long ago as the 6th millennium BC. The deposit contained cinnabar, from which liquid mercury was produced by a reduction process. The first recorded reference to Almadèn was in the 5th century BC, and during the Roman occupation of Spain the mines were extensively exploited. With the fall of the Roman Empire there was no further reference to mercury mining in Spain until 711, when the Arabs overran the country and operated the mines until their departure in the 12th century. The mine then came under the control of the Spanish Crown. In ancient times mercury was used as an elixir of life and even as a treatment for broken bones, later it was used as an ointment and in cosmetics. These are all uses that, bearing in mind the great toxicity of the metal, seem curious to modern minds.

The conquest by Spain of Latin America in the 16th and 17th centuries led to a massive boost in Almadèn's fortunes as the discovery of particularly silver, but also gold, deposits led to a major increase in demand for mercury as an amalgam to separate precious metals from their ore. Although presenting grave health risks to workers who operated the treatment process, the demand for Almadèn's mercury in Spain's New World colonies rose materially. By the 18th century, local mercury was being used to treat Latin American precious metals but Almadèn continued to prosper on the back of Spanish demand, indeed to such an extent that the Spanish government used the mine as a guarantee in receiving loans from, in particular, German financiers in the 16th and 17th centuries. Later the Rothschilds made similar loans to Spain backed by Almadèn.

With the advance of science, mercury found new uses in inventions such as thermometers, barometers and even in dentistry for fillings. In the 20th century it also was used in a wide range of chemical products and munitions. More recently its use in many of these areas has been phased out as governments have become concerned about its safety, but ironically the development of energy-saving fluorescent light bulbs has led to a recovery in mercury use.

As to Almadèn, it prospered in the 1940s when production reached over 80,000 flasks per year but things became more difficult after the Second World War and the mine was re-organised by the Spanish government in the 1980s, only for it to close in the early years of the new millennium as mercury demand fell away. It is calculated that over history the Almadèn mine has produced around one-third of all the mercury ever produced, and the Spanish government is seeking UNESCO World Heritage Site status for the mine.

7. THE LURE OF AFRICA

A considerable amount of the precious metal trade in the early Middle Ages between Europe and Africa occurred in the Mediterranean region, reducing the need for Europeans to travel to the African continent. However, although the European colonial surge into Africa still lay some centuries ahead, there was a thriving slave trade going on between West Africa and the Middle East at this time, and African riches, particularly in gold, were becoming quite widely known to the traders from the Middle East. The mines were primarily in the Sudan and West Africa and the trade routes, which carried the gold to Europe, crossed the Sahara. The routes going east to supply traders from Egypt went in the direction of what is now Somalia. During the Middle Ages the trading routes and thus the sources of gold were prone to change caused by material alteration in climate in the region due to temperature and rainfall fluctuations.

Many of the West African gold operations were alluvial, where indigenous miners panned the River Niger and its tributaries for gold as the waters swept down from the mountains of Sierra Leone. This area was a significant source of gold and some gold veins were sufficiently rich to warrant exploitation by shaft access. Other alluvial workings existed along the River Senegal as it wound its way to the Atlantic from its source in Mali. Mali is now one of Africa's largest gold producers and there is evidence from tools and pebbles discovered close to the operating Syama gold mine at Tembelini in the 1990s that there were gold diggings in the area in the 15th century. Although official figures for regional output do not exist, there is data relating to the value of gold trading, and in the 15th century gold passing through the Sudan amounted to between 30 and 50 tonnes per year.

In the 10th century copper mines were opened up in the Atlas Mountains area near Sijilmasa in Morocco, fairly close to what is now the border with Algeria, and the ore from those mines was treated at Igli in Algeria. That output was exported south to West Africa including Ghana – a rich kingdom and another major source of gold, known as the Gold Coast at the time – and east to the Sudan. Two of the bigger copper mines in the region were Tihammamin and Ten Oudaden. There was another important copper mine at Takkeda, situated in the Kingdom of Mali, whose product was also exported all over the region and to Egypt. One of the most important cities in Mali, which acted as a key entrepot for trans-Saharan trade, especially in gold, was the legendary Timbuktu, which as well as a commercial centre was also a centre of the arts and learning.

The main gold mining areas of Ghana were the Akan states in the southern half of the country, such as Asante, home of the modern Obuasi mine, and Wasso, now home of the Tarkwa mine. There is evidence that gold was being mined by indigenous Ghanaians at least as early as the 12th century and when the Europeans began to make landfall in West Africa from the 15th century onwards they found a thriving gold mining industry in Ghana. Over the next three centuries European traders regularly visited the Gold Coast to trade in the metal. With the coming of the Industrial Revolution and the colonial rush for Africa indigenous gold miners were slowly pushed aside as European miners began to apply new technology to developing mines, allowing them to mechanise their process and sink deep shafts to access rich ore that lay at depth.

Much of the gold and copper mined in the Middle Ages was used as coin, and supply levels could, of course, have a profound effect on the price of both the metals when traded for goods. Also, in stark contrast to today's world of money printing, coin in circulation could fluctuate materially over time, which could have a serious, usually uncontrolled, effect on local economies. When metals were in short supply coins either had to circulate faster, with inflationary implications, or if they circulated too slowly, sometimes due to hoarding, then economic activity dropped. As gold and silver flowed backwards and forwards between Europe and Africa during the Middle Ages the economic fortunes of trading countries such as Egypt, Syria, Hungary and Italy waxed and waned.

Producing regions were not spared in this process either. We can take an example from the mid-15th century when a shortage of mercury, which was key to the treatment process for gold, developed in Spain, causing severe economic pain as gold production slumped. This not only hit economic activity in both Europe and Africa but also led to debasement of the currency as precious metal content in coins was reduced to allow more coins to be minted.

8. THE OPENING UP OF NORTH AMERICA

Elsewhere we have talked about the discovery of ancient workings of copper in the Michigan Peninsular of the US and drew attention to the paucity of provenance as to who developed the workings and when. The earlier explorers who arrived in North America came across plenty of evidence of use of stone and metal for weapons and working implements such as axes and knives.

Copper artefacts and even jewellery were also found, suggesting that American Indians were able to work metal but that the raw material was native copper, mainly found lying on surface, suggesting no mining tradition amongst the continent's indigenous people. Indeed Sebastian Cabot and Giovanni Verazzano came across Indian women wearing copper when they landed in Newfoundland and Nantucket Island respectively around the turn of the 15th century. These sightings of copper continued as explorers continued to probe the east coast of North America over the following centuries.

THE SPANISH ARRIVE

The vastness of what is now the continental USA meant that without disturbing what was going on in the northeast explorers entered the dry wastelands of the southwest from the Caribbean and Mexico; these visitors were of course the rapacious Spaniards (of which more later), as ever searching for gold and silver, and later we will look at their activities in Latin America. However, in the 16th century from their base in Mexico, Francisco Coronado, Alvar Vaca, Antonio Espejo and Juan Onate all led expeditions north into what are now Texas, New Mexico and Arizona.

Rumours had circulated of gold in large quantities and Espejo headed into Arizona driven by a story of a lake of gold; whilst this proved far-fetched he did stumble across gold in rich veins near what is now Prescott, and this was followed by the arrival of Onate to stake claims in the area. This did not lead anywhere as it proved impossible to coerce the native Indians to assist the Spaniards, or Mexicans as they had by then become, into mining the veins. But though the evidence of gold in the region was patchy there was more substance to the belief that the region was prospective for copper, as stories abounded of copper items being found in the possession of indigenous Indians.

NEW MEXICO COPPER

One of the more substantial copper mines established in New Mexico was a mine at Santa Rita in the late 18th century. Over the years this mine was owned by Mexicans, Americans and even Frenchmen, the ownership regularly passing into new hands as the opposition to the operation, often violent, of local Apaches, which only ceased in the 1880s following the establishment of Apache reservations, rose and fell. By the start of the 20th century the ownership of the Santa Rita copper leases had passed through Anaconda Copper into the hands of Chino Copper, and the operating mine was now the El Chino open pit, which was taken over by Kennecott in 1933 where it remained until acquired by Phelps Dodge in 1986. El Chino still operates today and it is estimated that since the start of production in 1801 the mines developed around Santa Rita have produced $2 billion in value of copper and precious metal by-products.

Sadly this period of time was hardly a glorious one for the Mexican and American adventurers who penetrated the southwest in terms of their dealings with the Indians. Not only were the Indians usually reluctant to help the incomers over their mining needs, they resented the exploitation of raw materials from lands that they claimed were theirs, as we have seen above with the Santa Rita mine. Jesuit missionaries in Texas did not help foster relations with the indigenous Indians either and they were expelled by the Spanish crown in 1767, partly to encourage the Indians to help mine the rich silver lodes found in the area.

SEARCHING FOR GOLD AND FINDING BASE METALS

As with Latin America, the lure of North America for many explorers was the possibility of finding precious metals, but in this respect they were largely unlucky as in due course we know that gold was found out west, and in the east it was found over the Canadian border in central Ontario and in Quebec. It was copper that unsurprisingly kept on turning up – in Connecticut in 1707, in New Jersey in 1719 and again in New Jersey in 1750.

In colonial times metal discovered in the American colonies was usually shipped back to the UK if it needed smelting, which made for sometimes unpredictable economics. After independence one of the earlier copper mining and smelting complexes was established in Vermont in 1820 and in 1853 the Vermont Copper Mining Company was incorporated. It became one of the primary sources of copper for American users until the discoveries in Michigan set off a string of major discoveries across the Midwest, which we will come to later.

IRON ORE

As well as copper there were early discoveries of iron ore, the first being in North Carolina in 1585 by Thomas Heriot, who was on an expedition dispatched to the New World by Sir Walter Raleigh. The iron ore was useful to the growing colonies as their blacksmiths needed it to make tools and nails for construction. A small shipment of ore was sent back to England from Virginia in 1608 as ballast and was smelted in Bristol. A few years later an attempt was made to establish ironworks in Virginia for the manufacture of tools on a significant scale but local Indians destroyed the works established near Jamestown.

During the 17th century iron ore was found in a number of locations in the colonies – Massachusetts, Connecticut, Pennsylvania and Kentucky – and ironworks were established. One of the first was at Lynn, Massachusetts, built by John Winthrop Jr. who formed the Undertakers for the Iron-works Company in 1643. One environmental issue that had already raised concern in England with the growing iron industry was the rapid denuding of woodland to provide charcoal for the smelting process. The age of coal was about to begin, and so it was in the colonies as the cutting down of forests became equally unpopular.

The American colonies were settled with agriculture as their main economic activity, which perhaps explains why for many years the development of industrial enterprises, like ironworks, lagged behind Europe in technological terms, particularly behind the *old country*, as the Industrial Revolution loomed, and this continued beyond the achievement of independence.

In the early 18th century American iron making was based on blooms – wrought iron lumps – which could be beaten into implements on blacksmiths' forges. Cast iron working followed but operating problems due to the lack of waterpower in the summer meant that continuous working had to await the arrival of steam power at the start of the 19th century. There was also the problem that the British government sought to control iron production and encourage the export of iron ore to England for smelting; with American independence that practice ended.

The first iron ore used in Massachusetts came from bog ore – ore that had accumulated in ground hollows which had often then become ponds or small lakes. In due course larger accumulations of haematite and magnetic ores were discovered. At the end of the War of Independence every US state had some form of ironworks, although in terms of production they were still small and weekly output from each works was little more than 25 tonnes, an insignificant level compared to output in Europe. A century later the US was producing as much steel as Great Britain, Germany and France put together as huge iron ore and coal deposits in the Midwest were discovered and developed.

JOHN WINTHROP JR. (1606-1676)

John Winthrop Jr. was born in Suffolk in England, the son of John Winthrop who eventually became the Governor of Massachusetts. Winthrop Jr. was educated at Bury St Edmond's Grammar School and then at Trinity College, Dublin. On leaving Dublin he studied law in London for a short time in 1624, but then became something of an adventurer in the 1620s, joining the Duke of Buckingham's unsuccessful expedition to release the beleaguered Huguenots in La Rochelle on France's Atlantic coast. Thereafter he travelled around the Mediterranean.

On his return to England in 1629 he found his father had been appointed Governor of Massachusetts in the New World. Winthrop Jr. remained in England to look after his stepmother and siblings, and his father's business affairs. In 1631 he followed his father to America with the family and founded the town of Ipswich in Massachusetts, but was then encouraged by his father and interests in London to take up the position of Governor of Connecticut which led to his return to London to obtain patents from the Crown for the establishment of the colony of Connecticut.

He succeeded in obtaining the Crown's permission to establish the colony and on his return to America he became excited by the potential for minerals in both Connecticut and Massachusetts. He spent time in Massachusetts in the late 1630s persuading the primarily agrarian inhabitants of the colony to take an interest in developing the mineral potential of the area. Winthrop Jr. established two ironworks in Massachusetts, at Braintree in 1644, the first in the American Colonies, and then a bigger one at Lynn, both using locally mined bog iron. He then went back to Connecticut and continued his new role as a mining and metals entrepreneur building ironworks at North Branford. He also established a granite quarry on his land at Waterford on the Connecticut coast. Winthrop Jr. was known as America's first commercial iron maker, and due to his efforts Connecticut earned the title *cradle of American mining.*

But Winthrop Jr. was much more than a political figure and mining developer; he was a man of science also and very interested in alchemy. This interest meant he was curiously in demand over matters of witchcraft, something of an obsession in colonial New England. In the latter years of his life he spent a lot of time fulfilling the duties of Governor of Connecticut and in 1675 he became one of the

Commissioners of the United Colonies of New England, which merged Massachusetts, Plymouth, Connecticut and New Haven. He was also elected a Fellow of the Royal Society in London and contributed a paper on maize, another area of his expertise as he owned and operated a grist mill in New London, Connecticut, where he had a monopoly over the milling of maize.

Winthrop Jr. married a cousin, Mary Fones, in 1631 and she accompanied him and the other Winthrops to Massachusetts in 1633. Unfortunately Mary and their baby daughter died in the following year. The tragedy temporarily saw him back in England where he met and married his second wife, Elizabeth Reade, in 1635. They had nine children, one of which, Fitz-John, followed father and grandfather as a colonial governor.

Winthrop Jr. was a man of many parts, being interested in science, medicine and astronomy. Indeed he acted as a physician in Connecticut for a number of years and on the astronomical side claimed the sighting of a fifth moon around Jupiter many years before it was confirmed through the work of others. He also carried out experiments to mine salt from the ocean by evaporation. He died in 1676 of a cold caught in Boston where he was attending the New England Convention.

The Winthrop family is one of the exclusive group of rich upper class Americans who trace their heritage back to the founding days of the American colonies and beyond that back to England and even the Normans. Today the rich founding families are often seen as outposts of American snobbery whose past is perhaps more glorious than their present. John Winthrop Jr., dedicated administrator and politician, mining entrepreneur and scientist, a man of brilliance, underlines the difficulties for his ancestors in living up to the family's past glories.

MINING FOR LEAD

Further to the west in what was then French influenced territory, the search for minerals uncovered lead deposits. In the 17th century French traders such as Louis Joliet and Robert de La Salle made journeys from Quebec which took them down the Mississippi River, eventually as far south as the Gulf of Mexico, and in the process the French traders observed outcropping lead ore which resulted in the opening up of the lead mines of the Mississippi Valley. These were amongst the earlier fully documented mining developments in what was to become the USA.

The French explorer, Henri Joutel, recorded the existence of lead mines in the northwest part of Illinois in the late 17th century. It is also thought that local Indians had come across free lead and copper and not knowing how to work it had used it as adornment. With the coming of the French they soon picked up the idea of mining the lead ore and smelting it in hollowed-out tree stumps or in campfire hearths.

The reason for French interest in lead was its use as ammunition for the rifles of traders and trappers. One of the earlier Mississippi Valley lead mines was developed by Nicolas Perrot, a French explorer and trader, who was an important figure in the history of French colonies in North America as a diplomat working with the indigenous Red Indians. Perrot also had some knowledge of mining techniques and instructed Indian workers how to dig inclines into the orebody, thereby making it easier to set fires in order to fracture the ore and remove the richest material, a technique that should be very familiar to us by now.

Perrot was followed by Pierre le Sueur, another French explorer and trader, who had mistaken green sand for copper ore in an earlier loss-making deal. Le Sueur had hoped to further develop the lead mines in the area, including Perrot's, but after initial work they were deemed to be uneconomic. Other lead mines, however, were developed by both the French and the English, including Mine la Motte in Missouri's Ozark Hills. La Motte was eventually run by Philippe Renault, who came to America in 1719 and whose father was an iron founder back in France.

Mining in the Mississippi Valley was affected by Indian action from time to time, and for some decades lead mining was concentrated in Missouri, but by the end of the 18th century interest had returned to the region, including Illinois and Wisconsin. The treatment process remained fairly crude, using logs to shape a large fire and then melting the mined material in a large bowl within the structure. The resulting metal was then made into bars and often shipped to France.

Other French lead mine owners included Julien Dubuque, who operated in Iowa and Illinois and had a particularly close relationship with local Indians. Dubuque's mines were thought to produce around 1000 tonnes of lead a year, enough with his winter trapping activities to make him a very rich man. His Indian miners operated in the traditional manner, again using fire setting to crack the ore and then stag horn implements to break off the metal bearing parts of the ore. Smelting was done in a log furnace but with a stone base, and built into sloping ground. Lead recovery was satisfactory at around 65%, taking into account the fairly rudimentary technology used. Mining by the French continued into the early decades of the 19th century and when the French began to leave the US, Indians took over the mines for a short period before the expansionary and now independent Americans from the east moved in.

The pace of exploitation accelerated as the US pushed west and more lead mines were opened. The Indians lost out and the expansionary US government began to enclose land and then issue leases to interested parties with royalty requirements attached. The system was on the chaotic side and not well policed initially by the government, but in due course mining became more organised, and more efficient smelting technology was introduced. The displaced Indians, indignant at losing their land, created problems and security remained an issue for many years. One of the worst offenders in his ignoring of both US government instructions and Indian rights was Henry Dodge, who in 1827 was producing at least 1.5 tonnes of lead a day from land owned by the Winnebago Indians in the Wisconsin Valley. Dodge, having made his fortune, became Governor of Wisconsin, a post he held for many years, and then became a US senator. However, by the middle of the 19th century a recognisable and organised lead mining industry was beginning to emerge, as we will discover later.

9. GOLD, SILVER AND THE SPANISH CONQUISTADORS

One of the most wide-reaching events in the Middle Ages as far as metals and mining are concerned was the arrival of the Spanish Conquistadors in Latin America. If the Middle Ages ended with the Reformation and the dislocation that ensued, its golden time was surely the Renaissance when art and culture began to blossom – the first indications of an economic revolution to come were seen in the form of the futuristic sketchings of Leonardo de Vinci.

The Spanish certainly were inspired by the atmosphere of the Renaissance to sail out into the Atlantic, pushing west in the hope of finding a route to the east – in particular India and China – where riches in the form of spices, not to mention gold, beckoned. The Portuguese had earlier sailed into the Atlantic, also anxious to find a passage to India that did not take them through the lands of the Ottoman Empire; they first turned south to probe the west coast of Africa.

So it was that in 1492 Christopher Columbus, an Italian sea captain, supported and equipped by Queen Isabella and King Ferdinand of Spain, set sail for India going west across the Atlantic. Columbus made landfall in the Caribbean on the island of San Salvador, far from his target of India and the east. Over the years he made a number of trips to what became known as the West Indies, landing on Haiti, Jamaica and Cuba and establishing administrations there. Despite his pioneering success Columbus found little gold and certainly no spices and in due course he was recalled to Spain, having fallen out with the Spanish crown.

It was not, however, the case that the Spanish interest in Latin America's gold and silver came about as a result of any plan to explore for and then mine gold. Whilst Spain has a long history of mining, the Conquistadors were adventurers and though gold was their primary target they thought that their voyage across the Atlantic would bring them to India eventually, where it was rumoured that great quantities of gold, already mined, were to be found. When they reached the West Indies the Spanish did find some gold on Haiti, but it needed mining and the Conquistadors were really looking to acquire mined gold by trading other goods for it and then shipping it back to Spain for the Crown's treasury.

The Haitian gold operations were basically alluvial, mining river beds and streams, and the miners were indentured native Indians who the Spanish treated harshly with the consequence that many of them died, and were then replaced by slaves from west Africa. The Spanish in due course opened gold mines on Cuba. The gold mined was shipped back to Seville in Spain, which was the designated port where the gold had to be landed, and here it was delivered to the Casa de Contratacion which was the depot built to receive American gold.

Whilst the quantity of gold found on the Caribbean islands occupied by the Spanish was relatively modest, rumours abounded of great riches in a land called Mexico to the west. It was Hernan Cortez, an adventurer who had fallen out with the Cuban governor, who decided to flee Cuba before more trouble ensued and set sail for Mexico with a force of 600 men. There he encountered the Aztec civilisation, and gold.

Cortez found the rumours about gold in Mexico were correct and the natives were prepared to trade gold for beads and other trinkets, but the amount of gold that the natives had was not enough for Cortez. Fortunately for the Spanish, the natives knew where there was abundant gold – it was in the hands of the Aztecs, who were the most powerful civilisation in Central America at the time and who had a history of great cruelty in pursuing their geo-political aims. Unfortunately for the Aztecs the Conquistadors were even crueller and more ruthless than they were.

Initially the Spanish got on well with the Aztecs – the Aztecs showed great respect to their guests, being particularly impressed by their exotic white skin and formidable weapons and armour. The Aztecs were also happy to feed their gold lust, as they saw no value in the metal itself except as a raw material with which to make artefacts, and many of these artefacts were willingly given as presents to the Spanish. In the end gold lust drove Cortez to sack the Aztec capital of Tenochtitlan and, through a combination of the sword and European diseases introduced by the Spanish, he destroyed the Aztec civilisation.

HERNANDO CORTEZ (1485-1547)

Cortez was one of the earliest and most infamous of Spain's conquistadors who ravaged Central America in the early 16th century. His interest in Latin America was twofold – political and economic – reflecting his background as well-born but financially straitened. Like many Spaniards of that era Cortez was interested in gold and silver but unlike many of his contemporaries he eventually financed a number of precious metal mines in Cuba and Mexico.

Cortez was born in Medellin in Badajoz in northern Spain, his father Martin was an army officer and his mother Maria was a cousin of Francisco Pizarro, the conqueror of Peru. Cortez was educated at the University of Salamanca where it is believed he studied law. He did not complete his studies but at the age of 19 he sailed to Hispaniola (Haiti/Dominican Republic today) where he practiced as a notary. In 1509 he took part in the invasion of Cuba, where he materially enhanced his standing as a forceful leader; this led to administrative power within the new Cuban government and the allotment of land and other assets. It was at this stage that Cortez became involved in mining for the first time when he established an alluvial gold operation at Cuvanacan on the Duaban river in the east of the island, using a workforce of indigenous slaves that he had been allocated. The mining investment and his livestock operation generated great wealth for Cortez and he established a large estate and house near the mine workings.

By now he was a substantial figure in Spain's Caribbean colonies and his former supporter the governor, Diego Velazquez, became jealous of his success. This was despite the fact that he had married Velazquez's sister-in-law, Catalina Zuarez. Cortez was unfortunately a serial womaniser and was rumoured to have had a large number of children with both native and Spanish women. Catalina died childless in 1522 and in 1529 he married Juana Ramirez, with whom he had several children. Cortez, anxious for new challenges and perhaps realising that he had outgrown Cuba, raised an expedition to invade Mexico in 1519 where the Aztec empire was at its zenith, and sailed without final clearance from Velazquez. There he established a relationship with the Aztec king, Montezuma II, which Cortez turned to his considerable advantage as he acquired huge wealth in the form of gold and silver for the Spanish Crown.

Whilst Cortez much admired the advanced state of the Aztec civilisation he was ruthless in his treatment of both Montezuma and the

Aztec people, essentially ripping the heart out of the civilisation through a combination of imported diseases like smallpox and the slaughter of countless natives. He eventually toppled Montezuma, destroying the Aztec capital, Tenochtitlan, and establishing Mexico City in its place.

It was after this that Cortez turned once more to mining, discovering silver in the Taxco district and gold in the Oaxaco and Michoacan districts. As in Cuba, these mining activities earned Cortez further huge wealth but this was a wealth and influence that worried the authorities. For this reason, as negative comment rose about his burgeoning power, Cortez returned to Spain to plead his case with the king, Charles V, and the king gave him support. But when he returned to Mexico his powers, which had included the position of governor, were greatly reduced – though he was retained as captain general – and he turned instead to exploring Mexico.

In 1541 Cortez returned to Spain to try and resurrect his reputation but he was the subject of legal actions and was shunned by the king. His financial situation was weakened by his fight to save his reputation and after several frustrating years in Spain he eventually decided to return to Mexico. His health was now failing and he died of pleurisy before he could sail. Cortez remains a controversial character, clever and brave but also cruel and ruthless, a loving father but a deeply flawed human being. His economic success was overshadowed by his political activities, particularly as a conquistador, but his role as a developer of gold and silver mines in Mexico and Cuba makes him perhaps Spain's first known mining baron.

GOLD AND SILVER; THE DOWNFALL OF SPAIN

Mexico, however, was to yield the Spanish far more than just the gold artefacts of the Aztecs; in 1524 silver was discovered near Acapulco. Over the next three centuries before independence movements swept the Spanish from Latin America, Spain's colonies are thought to have delivered as much as $8 billion worth of silver to the Crown. Production in Mexico alone had reached 9 million ozs a year by the 18th century. During this period of Spanish hegemony the gold price was worth around 15 times that of silver and silver in volume was shipped to Spain in a ratio to gold of about 10 to 1. That means that in monetary terms silver was not far behind gold in value to the Spanish treasury.

Over 150 years from the mid-1400s to the early 1600s inflation in Western Europe saw consumer prices rise by 600%. The arrival of large quantities of gold and silver from Latin America, particularly Mexico and Peru, was thought to have been a major influence in this inflationary period. Certainly the Spanish economy became increasingly enfeebled due to inflation and the problems of its overseas possessions, and its European Empire was dismantled in 1714 following the Treaty of Utrecht.

Following Cortez, the Conquistadors had not finished with Latin America's great civilisations and within a few years Pizzaro led a small force of well-armed Spaniards into Peru where he confronted the Incas and their king, Atahualpa, in 1532. As with Mexico so with Peru – the Incas had extensively mined gold over the centuries and fashioned a large collection of artefacts. Atahualpa was prepared to gather these together as tribute for the invading Spaniards, but though a substantial quantity of gold artefacts were delivered to Pizarro he was not satisfied. He believed he had been duped and he had the Inca king executed. As with the Aztecs and other indigenous Latin American civilisations, the coming of the Spanish spelled the virtual extinction of the Incas.

It is fashionable today to decry the British for the establishment of their empire but the Spanish, with their gold lust and religious fervour, were equally as guilty – they destroyed many of those they had subjugated. When their American colonies wrested independence from Spain the new nations were culturally and demographically largely Spanish; Britain's colonies when granted independence in most cases returned to their original indigenous state. The great cultural treasures like the Taj Mahal in India survived intact, while the Inca cities were reduced to ruins and their fabulous gold artefacts melted down and sent back to Europe where the ensuing inflation reduced Spain to almost third-world status.

But how had the gold, which the Conquistadors looted, been mined?

MINING METHODS

The main mining method employed by the Spanish was alluvial panning and dredging of the rivers and streams of western South America, which yielded large quantities of free gold. There is little evidence of any widespread underground mining until later; as we will see further on with African copper mining, the native miner was clearly reluctant to dig deep in pursuit of the metals he sought. When silver was discovered in Mexico, however, the veins did often plunge and required the sinking of shafts to extract the metal. Also, the historic mining areas of Mexico, Peru, Chile, Bolivia and others do show evidence of open pit mining. This was something seen in central Europe as well, this region being a major producer of, particularly, precious metals in the Middle Ages.

CRAFTMANSHIP

As to the methods used by the Indians of South America to work the gold into the artefacts and jewellery which the Conquistadors accumulated, it appears that they were not so very different from those employed by the Egyptians and the Chinese centuries earlier. Whether there was any collaboration, as some observers have speculated, between the Chinese and the Indians on either side of the Pacific is unknown although it seems unlikely, if not impossible, due to the huge distances needed to travel and the lack of evidence.

Nonetheless some of the techniques the Indians used were not unique to them; in particular the lost-wax method where a craftsman carves his design onto a piece of wax, which is then covered by a layer of moist clay. The clay is then fired in a furnace. As the wax core melts and the clay layer with the design imprinted on it hardens, a hollow core emerges into which gold and copper are then poured. The clay is then removed from the furnace and the gold hardens and takes on the shape of the clay design. When it is finished the clay layer is removed to reveal the gold object. At this stage the copper is then removed by scouring and the gold artefact is then finished by the craftsman. Similar techniques were used to make silver objects.

Whilst the Indians' artefacts were in many cases both beautiful and complex the Conquistadors were not much interested in the quality of the workmanship but rather in the quantity of the gold bound for Spain – they melted down the jewellery and artefacts, thereby losing priceless objects. This turning of art into dross, even gold dross, was echoed several centuries later when many valuable silver artefacts were sold by their owners as scrap for cash as the underlying metal price soared crazily in the early 1980s; the value destroyed by this exercise was huge.

10. CHILE

During the 16th century, following the establishment of Santiago in 1541 – now capital of Chile and its largest city – gold mining was carried out in the south, with a number of rich but relatively small alluvial operations having been established by Spanish settlers. Output is thought to have been around 50,000 ozs annually but by the end of the century the indigenous Mapuche Indians had pushed the Spanish back towards the centre of the country and gold mining declined rapidly.

In the 18th century, after hostilities between the Spanish and the Mapuche had ended, gold mining began to revive, partly in response to the building of a royal mint in Santiago. New gold mines were established in the north of the country in the Atacama region. These were high-grade mines where the gold was found in vein deposits and annual output reached around 100,000 ozs by the beginning of the 19th century, from where, as we will discover later, production fell away as interest turned first to copper and then to nitrates.

11. THE EAST

Whilst the onset of the Dark Ages, as the Roman Empire collapsed, plunged Europe into social and economic depression and chaos, it did not materially affect countries in the East, although historically there had been trade and economic contact between the two regions.

CHINA

It is interesting how China always pops up in the forefront of developing technology, whatever the era. Its relative backwardness during most of the 20th century seems to be an aberration when viewed against the backdrop of recorded human history and the current revival in China's economic power could be viewed as a reversion to the norm.

So it was that as our Middle Ages period started in the 11th century and Europe was showing signs of having left the Dark Ages behind, China was way ahead. Its development of the coal firing of blast furnaces in the 10th century led to an annual iron output of perhaps as much as 125,000 tonnes in the first three quarters of the next century (though some estimates put the figure even higher), a rate of output that Europe did not achieve for almost another 500 years when blast furnace technology was finally mastered by European ironworks. It was also an early example of green environmental action as it ended the wholesale destruction of forests needed for the manufacture of charcoal, which had been the old fuel of choice.

We have earlier seen the problems that stalked Europe's recovering economies in the Middle Ages when supplies of silver became restricted leading to liquidity problems in trade settlements. This was a problem also in China around the end of the 10th century and it was at this time that the Chinese introduced paper money to try and address the liquidity issue. Up until that time China's silver and gold production, and that of its Central Asian neighbours, had been sufficient to finance trade flows in the broad region. But as the Middle Ages loomed many of these silver and gold mines – suffering from old age – experienced a sharp fall in output.

12. DIAMOND MINES IN INDIA AND BRAZIL

Later, in the next two parts of the book covering the Industrial Revolution and the Modern Age, we will look closely at the rise of Africa to its position as the largest supplier of diamonds in the world. But prior to this the main sources of diamonds were firstly India and then Brazil. Before the expansion of diamond usage as a result of the coming of the industrial age and then the mass marketing of diamonds by De Beers, diamonds were the exclusive preserve of the fabulously rich.

INDIA

Today diamond mining in India is very much a shadow of what it was, consisting now of one hard rock mine, the Panna mine in Madhya Pradesh in central India, which is controlled by state-owned National Mining Development Corp, and has a current capacity of 100,000 carats, small alluvial workings scattered around the country, and a new Rio Tinto project, the Bunder, also in Madhya Pradesh. However, going back many centuries to the time of the Phoenicians and the Romans there is evidence of trading in diamonds and many of them will have come from India, although there is a view that some African diamonds may have also been traded following incursions by the Phoenicians in the 5th century BC as they circumnavigated Africa.

Whatever the truth, the world both in ancient times and in the Middle Ages used diamonds for adornment – these came mainly from India and were highly prized, and therefore the property of the rich. The diamonds were largely found in alluvial settings clustered in a broad region with Madras as its southeast extension. Despite the alluvial nature of the host environment the diamonds quite possibly came originally from kimberlites formed in the Himalayas and over scores of millennia were lifted and then washed down to the central plains of India north of Madras, where they lodged in ancient river beds, and were then over many years covered by the movement of earth and silt.

Indian diamond deposits had been worked probably for four millennia at least and Roman writers like Pliny and Ptolemy who were interested in minerals had noted their existence, but it was around the 11th century that more frequent mention began to be made. Two of the most important producers of diamonds in the 16th century were the mines of Panna and the fabled mines of Golconda near Hyderabad, both in central India. Indian diamonds were usually found in alluvial settings, often laid down in terraces just a few feet below the surface. Mines were also established to the south near Madras (now Chennai) on the

Pannar River and there were other diamond workings to the west in the Anantapar district.

All these mines had been worked for some centuries but more extensive exploitation took place in the 19th century when a number of large stones were uncovered. Such stones were very old and hard, and had been weathered by movement to a smooth and brilliant finish. One of the problems facing early Indian stones was that cutting techniques, which were crude, were as likely to damage, even destroy, the stones as to shape and improve them. It was not until the 13th century that an advance in European technology allowed diamond cutters to cut rough diamonds, with increased confidence, to a basic number of shapes. Indian aristocrats before that, therefore, wore their diamonds largely in their natural state as crystals in a variety of jewellery and clothing settings.

In the 16th century the Antapar region yielded many diamonds of great value, which were claimed by the corrupt ruler of the State of Vijayanagara, Kama Rayar. However, his great wealth was unable to save him from defeat and death at the hands of the Muhammadans, who invaded from the north in 1565, at the Battle of Talikota.

Other diamond mines were worked in the region, including the famous mines of Wajrar Karur, and mines at Nandial and Kumal to the north, and also the nearby Ramulkota mines which were worked for a number of centuries in the Middle Ages. Wajirar Karur produced some large stones including the Gordonorr, a 67-carat stone, named with tongue in cheek after a Madras jeweller in the 1880s.

Historically the mines were worked as surface deposits and heavy monsoon rains helped wash the diamonds out into alluvial channels for the miners. More recently in the 19th century, miners, often returning to historic diamond-rich areas, dug pits into the diamond-bearing ground and then brought the ore to the surface to be washed and treated. This was a difficult task as the pits were often very wet at the deeper levels (50 to 80 feet) where diamonds had been laid down in extensive, if often thin, stratum. The water and the diamond-bearing earth had to be brought out by hand. Also, the earth was of a clay quality, which made the washing a more extended and expensive process. Fortunately the diamonds recovered were often of sufficient value to make the exercise worth doing.

In the 17th century, according to Jean-Baptiste Tavernier, the French traveller who brought a number of large diamonds back to France, there were as many as 60,000 Indians working in the diamond mining industry. Tavernier also may have been the first westerner to identify what became known as the Kohinoor diamond presented to Queen Victoria in 1849. It was a 280-carat diamond believed to have been discovered in the 4th century BC and over the centuries much fought over as a consequence of the frequently changing political face of the Indian sub-continent.

Its genesis was thought to be from the Golconda mines, where the 700-carat Great Mogul stone was also discovered, as was the famous Hope blue diamond. Historically the Golconda mines produced finer stones than the Panna mines but it is the latter area where, albeit in a small way, diamond mining continues today.

Spread as it was over many centuries, at a time when the keeping of statistical records was piecemeal at best and when the mass markets of today could not have been remotely envisaged, it is impossible to say how many carats of diamonds the Indian mining industry produced over this long period when it had almost a monopoly on world output. It is said that Muhammed Ghori, whose invasion of India in 1176 brought Islam to the country for the first time, pillaged almost 100,000 carats of fine diamonds in 30 years of tyranny before his death in 1206. It is also the case that at the turn of the 20th century Indian diamond output had fallen to less than 1,000 carats a year as Africa took up the reins of this industry.

BRAZIL

The predominance of India as the main source of diamonds began to flag in the 18th century, but fortunately in the early part of that century samples of diamonds began to reach Portugal from its South American colony, Brazil. The source of the first diamonds was the state of Minas Geraes, a region that will become very familiar when we look at gold and iron ore in Latin America later. The initial discovery was made in the area known as Serro do Frio or Diamantina, where substantial quantities of diamonds were found in rivers and streams, which led the Portuguese government to claim exclusive ownership of the area's diamond resources.

Originally diamond traders, afraid that Brazilian diamonds might lead to a sharp fall in the price of Indian diamonds, spread the rumour that the stones had actually reached Brazil via the Portuguese colony of Goa on India's west coast and were Indian and not indigenous Brazilian stones at all. Brazilian traders then turned the story around by taking advantage of these rumours and acquiring some Brazilian diamonds and selling them into the Indian market via Goa at premium prices, which Indian stones could command at that time.

However, Serro do Frio was not the only source of diamonds in Minas Geraes, and indeed the state itself had no monopoly on diamonds in Brazil once the *garimpeiros* – peasant diggers – inevitably joined the search. The state of Bahia was also an important diamond source and in the 19th century overtook Minas Geraes as the largest producer of Brazilian diamonds. There were also diamond diggings in the states of San Paulo, Mato Grosso and Parana. Stones from Bahia were not usually as fine or as large as those from Minas Geraes, but on occasion very high quality diamonds were found. The great proportion of Brazilian diamonds were below 1 carat, a handful were of greater size up to 20 carats, but very large stones were extremely rare. The largest found – *the Star of the South* – in Bogagem, Minas Geraes, was 254 carats, and there were two other recorded finds of plus 100-carat stones. Altogether this was a rather poor hoard compared with the output from the new diamond fields of South Africa.

When diamonds were first produced in Minas Geraes the cost of production was higher than in India, but despite that output rose and by the middle of the 19th century around 6 million carats of diamonds had been mined in the state, with a value of around $50 million, and around a further 4 million carats had come from other states. In Minas Geraes one of the key watercourses carrying the diamonds was the Rio Jequetinhonha; for Mato Grosso it was the Paraguay River and its tributaries. Digging took place during the dry season when the rivers could be temporarily diverted along a specially dug course and the mud beds of the river could then be dug down to around ten feet and the spoil be transported to treatment areas where, when the rainy season began, the dried mud could then be washed for diamonds. Within the mud there were occasionally large concentrations of diamonds found and sometimes even gold.

The heavy work was done by black slave labour, Brazil being one of the largest buyers of slaves. The black workers were incentivised not to secrete diamonds by being awarded their freedom if they found larger carat stones when washing the river mud, but despite that there was a high level of diamond pilfering. Nor did everyone see the discovery of diamonds in their area as unalloyed good fortune, indeed the social disruption was considerable as swathes of houses and huts with riverside locations were cleared away with the coming of the diamonds. Neither did the authorities always welcome diamond workings as these could lead to a dangerous fall-off in agricultural production.

In addition to gem diamonds, carbonado or black diamonds were also found in Brazil in the early 19th century. As described, these diamonds were black or grey in colour and internally flawed and therefore not desirable as jewellery. However, they were usually harder than diamond and as the Industrial Revolution gained traction the need for hard drill bits in industry as well as mining, and for effective abrasives, opened up a market for carbonado that Brazil was able to supply.

Brazil continues to mine and sell carbonado today. It also remains a diamond producer, but for a country that has produced in excess of 60 million carats of diamonds since mining started in the early 18th century, annual output of around 80,000 carats in 2008 constitutes a substantial fall from grace, although this figure may well understate real output levels due to the common incidence of diamond smuggling today. Having said that, Brazil does still remain a target for junior diamond exploration companies.

13. CONCLUSION

After an unpromising start in terms of human progress in the arts, science and economic activity – war and conflict was almost permanent and events such as the Black Death and the Great Plague provided major setbacks – the Middle Ages slowly began to witness a more promising environment. The Renaissance saw a flowering in education with exciting developments in art and the first signs of a broad revival in scientific activity for many centuries.

As the Industrial Revolution loomed the political and intellectual underpinning necessary for any great leap forward in human activity was in place. Such an advance also required human resources, capital and also raw materials backed by technology to be available, as they were. The mining industry was also ready and able to build upon the advances it had made in the 17th and 18th centuries, so all the blocks were in place for the greatest advance in human history.

THE INDUSTRIAL REVOLUTION (1800-1900)

1. INTRODUCTION

The 19th century felt the full force of the Industrial Revolution as scientific and technological advances accelerated and developments in a whole range of industries swept round the world. In earlier centuries restless European adventurers had launched expeditions to acquire essential goods, often agricultural commodities, and to chart the unknown or the barely known. In the 19th century the Industrial Revolution led to a more urgent need to find new sources of, particularly, industrial metals such as copper where local supplies were no longer adequate to fulfil the rapidly expanding needs of the newly industrialised powers such as Great Britain.

Once out of the bag, as we know so well today, the demands of technology which drive industrialisation are relentless. So it was in the 19th century, and the rapidly expanding appetite for raw materials led to a revolution in sea and land transport. It also provided the driving impetus for the great powers to colonise countries, which became key suppliers of raw materials, in order to guarantee them unfettered access to these metals and minerals.

2. DIAMONDS IN SOUTH AFRICA

South Africa was the most important source of diamonds for the worldwide diamond jewellery industry for over a century, its lead position only beginning to waver in the 1970s following the huge discoveries in neighbouring Botswana. Before the South African discoveries in the 19th century diamonds were found primarily in India in alluvial settings, until in the 1700s production fell away to be replaced by alluvial production from Brazil, as we learnt earlier.

The first diamond officially discovered in South Africa was the Eureka, a 21-carat stone found near Hopetown in the Cape Colony in 1867. It had been in the possession of Schalk van Niekerk, a farmer with an interest in gem stones, who enlisted the help of John O'Reilly, a trader, to determine whether the stone was a diamond. There was deep scepticism in Hopetown about the stone's provenance but in due course it was sent to a geologist in Grahamstown, Dr William Atherstone, who agreed with O'Reilly that it was a diamond; it was eventually valued at £500. Two years later another diamond came into the possession of van Niekerk, who bought it from a shepherd in exchange for livestock. This stone, at 83 carats, was much bigger than Eureka, and having been cut and polished was eventually sold for £30,000 in London to the Earl of Dudley. The stone was named The Star of South Africa and its discovery set off a diamond rush to the Cape.

Strangely, despite the original Hopetown discoveries the area failed to live up to its early promise and prospectors moved on to the far more prospective Vaal River to the north. Here the Cape Colony butted up to the Boer republics of the Orange Free State and the Transvaal; the local tribe, the Griquas, also had land claims in the area. The Vaal River diggings had been worked for a number of years and as discoveries were made at Klipdrift on the north side and at Pniel on the south side, potential political conflicts became inevitable.

The work was hard but straightforward – mud and gravel was removed from the river and riverbanks, and then sieved through a mesh placed over a cradle. Water was poured through the mesh to dissolve and wash away the loose mud, leaving only the stones and gravel. The contents were then removed from the cradle and placed on a table for examination and sorting. Initially the diggings were broadly peaceful, with behaviour well above the levels most mining rushes experience.

EARLY TENSIONS BETWEEN BRITISH AND BOERS

In due course the diamond discoveries led to land disputes between the two Boer republics, the Griquas and the British, which were finally settled by the annexation of the disputed areas by the British in 1870. Before that, outraged by manoeuvring by the Transvaal over concessions on the north side of the Vaal, the diggers had set up an independent republic, which lasted only until the British sent a magistrate to the Vaal River diggings to protect the rights of British diggers. But as so often happens in the mining industry, events were soon to overtake these Klipdrift disputes. Early in 1871 new discoveries were made little more than 20 miles away at the farms of Dorstfontein and Bultfontein and then at Vooruitzicht, which was renamed De Beer after the name of the farm's owner. Finally the discovery that was to ignite South Africa's diamond industry was made nearby at what was named New Rush. The discovery was in the form of a volcanic pipe, which eventually became known as the Big Hole, and around it grew up the town of Kimberley.

Diamonds come in two basic forms; they are either gem stones or industrial stones. Gem stones used in the making of jewellery are the most valuable, being able to be cut into a variety of shapes and then polished to a lustrous finish before being set in a jewellery piece, often made of gold. Diamonds are amongst the hardest minerals found and cutting them is a highly skilled art. The trick is to find flaws within the diamond that allow a cutter to saw or break the diamond as he shapes the stone. A large diamond may only be able to be cut into a much smaller stone but the diamond residue is valuable as it can usually, itself, be cut into smaller diamonds. A gem diamond has great value due to its rareness but in terms of usefulness, its beauty is its sole asset.

Industrial diamonds, in contrast, have great use as a cutting instrument in their own right and are used in particular in drill bits to cut through rock or as abrasives to cut, grind or polish other materials, including gem diamonds in the shaping and polishing process. Not only are they extremely hard, diamonds also do not heat up during cutting and metal can be cut extremely thin by a diamond cutter. Diamond can also be used to draw very fine wire. Industrial diamonds cannot be used to make jewellery stones as they lack the colour and clarity so important for cut and polished gems. Also, though diamonds are one of the hardest minerals known, the natural flaws and weaknesses in many stones mean that they can fracture and shatter.

For instance, the stones that lie off the coast of Namibia and South Africa are only a fraction in number of the stones from inland kimberlite pipes washed down the Orange River over many millennia. Some stones did not make the open sea but were caught in traps on the river bed and in many cases buried in the mud, many others were too flawed to survive such a journey and had been shattered into worthless shards and un-mineable dust as they bounced down the rocky river course.

THE RANDLORDS

The development of the South African mining industry in the 19th century led to the rise of a number of historic figures often referred to as The Randlords. Cecil Rhodes was perhaps the most well known of them but the group also contained huge figures like Barney Barnato, Joseph Robinson, Alfred Beit, Charles Rudd and the Joel brothers (nephews of Barnato). Their interest in diamonds predated their involvement in the goldfields of the Witwatersrand, but it was gold that eventually pitted Rhodes, the most political of the Randlords, against the Boer government in the Transvaal. Having said that, there seems little doubt that Rhodes's first love was diamonds and it was diamonds that saw him at his sharpest as he eventually battled Barnato for control of the Kimberley fields.

CECIL RHODES (1853-1902)

Once one of the British Empire's most revered but also widely hated figures, Cecil John Rhodes enjoyed a burst of popularity after his death when his legacy, the Rhodes Scholar awards to Oriel College, Oxford, in particular marked him as a great philanthropist and visionary. Since then his reputation seems to have been in almost permanent decline as first his admiration for the Anglo Saxon 'race' and then his vision of a worldwide British Empire poisoned his public esteem and earned the contempt, particularly, of modern historians.

Rhodes was born in Bishops Stortford 30 miles north of London, the son of a clergyman. Whilst his brothers attended England's great public schools and went up to Oxford, Rhodes was sickly and was educated locally. His health meant that he did not initially go to university but left school at 16. After being diagnosed with TB he went to South Africa to join his brother Herbert on his cotton farm in Natal. After a couple of years building up the farm Herbert, who had earlier spent a short time in the diamond fields near Kimberley in Griqualand West, returned to Kimberley and Rhodes followed. Herbert earlier had some success with his claims, but under Cecil's direction things took on a more organised and businesslike tone.

Although neither a geologist nor an engineer, Rhodes worked his brother's claims successfully as a digger before broadening his commercial interests to include a wide range of services to the Kimberley mining community. He also showed at least two characteristics of the traditional miner, being a man who enjoyed a drink and who used strong language, however he did not share that other vice of miners far from home; women.

Rhodes tended to have few real friends but those he did have were long lasting. As he expanded his Kimberley diamond interests he worked hand in hand with Charles Rudd and this enabled him to pursue his ambition of going up to Oxford, which he did in 1873. He divided his time between Oxford and Kimberley, not easy in those days, decades before overnight flights shrunk the journey time to a few hours. In Kimberley, Rhodes continued to build up his claims at the Big Hole where most of Kimberley's diamonds were to be found.

There were two things obvious to Rhodes: Kimberley's diamonds were high quality and the huge numbers of claims covering the Big Hole made mining inefficient and dangerous. Rhodes thus had two concomitant ambitions: to control the marketing of diamonds to obtain better prices from European jewellers and to create an integrated diamond mining operation at the Big Hole. Rhodes, however, was also drawn to the gold fields of the Witwatersrand in the late 1870s, when he floated Gold Fields of South Africa in London, and also to the mineral potential of what became Southern Rhodesia which was covered by the activities of The Chartered Company. His heart, though, was in the diamond fields around Kimberley and in 1889 with the purchase of the 'French Company' and Kimberley Central, Rhodes's De Beers company finally had control of Kimberley's diamond fields.

To all of this Rhodes added political ambition. He became an MP in the Cape Parliament in 1880 and ten years later he became Prime Minister of the Cape Colony. He represented a farming area with a considerable Boer vote whose interests he supported. However, he had begun to dream of a much expanded British Empire (from the Cape to Cairo) which hardly pleased the government in London at the time, and as far as it affected Africa did not please the Boers either. As a politician Rhodes began to tighten restrictions on the black and mixed race population in terms of voting, property rights and work. At the same time he introduced extensive and positive reforms in the agricultural sector and started a programme to renovate historic Cape Dutch houses that had fallen into disrepair.

As the Witwatersrand gold fields expanded, confrontation with the Boer government in Pretoria became inevitable. Rhodes was not always decisive – as conflict loomed in 1895 he backed Dr Jameson's infamous raid to lead an *uitlander* revolt in Johannesburg against perceived Boer oppression, but at the last moment his support wavered. The raid descended into farce with capture and humiliating trials for Jameson and his followers. Rhodes's political career then went into reverse. He was forced to resign as Cape Prime Minister but survived more serious personal damage due to his close relationship with Joe Chamberlain, the British Colonial Secretary.

He became leader of the Progressive Party, a capitalist-imperialist-liberal grouping, almost winning the 1897 Cape Colony election and still representing the Boer farmers of Barkly West. The following year the Anglo-Boer War began. Rhodes did not see the end of the war, his last days dogged by debilitating illness and the consequences of unwelcome attentions from Princess Catherine Radziwill, an Austrian aristocrat. He was buried in the Matopos Hills in Matabeleland in

Southern Rhodesia, a country which he loved and that he had hoped would eclipse South Africa as a source of mineral riches.

Modern historians have little good to say about Rhodes. His vision of a worldwide British Empire does not play well in the hindsight of a century passed, but there is little doubt that he remains a huge figure both as a politician and as a mining magnate whose name is synonymous with the rise of South Africa's diamond and gold industries. His grave, and those of other white and also indigenous leaders, lies in the beautiful and peaceful Matapos Hills in what is now Zimbabwe. Even today it is still keenly protected by the local Ndebele, a strange but touching fate for one of history's most notorious imperialists.

ROBINSON, THE LONER

Despite the role of Cecil Rhodes, it is Joseph Robinson who first became involved with diamonds when as a trader in the Orange Free State he came across them in the Orange River area in the late 1860s. He began to concentrate his diamond trading and exploration efforts on the Vaal River area and acquired a useful knowledge of alluvial diamond mining as well as a useful income from his exploration successes. With the discovery of what was originally called the New Rush mine and the rise of Kimberley, Robinson settled in the town and became a highly successful diamond buyer.

Robinson was an aloof and austere character who inspired respect as an upright citizen and a prosperous businessman. He had a dangerous temper though, which on occasions showed a different side and one that was not always to his advantage as he sought to build on his respectable image. He was a man who also sought to promote his own views and he used his growing wealth to purchase control of one of the Kimberley newspapers, *The Independent.* It is not surprising that such a prominent diamond buyer and mine owner made some enemies and his occasional indiscretions gave opponents plenty of opportunities to attack him. But Robinson's success ensured his reputation was known all over South Africa and his activities were always newsworthy. In due course in 1879 Robinson entered politics and got elected as Mayor of Kimberley.

PROBLEMS AT THE KIMBERLEY DIGGINGS

Up until 1880 the mining of the New Rush pit, and indeed the other pits around Kimberley, had been largely based on small diggers, some of them even black natives, although that was a situation that did not last long with ownership restrictions forcing out native claims holders; one of the earlier harbingers of the apartheid system.

As the diggers went deeper two problems arose. The first was that the rich near-surface material, which consisted of light, easily dug soil, gave way to what was called blue ground, which was much more compacted and needed increased effort to break up. Also, as digging descended into the blue ground the dug material appeared to be very low in diamonds. For a while despair reigned and many diggers gave up, selling their claims for virtually nothing. It is interesting to note that no exploration drilling took place in this era; when metals or minerals were found on the surface miners staked claims, set up a basic digging operation and continued to mine until the mineralised ore ran out. As the Kimberley diggers went into the blue ground they had no idea what lay beneath. As it was, the new deeper material became extremely productive. The scare was over but many small claims holders had already fled the diggings.

But now another major problem arose which created opportunities for the likes of Joseph Robinson. The deeper the diggings went the more costly was extraction and the more precarious the individual claims became, with water inrushes and earth collapses becoming common at greater depths. The era of amalgamation was looming and the future giants of the South African diamond mining industry were gathering; they included Robinson, Barnato and Rhodes.

THE RANDLORDS PURSUE AMALGAMATION

Robinson and Barnato had become enemies after clashes in the Cape Province Assembly in Cape Town, to which both had been elected in 1881. Robinson formed the Standard Company to expand his diamond interests and Barney and his brother Harry set up the Barnato Company. Both of these groups concentrated on the New Rush pit in Kimberley, but Rhodes's diamond interests were concentrated on the nearby De Beers mine and he incorporated De Beers Mining in 1881 to hold these interests. He also managed to merge the interests with those of the Stow, English and Compton company at the De Beers diggings.

RHODES SEEKS CONSOLIDATION AT THE DIGGINGS

Rhodes also attempted to acquire other De Beers operations to achieve one consolidated operation at the mine, but other independent operators, although happy to consider his plan, could not agree a price. However, Rhodes was already working on a much grander plan to amalgamate the whole of the Kimberley area diggings including De Beers and New Rush, better known as the Big Pit. At the same time very tough legislation was being pushed through the Cape Assembly to bring an end to illegal diamond buying and on this issue, if no other, Rhodes and Robinson were in agreement. The practice, which it was thought had been behind much of Barnato's success, had to be eradicated and with this aim in mind in 1882 the Diamond Trade Act was passed.

In 1884 the pace of change quickened. Rhodes had slowly but surely continued to work on the smaller holders at the De Beers mine and in 1884 merged his De Beers Mining with Baxter Gully and Independence, two of the larger groups still holding out. The New Rush mine in Kimberley continued to interest Rhodes in terms of amalgamation, and Rhodes was anxious to complete the amalgamation of all the mines in the area and to control them. The additional wealth an amalgamation would generate for him would allow him to pursue his political dream of an Africa annexed by Britain from the Cape to Cairo. Robinson and Barnato had less lofty reasons for supporting the concept of amalgamation; it would simply make them more money.

There were two key issues, which we have mentioned earlier, pushing the Kimberley area mines towards amalgamation – the problem of collapsing workings and the widespread practice of illegal diamond buying. There was also a continuing slump in diamond profits caused by overproduction as well as operating problems as the diggings went deeper. Another problem was one relating to the workers, who had already caused unrest in the 1870s with their Black Flag strike. To try and combat illegal diamond dealing a draconian search process had been instituted to cover all workers on the diggings, and the whites in particular objected to being searched so intimately. This unrest continued for some months and eventually led to a strike in 1883 that ended in a riot the following year. The police put the riot down with firearms, killing four protesters.

With all this unrest and the fall in diamond profits, share prices of the Kimberley mining companies fared badly. Central Company, which in 1881 had changed hands at £400, three years later had fallen to £25. Joseph Robinson's Rose-Innes had fallen over the same period from £53 to just £5. Rhodes's analysis, although he supported the war on illegal diamonds, was that the main problem facing Kimberley's diamond industry was that it was producing too many stones for the world market and this was undermining prices and demand. Amalgamation and production control was the only answer, and others agreed with him. There was, however, deep reluctance in the diggings to share the burden of reducing supply evenly amongst the still numerous operations. As if these problems were not enough, Kimberley, despite rigorous health restrictions, suffered a smallpox epidemic which further rocked the diamond industry.

The most forward plan for amalgamation was one hatched by J.X. Merriman, head of the Mining Commission in the Cape Assembly, which saw Joseph Robinson as the catalyst for change. Whilst Robinson was certainly a respected and powerful figure and very keen on mine amalgamation in Kimberley, he had a problem in that his diamond mining interests had been badly affected by ground collapse and he was deeply in debt. His control stake in the Standard Company was pledged as collateral against his bank borrowings and he could not raise money to support Merriman's plan and Merriman had to withdraw. The situation was opening up for Cecil Rhodes and in 1887 he and Beit successfully gained total control of the diggings at the De Beers mine and then turned their attention to the Big Pit, with Merriman and Robinson gone from the scene. However, someone else had been watching from the sidelines and now stepped forward – Barney Barnato.

BARNATO BATTLES RHODES FOR CONTROL AT KIMBERLEY

Barnato, who had always had exposure to the Big Pit, had gained control of the Central Company which had formerly been controlled by Francis Baring-Gould. He merged Central with Robinson's Standard Company in 1887 as Robinson divested himself of his shares to clear his bank debts. In summary, Rhodes had the De Beers Mine now sown up and Barnato had a very large stake in the Kimberley Mine, but there were still important stakeholders at Kimberley who Barnato needed to win over before he had complete control of the Big Pit.

One of these was Compagnie Francaise des Diamonds du Cap, popularly known as the French Company, which unsurprisingly was primarily owned by investors in Paris, including financier Jules Porges. Rhodes needed an interest in the Kimberley Mine if he was to achieve his dream of total amalgamation of all the Kimberley mines, which as well as De Beers and Kimberley included the Dutoitspan and Bulfontein mines. He decided to have a tilt at the French Company, but with much of his wealth tied up in the De Beers Mine he had to borrow over £1 million to have a chance of acquiring the French Company and thus getting himself into Kimberley.

BARNEY BARNATO (1852-1897)

If Rhodes was the great political figure in late 19th century South Africa, his business rival Barney Barnato was definitely the most colourful of the Randlords. Barney was born Barnett Isaacs in 1852 in the East End of London, one of five children of Isaac Isaacs, a Jewish shopkeeper. His early life was one of grinding poverty and earning money by his wits as a Petticoat Lane trader and also with his brother Harry, a magician, in a minor music hall act in London.

Hearing of the diamond rush in South Africa, Harry went to Kimberley and set himself up as a diamond buyer and also starred as a magician at the Mutual Hall. Barney soon followed Harry to Kimberley in 1873 and set up as a diamond dealer, and also not very successfully worked some claims. His talent as a stage actor led him to augment his income with regular performances at the Theatre Royal in Kimberley, where he was a popular turn.

Eventually Barney and his brother joined forces to trade diamonds at a difficult time for the market and in 1876 they bought a further number of claims. This proved a good deal for Barney – the claims were highly prospective and initially produced a weekly income of £1,800. Barney also decided in 1878 to enter politics and stood for the Kimberley Town Council. He won his election in circumstances that some observers thought dubious, but he proved both a colourful and diligent councillor, particularly when supporting the local market by using the old promotional charm he had learned in the East End as a youth.

In the early 1880s the amalgamation of some of the Kimberley diggings as a result of the increasing depth of the mining led to the formation of the Barnato Company, which became one of the most profitable operators for a time. The difficulties that many of the new larger organisations mining in Kimberley experienced did not seem to trouble Barnato's operations, which led to speculation that the company was deeply involved in dealing in illegal (stolen) diamonds, a common activity at the time. Indeed the popular Barney's reputation plunged when he manoeuvred to have one of Kimberley's leading policemen – who was responsible for cracking down on illegal diamonds – dismissed from the force. Barnato's operations eventually suffered ground collapse and mining on them ceased for three years.

When mining restarted in the late 1880s Barney was ready to take amalgamation at Kimberley to its logical end. This pitted him against Cecil Rhodes. He and Rhodes went head to head for control of the

Central Company and although Barney was far richer than Rhodes he lacked Rhodes's tactical skills. In the end Rhodes became the largest shareholder in Central, and he and Barnato became uneasy partners. Although Rhodes came out on top Barnato always believed that it was he who had triumphed, a view bolstered by his chirpy confidence. His self-bullishness was further supported by his election to the Cape Assembly in 1888.

The same year Barnato, intrigued by the gold discoveries in the Transvaal, went up to Johannesburg to see if this was an opportunity for him to make another fortune. Although sceptical at first he decided that the gold story was real and he enthusiastically began to buy up substantial leases on the Witwatersrand as well as other businesses, including the waterworks in Johannesburg. His nephew, Solly Joel, managed Barnato's Transvaal operations as Barney himself began to spend increasing amounts of time in London. His long-term lover, Fanny Bees, whom he had lived with discreetly for 15 years, became pregnant for the first time and so Barney finally married her in London in 1892. A daughter, Primrose, was born in 1893 and they had two other children later. Barney also built a mansion on Park Lane in London within walking distance of Joseph Robinson and Alfred Beit's mansions, all of them monuments to the wealth that could be made in the colonies.

For a few years Barney was intermittently feted by the establishment in London but was never really accepted due to his East End Jewish beginnings. Whether in South Africa or London he remained the jovial soul of the party, but the clouds were gathering over his business empire. He was overstretched and cracks began to appear. All was made worse by the infamous Jameson Raid which was to have supported an uprising of the *uitlanders* (foreigners who had gone to SA when the gold rush started) on the gold fields against the Kruger government. It was a disaster and many of the ringleaders were tried and sentenced to death.

Barnato aggressively attacked the Pretoria hierarchy and Kruger himself, demanding clemency and threatening to divest from the Transvaal completely. Whether his intervention was decisive or not, Kruger materially reduced the sentences but Barnato's days were numbered in any case as investors lost confidence in him. His penchant for seemingly divesting from businesses before they crashed and manipulating his own companies with no regard for the interests of his fellow shareholders has an echo of the more recent activities of assassinated *robber baron* Brett Kebble. It is eerie that Kebble's master company, JCI, was a successor company to Barnato's Johannesburg Consolidated.

In 1897 Barnato, who had become disturbed and erratic, and his family set sale for England from the Cape. During the voyage Barnato fell or jumped overboard and was dead when his body was recovered (a fate suffered more recently by fraudster Robert Maxwell). Although Barney had built and, in part, lost a great fortune from diamonds and gold in South Africa, and earned a place in history, to many in both South Africa and England he was just a spiv from the East End who eventually flew too close to the sun.

THE FRENCH COMPANY FALLS TO RHODES

Rhodes left for London in early 1887 and there got financial backing to buy out the French Company; he crossed to Paris and after much negotiation his offer for the company was accepted. But Barnato, alarmed by Rhodes's plan, moved to offer a higher price. This created problems for Rhodes, who was short of cash, but he had an alternative plan and he persuaded Barnato to allow him to complete his purchase of the French Company, which he would then sell on to Barnato's Central Company for cash and shares.

Barnato agreed and so Rhodes lost the French Company but gained a 20% stake in the newly expanded Central Company, which by now as good as controlled the Big Pit. So Rhodes controlled De Beers and he had a sizeable indirect stake in the Kimberley Mine. From this point Rhodes began to work on Barnato to persuade him to cooperate in the amalgamation of De Beers with Kimberley on a 50-50 basis. Barnato waxed hot and cold over the proposition but could not bring himself to agree to it as he believed that Kimberley was worth more than De Beers.

Whilst this was going on the situation in the diamond market was deteriorating. The De Beers and Kimberley mines were producing flat out, selling diamonds into a market that was, as it had been for some years, oversupplied. Diamond prices were weak and when Barnato proposed to float the Kimberley Mine on the London Stock Exchange – which would have intensified the price war and severely undermined Rhodes's plan for amalgamation of De Beers and Kimberley – Rhodes decided that a frontal assault on the Central Company to try and wrest control of it from Barnato was the only way to resolve things and so regain control of the diamond market.

In early 1888 Rhodes approached his loyal friend Alfred Beit to help him access the finance needed to enter the local stock market to buy Central shares. Beit responded positively and provided the required funds. Rhodes told Barnato what his plan was and battle commenced. Since both Barnato and Rhodes needed to acquire, either directly or for supporters, sufficient Central shares to take them above 50%, it was inevitable that the price of Central shares would rise sharply, and so they did from £14 to £49.

Whilst this was going on the underlying performance of the Central Company, and all other diamond miners in and around Kimberley, was deteriorating and most operations were barely profitable. Rhodes's plan, however, justified his tactics, for if he could gain control of all the Kimberley diamond operations, and Central was critical to him, then he would be able to cut overall production, bring the market back into balance and the mines would once more be profitable.

The battle for shares raged on but Barnato's supporters found themselves unable to resist taking profits as the share price of Central rose. Rhodes's supporters were more disciplined and in the end his holding in Central rose above 50% and Barnato threw in the towel. Though he was personally much wealthier than Rhodes he could not match Rhodes's tactical skill. The Kimberley diamond diggings were finally amalgamated and merged with De Beers Mining. The new company was called De Beers Consolidated Mines, a name that was to last until the end of the following century.

FINAL CONSOLIDATION OF THE DIAMOND FIELDS

In 1889 De Beers gained effective control of Duitoitspan and Bultfontein, and all the diamond diggings in and around Kimberley were finally under Rhodes's control. Barnato and Rhodes therefore became partners after years of rivalry, but their disagreements were not at an end. Rhodes had sought control of the diamond diggings, not only for the purpose of more efficient working and control of output levels and therefore prices, but also because he wanted to use it as his vehicle for extending British influence into the heart of Africa. Barnato strongly opposed Rhodes's imperial dream and it is unlikely that the government in London was keen either. But Rhodes was on a winning streak and the new deed of the merged company had at its centre the mission not only to mine diamonds but also to extend British influence throughout the continent using De Beers as the vehicle should the board (i.e. Rhodes) so decide.

The last years of the 19th century saw the newly amalgamated diamond fields bedded down and some recovery in the market as a more disciplined approach to mining and selling was implemented. At the same time the diamond czars themselves – Rhodes, Barnato, Beit and Robinson – were busy on new fronts in the goldfields of Johannesburg, Rhodesia and, for all but Rhodes, in the acquisition of mansions in London. The Boer War was also looming, which Rhodes welcomed as allowing his dreams of African hegemony to flourish in the wake of the inevitable British victory. The British did indeed defeat the Boers and annexed the independent Boer Republics but it was a very hard fight indeed, and Rhodes died before the victory. Barnato and Beit also died young, Beit made it past 50 but to just 53. Only Robinson survived into old age, dying in 1929 aged 89.

The 20th century saw the diamond industry go through, varyingly, periods of prosperity and struggle as economies grew and contracted. When the depression of the 1930s threatened the industry Sir Ernest Oppenheimer, building on Rhodes's amalgamation success, tightened market discipline by setting up the single marketing channel, the CSO. We will look at this in more detail later. It saved the industry and lasted over 70 years before new forces in diamonds led to a break up of what had come close to being a De Beers monopoly. Rhodes would have been highly satisfied that his brainchild had managed to survive so long.

3. THE GOLD RUSHES OF THE 19TH CENTURY

CALIFORNIA

THE FORERUNNER IN NORTH CAROLINA

Although the California Gold Rush, which began in 1848, holds its place in the consciousness of Americans as the country's ultimate mining event, it was most certainly not its first gold rush. That honour lies with the North Carolina gold discovery at Little Meadow Creek in Cabarrus County in 1799 by Conrad Reed, the 12-year-old son of John Reed, a soldier of the Crown who had deserted from the British during the War of Independence.

Gold mining in North Carolina lasted into the 20th century but after the Californian rush it struggled, and Canada's Klondike rush at the end of the 19th century all but killed it off. Whilst a number of very large nuggets were found, weighing several pounds, the mines established in Cabarrus County were small scale and the historic output of the area was insignificant compared to California and other great US goldfields to come, like Nevada.

FROM SAWMILL TO GOLD RUSH

The California Gold Rush started as a result of the discovery of a number of small gold nuggets in a ditch by James Marshall, a carpenter from New Jersey, who was building a sawmill near Sacramento, now California's capital, at Coloma. Within two years 90,000 prospectors had descended on California and four years later that number had swelled to 300,000, of which a quarter were thought to be foreigners. The gold fields stretched almost 150 miles, situated between the Sacramento and San Joaquin Rivers and the Sierra Nevada.

Many of the people who made the long journey either by sea to San Francisco or overland to Sacramento had not come to look for gold, but to provide the goods and services that are always in demand when a mining camp comes into being. As the number of camps increased so did the waves of people coming to California. In fact in the early days following Marshall's discovery the find was kept quiet as California was negotiating its separation from Mexico, which controlled the state at the time. Once the cession of California had been completed the news of Marshall's find began rapidly to spread, first across America and then round the world. The first prospectors who arrived called themselves the Argonauts after the ancient Greek heroes of mythology who sailed with Jason in search of the Golden Fleece. More popularly they were known as the Forty Niners (49ers) in recognition of the fact that many of them arrived in the gold fields in 1849.

What made the Californian gold rush unique at the time was the fact that there were no monopoly owners of the land, such as the state. Thus after staking a claim any gold found was the direct property of the claim holder. Anyone who could make it to California and, having staked a claim, find gold could become rich almost overnight. The Californian rush also led to the manufacturing of convenient sized unofficial gold coins to be used in commerce on the gold fields – this was preferable to the rather inconvenient use of gold dust as *coin*, which was to be a feature of the Klondike rush which lay ahead.

The mining methods used in California were primarily those appropriate to the alluvial nature of the gold deposits – pick, shovel and pan. Material was dug and scraped out of the rivers and streams and then washed in the pan, with the gold being simply picked out. This was relatively expensive in terms of labour costs but required very modest capital inputs. However, with literally thousands of miners working the streams the easily accessible gold was quickly found by the, as always, lucky few. After that, attention turned to the Sierra Nevada mountains to the east where the gold already found almost certainly had originated.

THE RUSH ACCELERATES

Mining now became more sophisticated as substantial sluices for washing the gravel and rock had to be built. This stage began in 1851 as the rush of eager prospectors continued. It is believed that once this more formal placer mining began, production ran for some years at the rate of 2.8 million ozs per year, with the gold price around $5 per oz. During this period the process of hydraulic mining was developed – this was the precursor of techniques used today to 'mine' the old dumps – particularly in South Africa – where power hoses or nozzles break up the old dumped tailings from past mining.

The expansion of mining into the Sierras did enormous damage when hydraulic techniques began to be employed. The need for hydraulic mining grew out of the fact that the surface gold gravels were soon worked out due to the huge number of miners working claims in the area. Prospectors discovered, though, that over time gold, having been forced to the surface by ancient geological movement, had been laid down on several levels. They had initially worked the surface deposits that had not, for a variety of reasons, been re-covered by the movement of earth and other materials. After that, deeper level mineralisation was worked but this was labour intensive, requiring large amounts of material to be moved by hand. Some gold had even made its way to bedrock and then been covered by moving material over eons, but it lay, sometimes, a few hundred feet below the surface and it required laborious digging and also tunnelling to extract it.

Faced with the prospect of mining huge quantities of dirt the invention of hydraulic mining was a godsend to the 49ers. It was not a godsend to the environment of the Sierra Nevada. Sluice boxes had littered the area for some years and the dams that each individual mining operation had built to store water for washing the dirt in the sluice boxes often flooded onto adjoining properties. The dams were also often filled by constructing flumes (open wooden viaducts) or cutting ditches to carry the water. By the late 1850s several thousand miles of these carry-ways had been built, causing huge environmental damage in the Sierras. Hydraulic mining also massively increased the scale of water use, although rainfall and therefore water runoff was fortunately more than adequate; the rights to this water were expensive and could be as high as $75 per day.

Although damaging, the effect of hydraulic mining on productivity and thus on mining costs (excluding water) was dramatic. Taking wages on the gold fields at $4/day, a cubic yard of dirt would cost $20 to process with an old fashioned pan, but just 20c to process by hydraulic nozzle.

HOW THE PROSPECTORS GOT TO CALIFORNIA

The experience of those taking part in the California Rush was easier than the privations that would be experienced by those making it to the Klondike in that later rush. However, getting to the Californian gold fields was no picnic. Whilst today there are many road and rail routes across and through the great Sierra Mountains into the Pacific USA, in the 19th century the overland crossing from the eastern and central states into California was fraught with danger.

The other routes used were sea routes, first the long voyage down to the tip of South America, round Cape Horn and up to San Francisco, before the relatively short land journey south to the gold fields. The time this route took and the considerable dangers posed by the huge storms that often blew around Cape Horn led to a third route being established, sailing to the Panama Isthmus and then going by land to the Pacific side to board ships to San Francisco at the Gulf of Panama. Whilst this latter route was obviously faster than the Cape Horn route it was not without danger as Panama's climate harboured diseases such as malaria and yellow fever, which led to many fatalities amongst the travellers.

THE ECONOMIC ASPECTS OF CALIFORNIA'S GOLD

The transfer of California from Mexican to American ownership, the gold rush and the US Civil War followed swiftly one after the other. The allegiance of California during the Civil War was to the Union side but there was some agonising over the issue as there were those in California who wanted the state to allow the ownership of slaves, but at the same time for California to be essentially a non-slave state. This meant that those already enslaved elsewhere could not escape to California to become free men but that no free black man could be enslaved in the state. Although this curious exercise in semantics led to a number of legal disputes, in due course California established itself as an anti-slavery state and a supporter of the Union during the Civil War. This may have been decisive in financial terms as Californian gold was shipped to New York and provided a backing for the Yankee dollar that the Confederate currency never had.

In fact, the importance of gaining some legitimacy for the Confederate currency led to attempts by Californian supporters of the South to waylay bullion ships bound for New York from California. Such attempts failed and as the Confederate side steadily lost ground to Unionist forces, the legitimacy (and value) of the Confederate dollar was severely undermined. The critical cotton trade between the southern states and the spinners of northwest England was also seriously damaged by the conflict, which all but cut off an important source of foreign currency earnings which could have provided some support for the Confederate dollar – the

British pound then being the world's most influential currency – although in those days gold was the sole reserve asset.

Metals, as we have seen, have played an important role in the economic and social evolution of civilisation with the role of precious metals, particularly gold, being generally, but not exclusively, to provide financial support for commercial activity and development. In the case of the Civil War it could be argued that gold was an important factor in the survival and triumph of the Union concept and the eventual advance of the US to superpower status.

Certainly a victory for the Confederates could have led to the establishment of close political ties between the South and the UK to match the already established commercial and trading ties. If such an alliance had prospered world history in the turbulent 20th century might have developed very differently, although one very large snag would have been British opposition to slavery. However, the ability of the Unionists to access strategic supplies to force home a military victory through the solvency of their gold-backed currency meant that speculation about an independent South remained just that.

We could also speculate about what might have happened if Mexico had kept hold of California and reaped the benefit of the gold discoveries. It is likely that much less of the gold produced would have found its way to New York and Washington and the Unionists might have found themselves forced to do deals with foreign interests to achieve their aims and bring to an end one of the most vicious wars of the 19th century.

THE RUSH MATURES

By the end of the 1860s, California's gold fields had increasingly matured into corporate entities. The individual prospectors and adventurers had slowed to a trickle and the industry had begun to organise itself. Many of the earlier mines had become worked out and had closed, fortunes had been made or lost and in most cases, as with all gold rushes, the dreams of the many had tarnished and died. Those who chose to follow the gold rush through the share market rather than at the diggings occasionally made money on their speculations but more often turned out losers.

Amongst those who won and then lost a fortune was John Fremont, whose earlier years had been spent in the US army where he had led a number of expeditions over the mountain ranges of the western US into California and Oregon, and was widely admired for his boldness and vision. In 1847 after effective, if at times illegal, actions against the Mexican government in California, Fremont became the military governor of California, but he unfortunately got on the wrong side of some army infighting and was court martialled in Washington on trumped up charges and dismissed both from his governor's post and the US army.

Before he was taken to Washington for the court martial, Fremont gave the American consul in Monterey money to purchase a ranch north of San Francisco. The consul, Thomas Larkin, instead purchased a large tract of land in Indian country in the Sierras named Mariposa, now Mariposa County which includes the Yosemite National Park. The purchase was to make Fremont a fortune as Mariposa turned out to be the southernmost end of the Californian gold mother lode, though at the time Fremont was furious with Larkin for his deception.

When Fremont returned to California to live at Mariposa he found himself in the middle of Indian unrest caused in part by the incursion of gold prospectors into the Sierras. Fremont's land was also the subject of long-running legal disputes over the ownership of mineral rights. Miners, who worked the gold deposits found at Mariposa before Fremont returned to California, claimed that he did not own the underlying mineral rights and it took Fremont six years to get a favourable legal ruling. In the end Mariposa made Fremont a very rich man but he was to lose most of it in due course, as he lost control of the Mariposa deposits – some of which he himself had developed – due to his frequent absence on political business in the East.

Fremont became one of the first two senators for California in 1850. Later, in 1856, Fremont became the first presidential candidate of the newly formed Republican Party and although he lost the election he continued in politics thereafter and returned to the army on the Union side when the Civil War broke out in 1861. After losing control of the Mariposa mines he invested the huge profits he had made from gold in other business enterprises, such as railroads, but they all failed, and politics and the army earned him very little. He died a poor man in Brooklyn at the age of 74.

But Fremont was just one of a number of key characters in the Californian Rush who made and lost fortunes. John Sutter, whose Coloma sawmill it was that James Marshall was working on when the first gold was discovered, was another. Sutter had no ability to manage the conflicts raised by the proximity of his land to the goldfields, the need for labour to operate and expand his sawmill business and his ambitions to build a major administrative centre in the area. His gold prospecting in the hills above his land, whilst successful, was poorly organised and he eventually abandoned the operation and returned to the sawmill and his other businesses. Unfortunately they had deteriorated in his absence and when miners diverted the American River the mill ceased working.

The golden age for Californian gold lasted 20 years until the 1870s when most of the 49ers had either made their fortune or lost their shirt, and all had grown old.

After this period California, which had attained statehood in 1850, continued to attract settlers in large numbers but gold was no longer the target, instead it was the state's verdant environment, fine climate and opportunities to prosper that attracted. As for gold mining, as with many other gold rushes around the world, the easy pickings were made early and in due course large amounts of capital, beyond the resources of prospectors, had to be raised to mine at depth and haul the large quantities of lower grade ore required to make a gold mine economic.

Today California is a difficult state in which to mine because of environmental regulations. Production at the state's largest mine, the Mesquite, is currently close to 200,000 ozs per year, which compares with 4 million ozs along the Mother Lode at the height of the Gold Rush in 1852. By 1865 output had collapsed to barely 800,000 ozs as placer mining was replaced with hard rock production. That figure had halved by 1929 but then the sharp increase in the official price of gold in the 1930s sent Californian production soaring to 1.4 million ozs by the end of the decade, before sinking back again.

OTHER AMERICAN GOLD RUSHES...

PIKES PEAK, COLORADO

Little more than ten years after the Californian discoveries, gold was found in 1858 in what became the state of Colorado. The location was in the Pikes Peak mountain area at the confluence of the South Platte River and Cherry Creek, in what is now the Denver suburb of Englewood. The area had been known to host small gold placers since the early 19th century and some of the prospectors bound for California found modest amounts of gold in the South Platte area as they passed through.

William Russell, who came from the southern state of Georgia and had been prospecting unsuccessfully in California, decided to try his luck in Colorado. With a growing party of curious bedfellows, including family, prospectors and Cherokee Indians, he made his way to the South Platte River. A small placer find of around 20 ozs was made but South Platte, although an intermittent producer, was never a prolific source of gold. However, prospectors widened their search and traced the placers to the Rocky Mountains to the west and also followed the Colorado mineral belt southwest, finding gold as they went.

IDAHO AND OREGON

Following the Californian rush gold was also found in the northwest in Idaho and in Oregon. Parts of this area had been the site for disputes between the US and Great Britain over the rights of the Hudson Bay Company's trappers and traders to operate around the 49th parallel. In due course these disputes were resolved and also treaties signed with local Indians to sanction prospecting, allowing miners to follow the fur trade into the area. A number of gold deposits were found and developed in Idaho including in 1860 the Cariboo placers on the Fraser River, the Clearwater gravels on the Snake and, in 1861, the Miller Creek gravels on the Salmon river. In 1863 the Boise Basin to the south was opened up as gravels and then quartz veins were discovered; silver mines were also developed in the nearby Owyhee valley.

CRIPPLE CREEK, COLORADO

Many of these early finds were small but in 1891 the persistence of the prospectors paid off when gold was discovered in Colorado at Cripple Creek, just a few miles from Pikes Peak. This find made by Robert Womak, a local rancher, was significant and eventually was developed as the Gold King mine. The same year Winfield Stratton discovered the Independence lode, a large, high-grade gold deposit where underground gold values ran at between 1 and 2 ozs per ton. In 1900 Stratton sold the Independence mine to London investors, who floated the project on the London Stock Exchange; the flotation was a failure when soon after listing reserves at Independence were significantly downgraded.

Despite that Womak and Stratton's discoveries set off another huge gold rush drawing thousands of hopeful prospectors to Cripple Creek; and many years after in 1990 Stratton was inducted into the National Mining Hall of Fame in recognition of his Cripple Creek discoveries. Such an honour would have surprised the early century London investors in Independence but they were long dead by then. Over the years the Cripple Creek area has produced around 700 tonnes of gold, half of Colorado's output, making it a major source of American gold, albeit much smaller than California where well over 3000 tonnes of gold have been produced since 1849.

LEAD IN SOUTH DAKOTA

Perhaps the greatest gold discovery, certainly in terms of mine size, made in the US in the 19th century was the Homestake mine at the town of Lead in South Dakota. The mine came into operation in 1878 under the control of George Hearst, a talented mining prospector and successful mine owner, who eventually founded the Hearst newspaper empire. The initial discoveries in Dakota's Black Hills predated Hearst's arrival by a couple of years and these discoveries were panned from creeks in the hills, and though rich were quickly worked out.

The Homestake mine and the adjoining Golden Star claim were staked by Moses Manual, a prospector, and were hard rock discoveries that contained ore of a lower gold grade than had been discovered in California, seldom more than 1 oz per ton. However, the width of the underlying orebody was very large, several hundreds of feet in places, making for efficient underground working with high-volume production possible.

Hearst was a ruthless man who had no conscience about the tactics he used to establish his hegemony in Lead and he employed enforcers to support his business strategy. By 1879 Homestake Mining had been listed on the New York Stock Exchange and it remained there until it merged with Barrick Gold in 2001. As an investment one of its claims was that it performed directly contra to the US stock market in the years of the Great Depression in the 1930s, making its shareholders large amounts of money while most around were losing their shirts as industrial stocks plunged in value.

GEORGE HEARST (1820-1891)

It is an achievement to found one business dynasty but George Hearst founded two – in mining and in publishing. He was born in Franklin County, Missouri, in 1820. His father was William Hearst, a pioneer farmer who married Elizabeth Collins in 1817. The family was poor and Hearst received only a sporadic and basic education. His father died in 1846, leaving the family with a large debt and Hearst had to look after his mother, sister and crippled brother as well as run the family farm and repay the debt. He had an aptitude for business, turning the farm into a profitable enterprise, opening a small store and leasing a couple of lead mines in the area, and in two years he had paid off the debt.

In 1850 Hearst, encouraged by his successful lead mine investments where he had immersed himself in understanding the mining business, decided to go to California with two of his cousins in the wake of the developing gold rush there. Hearst quickly discovered two rich gold-bearing lodes, firstly at Merrimac Hill and then at Potisi near to Nevada City, and with his efficient mining methods had amassed sufficient money within a year to invest in a theatre in Nevada City and a store in

Sacramento. These investments were less successful than gold mining and over the next few years Hearst and his cousins, having sold both Merrimac and Potisi, found additional gold deposits which they traded on, making handsome returns at each turn.

In 1859 Hearst trekked to Nevada intrigued by stories of rich silver potential and there he bought a 16% stake in the Ophir silver mine. Hearst worked out an efficient method of treating the metallurgically difficult ore and the world famous Comstock Lode was born. On top of his growing mining interests Hearst invested in land in California, some of it bought for ranching. He lived hard, gambling, drinking and caring little for his appearance; a dishevelled look belied both his financial success and his abilities as a mine owner.

In 1860 Hearst returned to Missouri to look after his ailing mother, who died the next year. Whilst he was home he met and fell in love with Phoebe Apperson. She was only 19 and her parents disapproved of the relationship so the couple eloped to California in 1862. A year later they had a son, William Randolph, who was destined to become one of America's most powerful newspaper magnates. In 1865 Hearst moved into politics as a Democrat, winning a senate seat in the California State Assembly. He also began to accumulate land near San Simeon, where the legendary Hearst Castle was eventually built by his son, William Randolph, with work starting in 1919.

Following his spell as a politician, with his family settled at San Simeon Hearst plunged back into mining and over the next few years, and with a new partner, James Haggin, he made investments in the Ontario silver mine at Park City, Utah and the Homestake gold mine in South Dakota. In 1881 he bought a 25% interest in the Anaconda gold/silver prospect in Montana, which eventually became the largest copper mine in the US. This string of investments crowned Hearst's career as a mining man and in 1886 he returned to politics becoming a US Senator for California. His son never showed interest in mining but Hearst had acquired the *San Francisco Examiner* in 1880 in payment of a gambling debt and it was this asset that his son used as his base for the building of the Hearst newspaper empire in the 20th century.

George Hearst's legacy is a formidable one, including the development of some of the US's greatest mines and the founding of the Hearst newspaper empire. Some were envious and portrayed him as a ruthless businessman set on personal aggrandisement, particularly in developing Lead, South Dakota, the town built around the Homestake mine.

Others saw Hearst somewhat differently, viewing him as a man of integrity and great business ability who was always courteous and kind.

Whatever the truth of these differing views Hearst was definitely an unpretentious man who lived simply; the development of the San Simeon estate into the lavish Hearst Castle was his son's doing, not his. His wife, who had played a key part in the building up of the education system in Lead, after his death became heavily involved in the expansion of the University of California, endowing the museum at Berkeley which bears her name. She died in 1919.

THE HOMESTAKE MINE

The genesis of the town of Lead and the Homestake mine are thought by some to go back to the 1830s when a group of prospectors are believed to have mined economic quantities of gold as a result of an exploration trip into South Dakota. Their fate was to be pursued and slaughtered by Indians, and their misadventures were briefly documented by the last of the hunted prospectors on a stone tablet found later. However, it was observations of gold shows coming from the unlikely direction of US army troops led by General Custer that started the early gold rush to the Black Hills; Custer, of course, was to perish along with hundreds of his men at the Battle of Little Bighorn in 1876. The US army considerably bolstered its forces in the region following the military disaster and mining development in the Black Hills was able to proceed in a secure environment.

The fortunes of the Homestake mine itself mirrored the improvement in technology over its life and never more so than in its early years when its low-hanging walls, timber supports and donkey based haulage was steadily replaced by much more capital intensive working methods in keeping with the size of the orebody. Over its life it produced almost 40 million ozs of gold and the workings reached a depth of 8,000 feet – not far short of some of the deeper South African gold mines – and a huge open cut operation was also established at the Homestake on the site of the original discovery made by Moses Manual. The mine was the largest US producer for a large part of its life and also the largest producer for much of Homestake Mining's long independent life until 2000.

Like many gold mining towns Lead at its inception was a rough place, but over the years, under the philanthropic hand of George Hearst's wife, Phoebe, a cultural glaze was applied to the town's activities, bringing music, theatre, film, sport and education into the lives of the isolated citizens. Whilst most mining towns have very short lives Lead was fortunately bound up with a mine that operated for well over a hundred years. It also has had the good fortune to have attracted a major scientific project to the closed depths of the old mine in the shape of the nuclear physics experiments and investigations of the Deep Underground Science and Engineering Laboratory. With its historic areas also preserved, Lead has outlived the mighty Homestake.

THE CAROLINAS

We have seen earlier that the first recorded discovery of gold in the US was in North Carolina and the state remained a small producer into the 20th century. South Carolina also hosted a number of gold mines and in the case of the Haile and Brewer mines they operated for over a hundred years. However, the scale of gold mines in the state was small. This was the case with many of the gold mines developed in the US. Another feature was the role of gold as a by-product and huge base metal mines such as Butte and Bingham Canyon have produced millions of ozs of gold over the decades even though their prime role was as copper mines.

...AND SILVER TOO

Whilst today we tend to view silver as very much the poor relation of gold (and we are far more respectful of platinum), for much of history silver was as sought after as the yellow metal, and the US had more than its share of world class silver discoveries. Perhaps the most famous was the Comstock Lode in Nevada, discovered in 1858 close to the Carson Trail, one of the routes taken by prospectors heading for the Californian gold fields.

THE COMSTOCK LODE

The Comstock had a relatively short life as a major producer (at 20 years), but during those years its output was huge, peaking in 1877 at around an annual 35 million ozs of silver and 400,000 ozs of by-product gold. The lode consisted of a number of separate operations and was named after Henry Comstock, a Canadian from Trenton, Ontario, who, although not the discoverer of the lode, bullied his way into the play and negotiated joint ventures with a number of the leaseholders on the grounds that he had a prior claim to the leases. This was Comstock's only trick; basically he was an uneducated, not very bright man and in due course sold out and then lost his money in poorly run business ventures, before committing suicide in 1870.

The credit for the Comstock discovery is usually shared amongst several men including Ethan and Hosiah Grosh, James Finney, John Bishop, Jack Yount, Aleck

Henderson, Peter O'Reilly and Patrick McClaughlin. In reality the Comstock ran for many miles and the competing claims could be rationalised as separate, coincident discoveries on the same geological phenomenon – the lode.

However, undoubtedly the key participants in developing the full potential of the lode and solving its considerable geological and operating problems were Adolph Sutro, a German, and William Sharon from San Francisco. The lode itself was very rich, running at over 3000 ozs per tonne in places, but it had two serious problems: the ore was very friable so workings tended to collapse; and water found at depth in the workings (the lode dipped to around 3000 feet) was scalding hot and even huge pumps were hard pressed to keep pace with the inrush. The solution to the water problem – which Sutro came up with – was to dig a tunnel eight miles in length from the mountain, which was up the valley from the Comstock mines, ending under the lode. This had the dual purpose of providing drainage and also ventilation to reduce temperatures in the deeper workings.

Sharon, who represented the Bank of California and other investors in Virginia City, the fast expanding town that had grown up round the Comstock mines, was opposed to the tunnel being built and he slowed its financing by lobbying eastern lenders. His plan was to take advantage of mining problems to acquire some of the financially failing smelters that had been built around the lode and thereby gain a stranglehold over the Comstock and the increasingly sophisticated urban services being offered to the mines and mining men by Virginia City. Sharon was successful but Sutro also, after eight years, got his tunnel built. The problem of collapsing workings was solved by building a wooden fretwork in the ore spaces to stabilise surrounding rock as the ore was mined and removed. The other problem that needed to be addressed was finding the most beneficial method of treating the ore.

In the early days of the Comstock, which was essentially a large and long orebody not a vein system as was most familiar to American miners, the mining and treatment process was extremely wasteful with much silver-bearing ore being washed away into the rivers with the waste rock. The problem was finding the right amalgamation process by which the silver could be separated from the ore in an economic manner. In the end the Comstock miners developed the Washoe process, a development of the earlier pan amalgamation process, where the ore was deposited in iron tanks, and mercury, salt and copper sulphate added to the mix to agitate and then separate the silver out.

Also, so important was the Comstock that direct rail links were laid to Reno to tie up with lines going to the east and to California, and a high-pressure city water system was installed in the form of a 12-mile pipe being constructed to bring water from the Sierra Nevada mountains to Virginia City. However, the Comstock's days of glory were short lived and by 1880 the mines were in rapid decline, though sporadic mining did continue into the 1920s.

OTHER SILVER STATES

Whilst silver was found and mined in many US states after the Comstock discovery, the two most substantial producers were arguably Idaho and Montana. One of the biggest silver producers was the Galena mine in the Coeur d'Alene district of Idaho, which began operation in 1887 and is still in production today, although it was halted in the 1930s and then again in the 1990s. The other giant silver producer in the district was the Sunshine Mine, which began operations in 1904 and, though currently closed, a plan to re-open it is being hatched by its new owners, Silver Opportunity Partners. Since 1887 the district has produced over 1 billion ozs of silver and substantial quantities of by-product copper and lead.

Second to Coeur d'Alene was Montana's Butte mining district, which started as a gold-producing area but in 1874 as the gold ran out rich silver veins were discovered. A key player in Butte was Anaconda, owned by miner and entrepreneur Marcus Daly, who bought a small silver producer in 1881 and turned it into a giant copper mine, a development we discuss later on. Because of Daly, who died in 1900, Butte became better known as a copper mining centre during the 20th century, although over its life it has produced around 700 million ozs of silver. The other main silver-producing area in Montana was Phillipsburg, where silver was first mined in 1864.

One of the problems that late 19th century silver miners in the US had to face was the repeal of the Sherman Silver Purchase Act in 1893, which had only been enacted three years previously. The Act ordered the US government to buy increased amounts of silver with Treasury Notes that could be exchanged either for silver or gold if the recipient so wished. Many holders exchanged their notes for gold-backed dollars, thus depleting the US gold reserves. The whole idea of the Act was to create inflation and so inflate away the value, and thus burden, of the substantial debts that the US farm sector owed and to give a boost to the silver mining industry, which was suffering low prices due to overproduction. Unsurprisingly the shenanigans caused by the enactment and then repeal only made matters more difficult for both farmers and miners.

Other major historic silver-producing states included Colorado and Utah, where silver was discovered in the latter half of the 19th century. The largest mine in Colorado was the Leadville mine, which over its life from 1877 until the 1980s produced around 240 million ozs of silver, although after the repeal of the Sherman Silver Purchase Act the main mining thrust switched to lead and zinc. Utah, as many states did, started as a gold state, quickly moved on to silver, then following Sherman continued to mine silver as a by-product to base metals. In the case of Utah, it ultimately became prominent in mining circles as the largest source of molybdenum in the world – this came from another Leadville giant, the Climax mine of Amax.

THE WITWATERSRAND

It is interesting that the first discovery of gold in South Africa's Transvaal, then the centre of the independent Boer South African Republic, was not on the Witwatersrand but at Eersteling in the Eastern Transvaal in 1871. Thereafter other discoveries were made at Barberton, Pilgrims Rest and Lydenberg.

The first gold discovery of significance on the Witwatersrand was made by an Australian, Henry Lewis, in 1874 at Blaauwbank, a farm 43 miles from where Johannesburg eventually sprung up. The first incorporated gold mining company was Nils Desperandum Cooperative Quartz, which, following its early liquidation, became Rustenburg Gold Mining when it raised new capital in 1876. Its efforts were greeted with success as samples from Blaauwbank were rumoured to have assayed between 4 and 8 ozs of gold per ton.

THE EARLY YEARS

Progress in prospecting and producing gold from these early diggings was hampered by the deteriorating political situation. By 1879 the British were embroiled in quick succession in two short wars, first with the Zulus and then with the Boers of the Transvaal, whose independent republic had been annexed by the British in 1877. The British did not acquit themselves very well, with the Zulus defeating them at Isandhlwana, but the Zulus were eventually suppressed with the help of the gold diggers from the Eastern Transvaal.

They fared no better against the Boers who were determined to re-establish their independent republic. Several small-scale clashes between the Boers and the British led to unacceptable reverses for the latter and London dispatched Lord Roberts with a force to South Africa to deal with the Boers. Before the force could be deployed a ceasefire was negotiated by President Brand of the Orange Free State. The Boers won back their right to self-government although under the overall protection of the Crown in 1881; in hindsight further trouble was probably inevitable.

However, for a time attention was able to return to gold. In 1882 another gold field was found in the Eastern Transvaal at what became Kaapsche Hoop near the Swaziland border. The following year a Frenchman, Auguste Robert, found the Pioneer Reef and this was quickly followed by the discoveries of the Barber and Sheba reefs. The latter hosted the Golden Quarry, found by Yorkshire coal miner Edwin Bray, where the ore ran 8 ozs to the ton. The rising excitement in the Transvaal began to attract the attention of South African mining's great and good, who were then headquartered in Kimberley near the diamond fields of the Free State, and Cecil Rhodes and Alfred Beit, amongst others, visited the diggings.

The attitude of the legendary Rhodes to the gold discoveries in the Transvaal, both Eastern Transvaal and later the Witwatersrand, has puzzled historians who believed that he never took gold as seriously as he took diamonds, although he did form the Gold Fields of South Africa company in 1887. On the other hand perhaps he was looking for Eldorado and doubted that it was to be found in the Transvaal. Certainly he pursued the search for gold in Rhodesia with enthusiasm, although it was enthusiasm that proved ultimately to be misplaced.

Whilst at this time gold production was mostly coming from the Eastern Transvaal, and turning the new town of Barberton into a thriving commercial centre, the most interesting prospecting was further west on the Witwatersrand. Indeed the Eastern Transvaal, though it spawned many hundreds of incorporated exploration companies, ultimately was a disappointingly small gold producer and hundreds of investors ended up with worthless paper. The experience soured investors both in Kimberley, then the centre of South African mining, and in London, and confidence did not return easily.

THE RISE OF THE WITWATERSRAND

At the time of the South African discoveries the vast proportion of gold historically mined had come from placer or alluvial deposits in sand, gravel and river beds where the gold was believed to have been washed down from much higher ground. On the Witwatersrand the gold was found in reefs of conglomerates, although the discovery that bolstered the spirits of the financially beleaguered Transvaal government was found in the more common quartz bed from the Confidence Reef of a prospector, Fred Struben, in June 1885. The samples sent to the government analyst in Pretoria contained results per tonne of astonishing richness: 913 ozs, 362 ozs, 301 ozs and 123 ozs. The results were unfortunately simply a flash in the pan as follow-up work revealed no sign of any more bonanza ore. However, prospecting over the Witwatersrand was accelerating and the men from Kimberley once more began to take notice.

Joseph Robinson was the first of the Kimberley bandits to go up to the Witwatersrand but his journey and his staking, once there, relied entirely on the shrewdness and generosity of Alfred Beit, who funded the financially embattled Robinson. Cecil Rhodes and Charles Rudd, who had between them tied up a large part of the Kimberley diamond fields, encouraged by another Kimberley bandit, Hans Sauer, who had already been to the Witwatersrand and fallen out with Robinson, travelled to the Witwatersrand as well.

It was Robinson, however, who managed to stake the most ground. Indeed he did deals right along the 30-mile stretch of the Witwatersrand. He also formed the

Robinson Gold Mining Company from a group of claims that were offered by Hans Sauer to both Rudd, who did not believe there was much gold to be found on the Witwatersrand, and to Rhodes, who also refused. Beit bought them instead for Robinson. The shares of Robinson Gold eventually rose to £80 from £1.

On 20 September 1886, Carl Von Brandis, Gold Commissioner for the Witwatersrand, on behalf of Transvaal President Stephanus Kruger, proclaimed the Witwatersrand a public digging. What was to become the world's most prolific gold field had been officially born. The proclamation more sinisterly also sowed the seeds that eventually led to the savage Anglo-Boer War.

THE WITWATERSRAND GOLD FIELD OPENS UP

With the Witwatersrand gold field now officially proclaimed, Von Brandis proceeded with laying out a town, in effect the birth of what is now Johannesburg. The leases offered were initially only for five years and this caused a storm of protest that led to one, Henry Marshall, buying land nearby and offering leases on more favourable terms. Although illegal this was quickly incorporated into Johannesburg in a deal with the Transvaal government that increased the lease period from 5 to 100 years.

Marshall's town is now the site of Johannesburg's main Post Office and the old financial district. The framework for what was to become Africa's premier financial and commercial city had been laid. And although in its early years it was more like a short-life mining encampment it was not long before more substantial buildings appeared, some even made of stone and brick. Unusually, perhaps uniquely, over the decades the mining city of Johannesburg grew to a size that meant that it has been able to outlive the industry that spawned it and prosper despite the material contraction in gold mining on the Witwatersrand experienced over the last 20 years.

Johannesburg was also very easy to get to when compared with the travails that faced diggers trying to get to the Klondike in Canada a few years later. Regular and competitive coach services from Kimberley and also Durban began with an eye on commercial business, particularly the carrying of mail as well as passengers. However, the main method of travel for prospective diggers was by foot or horse, a gruelling journey of some weeks from the coast, but a walk in the park compared with Canada.

Along with the diggers came the essential service providers for any gold camp; storemen, hoteliers, engineers, mechanics, entertainers and even bankers. In due course, because the Witwatersrand, unlike many gold fields whose lives were often limited, just grew and grew, many of the initial cottage businesses set up to service the diggers and their activities became huge enterprises, even world class, like South African Breweries.

The initial mining operations on the Witwatersrand were those appropriate to prospecting, consisting largely of surface digging, relatively short and wide and shallow, targeting the outcrop of the gold-bearing veins. In the odd case the trench may have gone down 50 feet, deep by the early standards of the gold fields but hardly a dent when compared to the depths finally reached on the Witwatersrand where mines have gone down more than two miles.

What was required was a proper capitalisation of the exploration effort. This would enable groups to bring equipment onto the gold fields to mine at greater depths and also more productively. So in 1887 began a process of incorporation of gold mining companies that was to continue almost up to the present. Although the windy wastes of the Witwatersrand were unsurprisingly unable to provide the capital at this stage of its development, South Africa's other urban centres such as Kimberley, Durban, Cape Town and even sleepy Pretoria had the capital resources, and overseas London was also involved from an early stage.

Amongst the mining companies established were Black Reef, Paarl-Ophir, National Gold, Grahamstown Gold, Primrose, Vierfontein Farm, Kromdraai and Leeuwpoort, and it was also in 1887 that Rhodes floated his Gold Fields of South Africa company in London, which in several guises still prospers to this day. American investment was also involved but mainly in the area of engineering companies such as Fraser and Chalmers and Studebaker Bros, who had built up unparalleled expertise in gold mining in North America. Also in 1887 the Simmer and Jack company, which is still in existence today, was formed.

The impact of the Witwatersrand as a gold field and as a financial phenomenon dwarfed the earlier gold fields of the Eastern Transvaal. A stock exchange was established at the end of 1887 and there was a total of 68 gold companies formed with a combined capital of $US15 million, operating over 1,000 stamp mills which were used for processing the ore. These stamps had the capacity to produce a quarter of a million ozs of gold. Not bad for a standing start, but a mere flicker of what was to come. In terms of their impact on the Transvaal these new gold mines had a massive effect. In 1884 the Transvaal's national revenue was under $1 million, by 1889 gold output on the Witwatersrand had topped $7 million in value and the number of incorporated companies had reached 450. The stock exchange boomed and so did Johannesburg.

The issue of fuel to power steam engines on the gold workings quickly arose. Fortunately, low-grade coal seams were discovered near Boksburg, just 13 miles from the gold fields; it was cheap and accessible, and a short train track was laid to carry the coal to the mines. At the same time, coal was also discovered at Springs to the west of Johannesburg, easing the development of the gold fields as they spread in that direction. The ability to bring in fuel readily was also a boon to the expansion of Johannesburg, which was growing exponentially. In 1888 a small gas plant was built, in the following year a dynamo to generate electric power was installed, in 1891 a horse drawn tram system was laid out and in 1894 a telephone exchange was built.

POLITICAL EVENTS SURROUNDING WITWATERSRAND GOLD

There is much irony in the story of the development of the Witwatersrand as the world's largest gold field and its central role in the growing dispute in the late 19th century between the *uitlanders* in Johannesburg and the Boer government in Pretoria. The dispute ultimately led to the Anglo-Boer War and the growing gold mines of the Witwatersrand were a catalyst for a long-running dispute over representation and taxation.

It is particularly ironic that the US public, keen at that time to twist the lion's (Great Britain) tail at any opportunity, and therefore sympathetic to the Boer cause, should have taken this stance when one of the biggest groupings of *uitlanders* in South Africa were Americans. The American public's sympathy for the Boer cause reached a crescendo during the war and after it many US states offered to take in thousands of the defeated Boers to spare them the humiliation of living under the British yoke. Probably this generosity was born of America's own war against the British.

However, one of the key *uitlander* issues with the Boers was that they had no vote in Transvaal even though their mines provided huge tax revenues to the government in Pretoria, a sentiment that bore more than a passing resemblance to the complaints of the American colonies about the British a century before, and which one might have thought Americans would have had some sympathy with. American support for the Boers also looks odd in the light of more recent events, when in the 1980s the US decided apartheid was a big international political and moral issue, and applied huge pressure on South Africa to abandon the system. This is particularly ironic as many considered the US to be a deeply racist society until the civil rights movement in the 1960s won fundamental rights for American blacks.

The policy of apartheid, which really came to the fore with the attainment of power by the Afrikaaner National Party in 1948, had in fact been around for decades before that, although the British-inclined United Party, who were in power until 1948, practised it with a light touch, with people of mixed race being able to vote in SA elections until 1956. This was also an era when racial discrimination simply was not an issue. Foreign interest in the social structure of South Africa was negligible, and indeed the gold mines eventually developed in the West

Witwatersrand in the 1930s and in the Orange Free State after the Second World War were substantially financed by disinterested foreign capital looking for a good return.

SOUTH AFRICAN GOLD SHARES IN THE 19TH CENTURY

It was inevitable that the discovery of the largest gold field in history should spark off financial excitement as well as political controversy. The requirement for capital to develop even the earlier shallow, high-grade gold deposits made many rich but also, as is usually the case, impoverished many more as eager investors bid up the value of the expanding pool of securities listed in both the goldfields and London.

The first gold company incorporated for the purpose of exploring and exploiting the Witwatersrand, as the Blaauwbank area eventually became known, was the Nil Desperandum Company. There followed over the years a river of company formations, some of which – like Durban Roodepoort (DRD) and Simmer and Jack – are still in existence today over 120 years later. Share price movements were often dramatic, and no wonder when information was so imprecisely circulated.

One example will suffice. Turffontein Gold Mining, formed in June 1887, had sunk a 200-foot shaft in three months but provided no gold assays and merely stated that within two weeks a ten-battery stamp mill would be in operation. The implication of such an announcement was that sufficiently good gold values had been found as to warrant the establishment of such a crushing plant, but the announcement was very short on detail and thus highly misleading. Unfortunately the Turffontein had very low gold values at depth and the mine was closed when the company ran out of capital.

The volatility of gold share markets the world over has always been acknowledged by investors, but give the market a scare like the 1890 Johannesburg crash and panic can destroy fortunes. This collapse in gold shares followed the discovery of pyritic ore at depth on the Witwatersrand – this type of ore did not respond to the conventional treatment process of separating gold from ore by using mercury. The share price movements were huge. Barney Barnato's Moss Rose, for instance, fell 96% from £5 6s to 4s (536p to 20p), Spes Bona fell 99% from £5 15s to 1s 4d (577p to 7p) and even Cecil Rhodes's holding company Gold Fields of South Africa fell 75% from £5 15s to £1 9s (577p to 144p). Following the crash those who trusted the industry to find a solution, which it did in two years, made a fortune; the famous City Deep for example, after a 70% fall, recovered from 1s 6d to reach £17 8s in the 1895 boom (this was the equivalent of a rise from 7.5p to £17.40).

The recovery in the industry's fortunes in 1895, which lifted City Deep, also led to huge rises in other gold mine share prices as the travails of 1890 were completely forgotten. Speculation was also in the air as many companies were carrying out drilling programmes and these were indicating that the Main Reef series, which production was centred on, had real legs, with a depth extension down to 10,000 feet. This suggested, if technology allowed, a life of many decades. It was no wonder that Johannesburg had already, as a city, far exceeded the size of other gold mine towns like Dawson City and Kalgoorlie. Its stock exchange was also far larger than those which grew up round the 19th century gold rushes, and it was further bolstered by the London Stock Exchange where large amounts of the development capital had been raised and where SA gold shares were also traded.

As the boom raged investors were attracted to the dividends that many mines were paying. For example, Crown Reef in 1894, as the boom began to take off, paid out £50,000 in dividends (its initial capital had been £125,000) and Ferreira paid £92,000 on initial capital of £138,000. The share price rises were also equally juicy. Between 1893 and 1895 Simmer and Jack rose from £2 10s to £23 (247p to £23), Rand Mines rose from £5 3s to £37 10s (515p to £37.50) and Durban Roodepoort Deeps from £2 10s to £10 10s (247p to £10.50). In 1894 alone the aggregate stockmarket value of Witwatersrand gold shares increased from £31 million to £155 million. But clouds were building over the Highveld, with many of the residents becoming increasingly angry about their political status because, as we have mentioned before, Johannesburg and the gold mines provided huge tax revenues to the Boer government in Pretoria but citizens had no voting rights in elections.

At the end of 1895 the failed Jameson Raid plunged the city and the gold mines into chaos, although this was short-lived as compromises were struck both locally and internationally to defuse the very difficult political situation which the raid had thrown up. However, the collapse in gold share prices could not be put down solely to Dr Jameson as the previous three years had taken gold shares to exceptional heights and they had begun to buckle some months before the raid. Simmer and Jack fell an astonishing 93% from £23 to £1 13s (£23 to 165p) and Rand Mines fell 67% from £37 10s to £12 7s (£37.50 to £12.35).

The boom was over and the impending Anglo-Boer War created further problems for the mines as output stagnated and then, for the last two years of the War, stopped completely. However, for all the problems that the end of the century had brought the gold mining industry, it entered the 20th century as the largest in the world, a position it was to hold for the whole of the century. As for the gold share market, boom and bust would always be an unavoidable hazard for those participating, partly mitigated by the huge flow of dividends during the boom years.

WESTERN AUSTRALIA

The second half of the 19th century saw a surge in the search for gold, driven by the Industrial Revolution and the huge expansion in world trade that required a central asset for settling accounts that held the confidence of bankers, merchants and nations alike – that asset was gold. So it was that following the Californian rush and the discovery of gold in South Africa's Witwatersrand, Paddy Hannan found gold near Kalgoorlie in Western Australia, thereby starting another gold rush in 1893.

Australia's search for gold had, however, started well before Hannan's find. Indeed the first official recording of gold was in 1823 near Bathurst in Victoria by James McBrian, a surveyor. Although when these gold shows were eventually worked they yielded only small quantities of the yellow metal, they constituted Australia's first gold discovery. Whilst many, particularly amongst the pastoralist community, looked for and found gold, a quirk in the law meant that initially prospectors were more interested in finding base metals.

The quirk came about as a result of Australia being a crown colony and in English law the crown having all rights to any gold found. The crown claimed no such rights over other metals, which was one of the reasons for a surge of activity in South Australia, where in 1844 a copper mine at Kapunda was developed. But the lure of gold could not be held at bay forever by such technicalities, especially when Edward Hargraves, an Australian who had been working in the Californian gold fields, arrived back in New South Wales (NSW) determined to prove that similar geology meant that NSW was as prospective for gold as California.

DEVELOPMENT OF AUSTRALIAN GOLD MINING

The structure of early gold mining in Australia was basically small scale, with miners being awarded very small-sized claims. On these early gold fields many shafts were sunk by individual claim holders to access the deep gravels where the underground water courses ran and where the gold lay. This meant that mining was inefficient and also encouraged illegal claim infiltration underground. On top of this there were constant legal and political battles with the authorities over mining licences and rights over the mined gold. Most of the mining was done laboriously by hand, as was the shaft digging and de-watering. The introduction of mechanical digging and mining methods was often met with hostility by the small claims holders, especially when larger-scale operations were finally permitted. In these early days gold mining in Australia was about individual rights, rather than efficient mining.

Over the next few decades gold was discovered in NSW, Victoria, Queensland and the Northern Territory. Amongst the famous names which left their mark on Australia's gold mining history were Bendigo, Ballarat and Mount Alexander in Victoria; Turon and Ophir in NSW; and Charter Towers in Queensland. Mount Alexander in 1853 was thought to have produced well over 600,000 ozs, a phenomenal figure for that day. At its peak in the late 1850s Victoria produced as much as 3 million ozs of gold per year and employed over 140,000 men mining gold. Such was the wealth Australia's gold generated that it led to the country consuming 15% of all Britain's exports, then the most powerful economy on earth. The development of Charter Towers also introduced British capital into Australian mining for the first time; a trend that was to last for over a century and reach its zenith with Poseidon in the 1960s, of which more later.

By the 1880s gold mining had moved west to the Kimberleys in the northern part of Western Australia after mining in Queensland's Charter Towers district had begun to peak. The great Western Australian gold rush was looming, but in advance of this historic event Australian gold mining actually appeared to have begun to decline after decades of discovery and growth, and interest had begun to dwindle. From the Kimberleys miners pushed south and west to the Pilbara and then to Murchison and Southern Cross.

Western Australia was a forbidding place to prospect for gold compared with the east. It was hot and dry and surface gold showings were rare, as the gold bearing reefs were usually hidden below the area's Pre-Cambrian shield. Though thousands of prospectors flocked to the state, they were spread thinly over this vast and forbidding land and the journey to the gold fields was often as hazardous as those taken by prospectors in the Canadian and Californian rushes.

GOLD AT KALGOORLIE

From Southern Cross prospectors travelled east and found rich gold deposits at Coolgardie, where a small town quickly grew up. In 1893 Paddy Hannon with Tom Flanagan and Dan Shea drove north east and in early June found gold. Hannan's gold haul on that first day was 130 ozs, and Australia's greatest gold field at Kalgoorlie was born. On 17 June Hannan returned to Coolgardie to register his claim and re-supply.

Gold at Coolgardie had been found a year earlier when Arthur Bayley and William Ford had found 554 ozs on their lease. When the residents of Coolgardie heard of the discovery at Kalgoorlie, what became known as Hannan's Rush was on. Hannan himself concentrated on his 20-acre claim with its gold in quartz veins and did well from his efforts. He ignored the ironstone hills to the south where, unbeknownst to him and others, lurked what became the richest part of the Kalgoorlie gold fields – the Golden Mile.

These ironstone hills covered with quartz dolerite greenstone are now a highly inviting setting for gold explorers but in Australia in the 19th century no gold had been found in greenstone. But that changed with the arrival in Kalgoorlie of two South Australian latecomers, William Brookman and Sam Pearce, acting for an Adelaide syndicate.

Disappointed that all the claims around Hannan's Find had gone the two men headed south and began to peg the ironstone hills. An interested but sceptical Paddy Hannan visited their claims and was surprised to see that gold was visible in the crushed rock from the greenstone. He had found the Kalgoorlie gold field but missed the richest part of it. Brockman and Pearce were acting for the Coolgardie Gold Mining and Prospecting Syndicate and within months the Syndicate floated Ivanhoe, Great Boulder and Lake View on the Adelaide Stock Exchange. Huge profits accrued to the Syndicate members.

PADDY HANNAN (1840-1925)

The history of gold prospecting in Australia is populated by countless thousands of mostly unlucky and long forgotten men. One of the few whose name still survives is Paddy Hannan, who found the fabulous Kalgoorlie gold field in Western Australia in 1893 and whose statue is still to be seen in the centre of Kalgoorlie today.

Hannan was born in Ireland in 1840, one of five brothers and six sisters. He travelled to Australia in 1862 and worked for several years as a miner in the gold fields of Ballarat in Victoria where his uncle, William Lynch, was a miner. After that he went to work in the gold fields of New Zealand for several years and returned to prospect in New South Wales and then South Australia. He later crossed Australia to prospect for gold in Western Australia around Southern Cross. Hannan was a careful man with an ability to find water as well as gold, something that stood him in good stead in parched Western Australia. It was this skill at finding water that led to Hannan's discovery of the famous Kalgoorlie gold field in 1892, for it was while he was looking for sources to fill his waterbags that he stumbled over surface gold.

Hannan did well from his claims, but Kalgoorlie's gold ore was complex and really required sophisticated equipment to treat it and release the gold. As this kind of mining was capital intensive, it was beyond the financial means of small-scale prospectors and Hannan left Kalgoorlie in 1894 for the Victorian coast, leaving the development of the Golden Mile to the capital markets of eastern Australia and overseas, particularly London. After several months of holiday he returned to Kalgoorlie and then struck out north to Menzies where quietly on his own he prospected with little success. In 1897 he returned to a Kalgoorlie transformed into a growing town of increasingly permanent structures and now the centre of one of the world's largest gold fields. He gave an interview to the *Kalgoorlie Miner* whose editor called for the Western Australian government to grant Hannan a pension in recognition of him having effectively found the Golden Mile, which even in those early years had yielded around £100 million of gold, enriching the state in the process.

It was ten years before the pension of first £100 and later £150 a year was granted. Then Hannan, at the age of 61, decided to leave WA and return to Melbourne where, a bachelor all his life, he lived with one of his sisters and two nieces. But he was not done with prospecting. In 1910 a gold rush at Bullfinch, just north of Southern Cross, had Hannan sailing to Perth and then taking the train to Coolgardie having wired money ahead. Hannan found nothing and returned to his nieces in Melbourne with his prospecting days finally over. He lived there quietly until his death at the age of 81 in 1925, active and respected and liked by all who knew him, a man whose modesty was in stark contrast to many long-forgotten gold prospectors who talked a better game than they played.

A man of great integrity, he did not believe that he had achieved anything special, although many Western Australians, then and now, would disagree with him. It is therefore a fitting monument to Paddy Hannan's memory that when the history of Kalgoorlie's gold is told it is his name that we remember and his statue that we see in Kalgoorlie.

IMPORTANCE OF OUTSIDE INVESTMENT

But huge though Kalgoorlie was, it was not the ideal geological setting for individual prospectors; the deep lodes in which the gold was found demanded mechanical mining methods and substantial capital investment, which was beyond the pockets of the prospectors. Australia, along with the rest of the world, was also in the midst of an economic slump at the end of the 19th century due to the worldwide recession. Since there were limited opportunities for European or North American investors seeking profitable investments in their own markets, promoters looking for development capital for Western Australian gold prospects turned to London where, as a consequence of the slump, there was plenty of capital sloshing about and looking for a home.

The Australian banking system was also under severe pressure, something felt sorely in the gold fields of Western Australia where bank failures often led to the loss of miners' savings. However, clouds often have silver, or in this case gold, linings and the depression, which badly hurt more conventional industrial and financial assets, pushed investors towards gold shares in a manner that was repeated 80 years later, in the 1970s. The appetite in London for gold mining paper was immense, gold being seen as a safe port in a storm, and Australian and British promoters were only too happy to supply the market with product.

In a short space of time, from 1894 to 1896, almost 700 Western Australian gold mines, most of them of dubious quality, were floated on the London Stock Exchange with enthusiastic British support – not the first time or indeed the last when mining ventures have turned to London for capital support. Amongst the companies floated were Londonderry Gold Mine, a promotion of the Irish peer the Earl of Fingall and the Chilean nitrate magnate John North. The Londonderry prospect, near Coolgardie, was based upon a very rich surface deposit that had no legs at depth. It led to heavy losses amongst British investors and consternation sufficient to drive the infamous Oscar Wilde/Marquess of Queensbury libel case from the front pages of London's newspapers.

John North was also behind another offering, Wealth of Nations, a company with a promising gold show at surface that died as quickly at depth as had Londonderry Gold. These failures and many similar ones did not put everyone in London off; many early investors did well selling their shares to latecomers, who were the ones who took the knocks.

Another English promoter was Albert Calvert whose Mallina Gold, formed to exploit reefs near Roebourne in the Pilbara, was such a successful flotation that half a dozen followed in quick succession and happy London investors mounted a lavish banquet in 1895 for their benefactor as a mark of their gratitude. Most of them saw their profits melt away in due course.

PROBLEMS IN THE GOLDFIELDS

The main show, however, remained Kalgoorlie, but it was not always gold prospecting that caught the headlines. Health issues abounded in the camp and typhoid was rife in the last years of the 19th century. Another key issue was water – the towns of the goldfields were by then finding their main source, the condensing of water from the area salt lakes, inadequate for their fast growing needs. A 350 mile pipeline, a huge construction project in any age let alone the late 19th century, was therefore built from Perth to Kalgoorlie, a city serving a broad population of 35,000 people, to bring water to the towns and mines. The first water flowed along the pipeline in 1903.

The gold boom brought a thirst for more than just water; information and, of course, rumour were the staple diet of the goldfield dwellers and six of the eight newspapers published in Western Australia around this time were produced in the goldfields, with the *Kalgoorlie Miner* perhaps the best known and which is still in vigorous health today, over 100 years later.

Kalgoorlie itself had grown quickly and due to almost non-existent planning was subject to frequent fires in both commercial and residential districts, a problem faced by gold mining camps everywhere. Nevertheless, the attractions of Kalgoorlie's rich mines drew people from all over the world who were anxious to provide services and to invest in businesses supporting the mining community. So hotels, breweries, brothels, sophisticated shopping and up-to-date transport links by railways going both east and west, and even tram systems between goldfield towns, were established.

Gold has always been a hard master and in the case of the mines of the Golden Mile situated just outside Kalgoorlie in Boulder City, the problem was metallurgical. The crush of mines on the Mile where leases were seldom more than 25 acres in extent and which included legendary names such as Golden Horseshoe, Ivanhoe and Paddy Hannan's Brown Hill, led to many of them having to go deep. Fortunately the gold continued at depth to below 1000 feet and grades were rich; in some cases more than 5 ozs per ton. There was a problem, however, in that recoveries were poor and this meant that a lot of gold escaped into the dumps or remained locked up in the plant.

The old techniques used on the shallow oxidised ores, crushing the ore in the stamp mills and then recovering the gold by amalgamation, did not work efficiently on the sulphide-type ores known as telluride, only found elsewhere in a gold setting in Colorado and Russia. To make Kalgoorlie's deep levels pay, the problem of poor recovery had to be addressed. A number of different techniques were tried, but in the end success followed laboratory work done by German chemists in Kalgoorlie led by Dr Ludwig Diehl, who found that using bromo cyanide and crushing the ore into a slime in a tube mill did the trick. And so Kalgoorlie's growth could continue.

The Kalgoorlie prospecting boom came to an end in the first decade of the 20th century. Gold mining, of course, continued for decades but it became the preserve of bigger, well-capitalised companies able to invest in deep-level operations. Aggregate production was huge, with 26 million ozs produced between 1893 and 1943 and an additional 13 million ozs produced between 1943 and 1983 when development planning started for the conversion of the ageing Golden Mile underground mines into what is now called the Superpit, which since 1989 has produced around 23 million ozs.

When the Superpit closes in 2018, it should have yielded over 1.85 billion tonnes of ore and extend to around 3.6 kms by 1.6 kms with a depth of 0.65 kms – over 1,800 feet. Grades at the Superpit now run at around 1.8 gms gold per tonne against 41.7 gms on average on the Golden Mile a century before. However, one of the largest nuggets found in the region was the Gold Eagle at Larkinville, south of Coolgardie in 1931, long after the end of the prospecting boom, which weighed an enormous 1,135 ozs. Overall, the value of gold output in today's money since Hannan's find, in what was formerly called the East Coolgardie Goldfield, is at least $20 billion.

SPECULATION OVER KALGOORLIE GOLD

The final years of the 19th century saw speculation over Kalgoorlie's gold at its height. Later on we will look at a few of the mining promoters and crooks, long dead both physically and litigiously, whose activities in a disparate number of locations around the world separated, sometimes skilfully, investors from their capital. But not all promoters were bad and perhaps the peak of Kalgoorlie's powers in those closing years of the century was signalled by the rise and fall of two of them, Whitaker Wright, an Englishman who it is believed never set foot in Western Australia and Frank Gardner, a dour American based in London.

Wright controlled Lake View Consols and Ivanhoe, two of Kalgoorlie's blue chip mines, but he was also a speculator who had earlier made and then lost large amounts of money in the US. Back in London his reputation was revived as his new Western Australian projects prospered and he began to milk Lake View as its exceptionally rich Duck Pond gold shoot began to throw off huge amounts of cash. However, in a repeat of his actions in the US, Wright began to speculate that Duck Pond could continue to provide large profits almost indefinitely. In an unusual version of postman bites dog, smaller Lake View investors did not agree, thinking Duck Pond would soon peter out, and began to sell their shares which Wright bought energetically through his holding company London and Globe Finance. The smaller investors were right and Wright, finding that his buying could not hold Lake View's share price up, began to manipulate the London and Globe's accounts to stave off insolvency. He failed and was eventually tried for fraud in London and found guilty on Australia Day, 1904. He died of a heart attack on the same day.

Frank Gardner's fraud was in mining terms the more traditional one of salting ore before assay to boost the gold content. Gardner was Chairman of two Kalgoorlie gold operations, the high-quality Great Boulder Perserverance and the much smaller Boulder Deeps. The latter, in 1903, reported sensational drilling results to Gardner, with a new lode found running, so it was said, 7 ozs of gold per ton. Gardner, as Chairman, sat on the story with the intention of quietly adding to his shareholding before making a public announcement of the sensational result, but others were in on the news and the shares started to soar on their buying. Gardner had been outflanked and his attempted news manipulation was exposed when the ore was assayed again and found to contain little gold.

Kalgoorlie survived these and other scams and prospered well into the 20th century. But the discovery of nickel south of the city at Kambalda in 1966 seemed to seal Kalgoorlie's fate, as did the continuance of a fixed gold price, which squeezed margins almost to extinction. However, Kalgoorlie's continued existence was ultimately secured by these nickel discoveries, with gold mining in the region having been reduced to a trickle by the mid-1960s.

The city's current prosperity, though, has been based on a revival of gold mining thanks to a gold price that over 40 years since the nickel discoveries has climbed from $35 to well over $1,000 per oz. This revival of Kalgoorlie's gold fortunes underscores the old mining adage that the best place to find a mine is where one existed before, but it needed the freeing of the old fixed gold price to encourage the return of gold mining capital, initially led by one-time America's Cup hero and ultimately disgraced entrepreneur, Alan Bond.

The second half of the 19th century saw aggregate gold production in Australia reach almost 90 million ozs by mid-1897. This was made up as shown by the figures in the table.

AUSTRALIAN GOLD PRODUCTION BY STATE (1851-1886)

Start of production	State	Production (m ozs)
1851	Victoria	61
1851	New South Wales	12
1858	Queensland	12
1870	Tasmania	1
1886	Western Australia	1
	Total	**87**

Source: Mining Hall Of Fame, Kalgoorlie

The figure for Western Australia is depressed because Kalgoorlie, the biggest WA producer, only got going in the mid-1890s so was still gearing up as the century drew to its close. By the end of the 20th century Kalgoorlie had produced around 60 million ozs, with the Golden Mile (now the Superpit) responsible for much of that.

THE KLONDIKE

The Klondike rush in Canada's Yukon Territory, which occurred between 1896 and 1899, was arguably the most famous of all the 19th century gold rushes. It crowned a century that had seen a number of similar excitements in the US, Australia and South Africa. Canada itself had experienced a number of gold rushes before Klondike, particularly the Cariboo Gold Rush which attracted would-be prospectors from around the world. Strangely, despite its notoriety, Klondike's glory lasted only a few years; rich deposits were discovered and large personal fortunes made, but the richer mines were quickly worked out and the prospectors moved on, leaving behind a gold camp whose future was to depend on more capital-intensive operations.

BEFORE THE KLONDIKE RUSH

The Klondike rush had been preceded by a couple of decades of low level exploration in the Yukon by a handful of prospectors who found gold in the riverbeds of the Yukon River and its tributaries. Some of the finds were of significance and triggered mini-rushes, like the 1886 discovery of placer gold (free running gold in the form of dust and nuggets) on the sandbars of the Stewart River where it flowed into the Yukon River. Here $100,000 worth of gold (around 5,000 ozs) was found in a year. This stimulated further prospecting in the region, both in the Yukon and in Alaska on the American side of the border.

Although the Royal Mounted Police had had a presence for some years in the Yukon, the region was very sparsely populated and difficult to access. There were three main ways into the developing placer gold fields of Alaska and the Yukon – a route west from the Mackenzie and Peace River valleys, a long route down the Yukon River from Alaska's west coast and the infamous route through Alaska's Chilkoot Pass. In time, a deadly overland route from Edmonton was added.

If prospecting for gold was a hard existence near the Arctic Circle, mining the gold only exacerbated the experience. With the permafrost solid all year round, a technique was developed that used a wood burning stove to thaw out the ground. When the stove burned itself out, the miners would lift the stove away and dig out the thawed dirt and then re-light the stove and place it back on the exposed permafrost below; this would continue in sequence and slowly a shaft would be sunk. When bedrock – where the gold was to be found – was reached, the miners would then tunnel off in search of ancient creek channels where gold could have been deposited.

This ore, or paydirt as it was then called, was hauled up to the surface and placed in dumps to await treatment in the spring. When the creeks burst and water started to flow again, the water was diverted into long sluiceboxes and the material from the dumps thrown into the boxes. The flow of the water would take all the waste dirt away but due to its weight gold in the ore would fall to the bottom and get caught on the floor of the sluiceboxes. Every few days the boxes would be closed and as they were emptied the miners would pan the residue left behind, the lucky ones finding gold at the bottom of their pans.

In the years before Klondike, gold was found in both Alaska and the Yukon and two sizeable communities grew up at Circle City and Fortymile, the former in Alaska and the latter in Canada's Yukon. Whilst this earlier gold mining was important, it was not on the scale of the Klondike rush. Nonetheless, the services that grew up in these cities, or towns as they really were in terms of size, were adequate for the needs of the miners, with Circle City rather more sophisticated than its Canadian counterpart.

Although the initial Alaskan and Yukon diggings provided substantial rewards in terms of scale, they were by no means bonanza discoveries; the restlessness of prospectors looking for giant finds drove on some to extend their efforts further into the Yukon. In 1896 these efforts were rewarded with such a success that it reverberates even today when mining men meet to discuss the great events of what is historically one of mankind's oldest industries. But mining has always been an industry that generates huge controversy and there is no more sensitive subject than the genesis of mining discoveries, with the Klondike being no exception. Two men, George Carmack and Robert Henderson, the former a Californian and the latter a Nova Scotian of Scottish descent, are credited with setting off one of history's greatest gold rushes.

THE SETTING OFF OF THE KLONDIKE RUSH

Robert Henderson was a prospector who had almost circled the globe in his search for gold. He had been drawn to the northern wastes by stories of great potential in the Yukon, but had ignored the conventional targets around Fortymile and pressed on down the Yukon to access the rivers and creeks around what was to become Dawson City; perhaps, after Johannesburg, the 19th century's premier gold city. There he ran into French Huguenot prospector Joe Ladue, who made a fortune developing Dawson City when the gold rush exploded. Henderson teamed up with Ladue who provided Henderson with financial backing (a grubstake). At Gold Bottom Creek, off one of the Klondike River's small tributaries called the Hunker Creek, Henderson struck gold; the birth of the Klondike Rush was now close. However Henderson was a restless gold panner and though he had found useful amounts of gold at Gold Bottom he kept moving. This led him over to Rabbit Creek, later renamed Bonanza Creek. There he ran into George Carmack and two Indian friends who were not as preoccupied with gold as was Henderson, but who were nonetheless interested enough in making money.

There was a code of conduct among prospectors that laid down that they should share knowledge of their discoveries with fellow prospectors, after of course having staked their own particular claims. Henderson followed the convention telling Carmack about Gold Bottom. In due course Carmack decided to have a look at Gold Bottom but decided not to stake, and struck up Rabbit Creek which had interested him as a possible site for cutting logs to float down to a timber mill at Fortymile. It was then that Carmack and his two Indian colleagues began to pan and a new gold province was born.

Over a century later it is difficult to calculate how much gold the four peak years of the Klondike Rush saw mined; government attempts to rake in royalties and apply other imposts led to misreporting, probably on a grand scale. However, the nugget found by Carmack at Rabbit Creek was significant in terms of prospecting standards, although as it was the size of a male thumb it may not have seemed huge to everyone. The fact was that further panning by Carmack and his friends yielded a panful of gold measuring a quarter of an ounce in weight, valued at the end of the 19th century at around $5/oz. Following this discovery first Carmack and then, as the news spread, others staked claims in the area and began to mine.

Stories of huge and valuable discoveries began to reach the outside world, and, as with other gold rushes before, interest was ignited and thousands around the world began to consider whether they should join the rush. In the end it was the ships arriving in San Francisco from Alaska which sparked the most interest, with early prospectors from the Klondike unloading thousands of dollars of gold dust from the boats and telling stories that grew larger and larger the more liquor eager listeners poured down their throats in the city's bars.

As far as action was concerned, the thousands who eventually travelled to the Klondike had many motivations for their participation in the great rush. Some went as representatives of syndicates who financed them to exploit some part of the Klondike project. The US, for example, by August 1897 sported over 80 syndicates spread across the country with a capitalisation of over $160 million. Although many of these syndicates were expecting to find gold, many of them aimed to make money by providing services to Klondike miners. Indeed, as with most mining booms, the men who in the end made the most money were often those who sold shovels to the miners rather than those who used the shovels.

THE HARD REALITIES OF THE RUSH

The conditions that those travelling to Dawson City overland had to endure were often horrendous, especially for those who took the sea route to Alaska and then went over the Chilkoot Pass in winter. The Canadian authorities laid down the supplies that prospectors aiming for Dawson City and the gold diggings had to bring with them. These made up a material weight of around a ton, consisting of food and equipment to support them on their journey and for when they arrived at the gold diggings; the supplies were supposed to last for a year.

These were usually carried on sledges or even using pack animals, but those taking the Chilkoot route had first to carry them up the famous pass in lots, sometimes having to make forty trips until all the supplies were on ground at the top of the pass from whence the long journey to Dawson City could begin. Although many heading for the Klondike tried to time their travel to coincide with the short summer months which enabled them to make the journey by river, for many, particularly those coming from distant countries, this was often not practical due to timing problems and so they had to travel through the grim winter months. Needless to say loss of life was not uncommon and certainly loss of provisions, often due to exhaustion and inability to transport the full load, was fairly common.

At its peak Dawson City, the centre of the Klondike rush, offered a substantial array of services to the miners, prospectors and others who had followed the gold rush. There were hotels, bars, brothels, theatres and music halls and a whole array of stores and businesses both serving the prospectors and miners and the society that had grown up around the gold rush. The old adage *all human life was there* applied in spades to Dawson City.

With Fahrenheit temperatures ranging from the low 90s in the summer to tens of degrees below freezing in the winter, Dawson City was a difficult place to live, with supplies often uncertain in the winter and the summer melt turning streets to rivers of mud on occasions. The summer also brought midges and mosquitoes in their millions, and disease in such a cramped environment was never far away. The Canadian authorities did a fair job in trying to bring some order to Dawson, but it was always a frontier town and public order did break down from time to time as happened in most mining camps. Fires were also a problem, but such was the energy and optimism of Dawson's residents that re-building started as soon as the last ember had been doused.

At the height of the rush in 1898 the town's population reached 40,000 but when the rush ended the next year the population fell rapidly, reaching 5,000 in 1902 when Dawson was incorporated as a city. In 1898 around $19 million worth of gold was produced from the diggings of the Klondike; gold traded at around $21/oz so this added up to 900,000 ozs or 28 tonnes. The year of peak production was 1900 when, with the Klondike fields fully pegged and a new gold rush starting in Nome, Alaska, which had drawn thousands of people away from Dawson City, gold output reached $22 million or just over 1 million ozs. Overall it is believed that around 12 million ozs of gold has been won from the Klondike's alluvial mining operations but only a few thousand ozs has been mined from hard rock sources. The source of the Klondike's alluvial gold remains a mystery and still remains a target for explorers to this day.

INCREASED MECHANISATION AND THE END OF THE RUSH

The Klondike rush ended when the easy surface gold in the creeks was worked out and mining had to turn to mechanisation, primarily dredges, which required substantial capital input in order to get at the gold lying at depth. This, of course, was the fate of other rushes such as the Witwatersrand and in Western Australia, where the need for capital-intensive investment drove out the prospectors and ushered in organised mining companies.

Two major companies worked the Klondike after the rush ended in 1899 – the Canadian Klondike Mining Company and Yukon Consolidated Gold. Canadian

Klondike went bust in 1921 but Yukon Gold continued dredging operations until 1966. Dawson City, although a shadow of what it had been during the rush years, continued to service mining activities in the old gold fields and the prospectors and trappers who operated in that part of the Yukon. After 1966 things became particularly difficult for the town, but in more recent years it has enjoyed something of a revival as a tourist destination and gold exploration companies such as Klondike Gold Corp have also arrived as the gold price has risen.

4. RUSSIA'S GOLD

Gold rushes in the Anglo Saxon world were in spirit and development very much mirrors of the entrepreneurial, free market system that the Industrial Revolution had unleashed. In autocratic Russia, the government had also begun to encourage the private gold prospector in May 1812 when it allowed Russians to search for and mine gold subject to a state royalty. This was interesting timing, as within a month the French began their ultimately disastrous invasion of Russia.

Before that, gold mining was the monopoly of the state and in addition there were only limited numbers of free men to work in mines, as so many of the population were indentured serfs largely working on the land. Indeed before Peter the Great travelled to Europe in 1697/98 and was exposed to advances made in science and technology, Russia was a backward nation. On his return he used his new knowledge to modernise Russia and built the country's first navy. Peter had also made a close study of mining and metallurgy and with the help of European expertise organised a programme of exploration. Up until then most of the metals used by Russia were imported from nearby neighbours such as Sweden.

GOLD EXPLORATION SUCCESS

The programme was successful in finding precious metals and in 1719 the first gold was produced in Russia from the Nerchinsky Mine in Transbaikalia; the same year that Peter issued his Privilege of the Collegium of Mines to stimulate private exploration (although any gold found still remained the property of the state). Nonetheless, gold and silver was found in the eastern Urals, the Altay Mountains and the White Sea.

From the time when the first gold was found in 1719 until the beginning of the 19th century around 25 tonnes of gold was mined in Russia, with the mines in the Altay Mountains being the largest source. This made Russia perhaps the largest gold producer in the world at the time. Rather strangely, possibly as a result of the metallurgical skills and treatment technology gained from Europe, Russian gold mines were initially underground, hard rock operations. In 1813 the first alluvial gold deposit was found at Neiva River in the middle Urals and within ten years around 200 alluvial operations had been established. The Neiva River discoveries could have set off Russia's first gold rush, bearing in mind the change in the law, but the authorities were alert to that possibility and ruthlessly kept the lid on the initial discovery. Gold fever however, even in an autocratic country like Russia, could not be held down forever.

As well as the Urals, gold alluvials were being uncovered in the Altay Mountains – previously host to hard rock operations – and in the Sayan Mountains that run on into southern Siberia. In the first quarter of the 19th century, this push towards the mountainous south of Siberia led to large numbers of serfs running away from land and factories of their 'owners', to seek their fortune, thereby initiating Russia's first gold rush. Much of the prospecting was done, as in California and Australia, by sight, with pits being dug to extract the gravel for washing in sluices to extract the gold.

With the abolition of serfdom in 1861 gold mining in the Urals and Siberia expanded rapidly, with output running at around 9 tonnes per year in the 1870s in the Urals, with a similar total coming from Siberia. These output figures made Russia one of the largest gold producers in the 19th century, a position it held for much of the 20th century as well. To start with, the mining methods were crude, as they were in California, but technology and the importation of western expertise saw more modern methods introduced.

THE AMUR REGION

As the 19th century rolled on, the successes in the Urals and Siberia led to gold prospecting moving further east to the Amur region where gold was first discovered in the basin of the Amur River which runs from China north into Russia and then east into the Pacific Ocean. This area was called Primorie and attracted Chinese as well as Russian prospectors, and although a substantial producer, with output of around 60 tonnes between 1871 and 1918, it was basically un-mechanised. A similar amount of gold was produced over the same period in the Prickhotie region in the hostile mountains in the southeast of Siberia, and foreign capital played a big role in the opening up of this gold province. Russia's first gold dredging operation was established in the Upper and Mid Amur region in 1894 and in 1910, decades after it was used in California, Russia saw the start of hydraulic mining in Amur province. From the start of gold mining in Amur, around 1870, until 1900, gold output was in the region of 170 tonnes with a total between 1870 and 1922 (when the Soviet Union came into being) of around 300 tonnes. This figure does not take into account gold smuggled into China and illegally mined gold, which suggests a total production closer to 1,000 tonnes from Amur.

THE LENSKY DISTRICT

The Lensky gold district, which is part of Transbaikalia where gold in Russia was first mined, began producing alluvial gold in 1843 and continues as a major producer to this day. The dominant operations were owned by the Lenzoloto Company, which in the guise now of Polyus Gold remains the biggest gold mining company in Russia. Lenzoloto produced at least 600 tonnes of gold in the years up to the establishment of the Soviet Union in 1922 and the region, with its unregistered mines and Chinese smuggling, probably produced a good deal more in unrecorded form. Lenzoloto was progressive in terms of technology and introduced steam powered water pumps and ventilation systems in the late 19th century. It also increased the scope of electric power within the production process and embraced hydraulic mining after early problems.

AFTER 1922

In 1922, with the forming of the Soviet Union and the nationalisation of the economy, gold mining entered a new age, driven by resource requirements dictated by central planners, and we will return to this later. However, the Soviet gold industry retained an important position in the mining industry and indeed on the world stage where decade after decade it was the second largest gold producer in the world, albeit well behind the largest producer, South Africa. For many years in the 1970s the prestigious Platts Metal Guide showed Soviet gold output just behind South Africa, but when the Russian archives were opened after the fall of the Berlin Wall, official figures showed that the Platts figures had been very optimistic.

5. SPAIN - EUROPE'S MINING LEADER

The 19th century witnessed a further attempt by the great European powers to extend their overseas possessions, but the motivation this time was less a case of trade and geopolitics and much more a case of obtaining control of natural resources to feed the industrial revolution that was in danger of outgrowing indigenous supplies of raw materials. Such a strategy is mirrored today in the 21st century by China's push to secure raw materials for its own industrial expansion. Over the centuries one of Europe's leading mining countries has been Spain, with a history going back beyond the Roman Empire to the time of the Phoenicians. Earlier we looked at Pliny's observations on gold mining under the Romans, but Spain has also been a major miner of copper, zinc and lead for centuries.

LEAD

In the 19th century the UK not only sought raw materials from its Empire to fuel its industrial expansion, it also looked nearer to home on the European continent, particularly to Spain. Spain's ancient lead mining industry had ironically declined in the face of rising UK output in the 18th century and early years of the 19th century but it began to stabilise after that and when the Spanish government allowed private capital into an industry that before had been largely state owned, British finance became an important component in the industry's revival.

This revival in lead mining began in the Sierra de Gador, not far from Almeria on the southeast coast and this made transport to foreign markets relatively straightforward. Two of the new mines established were the Linares and Fortuna lead mines in Jaen province in Andalucia, financed and operated by the British, which were opened in 1852 and 1854 respectively. Skilled workers and management at the resurgent lead mines had a substantial Cornish flavour, as happened in other parts of the world. This revival in Spanish lead mining saw output once more climb above UK production levels.

By law the lead had to be processed in Spain before export, so a large number of small, rudimentary, inefficient and polluting treatment plants were established in the area. In due course, because of weak lead prices, much more efficient smelter plants with new technology were built, cutting waste and improving recoveries – formerly only 50% of lead in the ore was extracted, leaving over 30% to go into the tailings ponds.

Although many of the smaller lead smelting plants were inefficient, the San Anares plant between Malaga and Almeria at Adra utilised modern English technology and equipment, including the use of waste heat for the steam boilers, and condensation chambers were installed to capture lead vapours and fumes. The man who eventually put the San Anares smelter on an economic footing was Manuel Heredia, who founded Andalucia's iron ore mines and smelters – we will discuss him further below. San Andres was also a major producer of silver from the lead ores mined and it kept at the forefront of technological developments. However, by the end of the 19th century lead mines in the Sierra de Gador were substantially exhausted, and, despite the ability to bring in lead ore from elsewhere in Spain, the economics of this were poor and treatment plants in the region began to close. By the end of the 19th century the lead industry was essentially dead.

COPPER

Another highly influential British mining company, which began to operate in Spain in 1869, was Tharsis Sulphur and Copper from Glasgow. Its mine was in Huelva province in the southwest part of Andalucia and was a copper pyrites operation. Unfortunately the pyrites in the ore threw off large amounts of sulphuric acid, polluting surrounding vegetation using the old smelting process; the process also retained some of the copper in the burnt waste – as the copper could not be recovered this undermined the financial return from the operation. Tharsis's original ownership was French – Mines de Cuivre de Huelva – but they sold out to Scottish interests when a new treatment process for copper in pyrites, the Henderson process, was developed but was beyond the resources of the French owners to introduce themselves. Like other British-controlled mining companies in Spain, its key personnel were not locals; in Tharsis's case they were mainly Scottish and were specialists in the new treatment technology.

Many other Spanish copper mines, some of which had operated for centuries, were to be found in Huelva province. Perhaps the best known of these was the Rio Tinto mine, which was for many years under the control of the British mining house of the same name. Copper had been mined around Rio Tinto since the 3rd millennium BC, but the mines closed down in the Middle Ages and were only re-opened in 1724. A hundred and fifty years later the British, as they did with lead, took control following the sale of the leases by the Spanish government to a group of London investors which included Hugh Matheson. Matheson's family company, Jardine Matheson, was at the time also negotiating a deal to market Japanese copper worldwide.

With its new investors, the Rio Tinto mine became one of the largest copper mines in the world at that time, and shipped its product using its own facilities at the port of Huelva, where it constructed an unusually long pier to allow continuous mechanical loading of multiple vessels. The mine itself was re-designed as a large open pit operation, as were other mines in the area; this form of mining, although necessary to provide an economic return for investors, increasingly turned the terrain of the mining area into a moonscape. The re-opening of the ancient mine was not without its problems, caused by much harder rock than expected and also water inrushes into the pit; this caused development costs to go over budget. However, once the mine was operating at full tilt annual profits came in at £1.6 million, not an inconsiderable sum in the 1880s.

The grade of the ore first mined by Rio Tinto was between 2% and 3% copper and around 200 million tonnes of ore were estimated to lie in the area; in the hundred years from when it took control of the mine, Rio Tinto mined over 2 million tonnes of copper. The mine was closed in the late 1990s due to persistently low metal prices and a rather inefficient local cooperative ownership structure. A junior Cyprus-based group, EMED, is hoping to restart mining in 2013 to produce around 37,000 tonnes of copper annually, which is not bad for an area that has been producing copper for over 5,000 years. Another Huelva copper mine currently receiving renewed attention is the La Zarza mine, which was opened in 1867 and was finally closed in 1991. It was a material producer in the 19th century and after, with a peak annual production of just under 14,000 tonnes of copper in the early 1930s.

IRON

Andalucia was also an important iron ore mining area in the early 19th century, with deposits near Malaga and Marbella. The man who developed the industry was Manuel Heredia, who made his fortune in wine and olives, and had cut his mining teeth on developing graphite deposits further inland in the mountains near Ronda and also, as mentioned above, in the operation of the San Andres lead smelter. Heredia registered the La Concepción company to mine the iron deposits at Ojén near Marbella in 1828 and built two plants to treat the ore, the first near Marbella and then another in Malaga, on what is now the Costa del Sol, one of Spain's prime tourist areas. In 1832 Heredia established a third works in Seville, but already competition was beginning to loom as iron ore operations were being established in the Basque region in the north in the province of Asturias and later in Viscaya.

Both areas suffered from political upheaval in the early decades of the 19th century, but despite this they established a technological superiority in terms of the treatment process they used. However, it was not until the latter years of the century that they realised their full potential. Bilbao became a key centre for iron ore mining and for the production of iron and steel, and attracted substantial foreign investment interested in securing supplies of both iron ore and finished metal. Two of the largest operations were Orconera Iron Ore and Mines de Somorrostro, a French/Belgium company. The establishment and expansion of the iron ore industry in the Basque area also led to the development of coal mining in the region, although Welsh coke for Spanish furnaces was still imported as ballast in ore carriers returning from export voyages to Britain.

6. THE EMERGENCE OF CANADA

We have already had a look at the great Canadian Klondike gold rush but Canada as a mining nation is, and always was, much more than just the Klondike. Whilst the early settlers arrived in Canada in the first decade of the 17th century, the country only really began to grow following the American War of Independence and the trek north by the defeated United Empire Loyalists who sought the protection of the British Crown in Upper Canada, now the southern part of the province of Ontario. Throughout the 19th century Canada steadily integrated itself into the vast landmass it is today, separated from the US by the 49th parallel from the Great Lakes to the Pacific, with the eastern part of Canada broadly separated from the US by the Great Lakes and the St Lawrence Seaway, although part of Quebec and the Maritime Provinces lie south of the St Lawrence.

EARLY MINERAL DISCOVERIES

It was into this eastern part of Canada that the English and Scottish immigrants of the 19th century flowed and it was here that many of the earlier mineral discoveries were made. Coal was found in New Brunswick in 1782 and mined by settlers. Iron ore was discovered in Quebec near to the confluence of the St Lawrence and St Maurice rivers in 1737 and was smelted on site. In the first quarter of the 19th century further iron ore finds were made in southern Ontario and iron furnaces were built.

The first base metal mines were developed in 1847, around the town of Bruce Mines in Ontario, named for the eponymous Governor General of Canada at the time, James Bruce, on the north shore of Lake Huron. These copper operations, collectively known by the town's name, consisted of up to ten shafts or openings and were mined by a number of companies including Huron and St Mary's Copper, Montreal Mining and Wellington Copper. These were the first copper mines opened in North America and shafts went down to around 45 feet from whence the miners chased the copper veins in tunnels and drives. Ore was hoisted to the surface using a basic pulley system. Working conditions were not particularly good, although the pay was attractive enough to lure the ubiquitous Cornish miners from England. The mines operated until 1876 when flooding led to their closure. They re-opened during the First World War but were closed in the early 1920s. Another copper mine at Rock Lake and a gold mine at Ophir, both close to Bruce Mines, were opened towards the end of the 19th century but closed in the early 1900s.

The first gold discovery in Canada was made at Chaudière River in Quebec in 1823 but production did not begin until 1862, two years after production started on Queen Charlotte Islands in British Columbia (BC), just after BC became a British colony in 1858. Around the same time placer gold mining started in the Cariboo region of BC and over the years more than 100 tonnes (3.2 million ozs) was mined from there. Gold was also found in 1860 in Nova Scotia in the east and although the mines were always small, around 45 tonnes of gold has been mined in the province.

Coal was another increasingly important mineral where the new Canada had potential and indeed the mines of Cape Breton, Nova Scotia had been originally worked by the French in 1672. A hundred years later coal began to be mined in New Brunswick and in 1800 it was found in Alberta, in the heart of Canada's prairie lands. By the mid-19th century coal was being mined on Vancouver Island in BC, which was fast becoming a serious mineral producer. Although in no way fully compensating for the loss of its American colonies, Great Britain had in the shape of Canada a consolation prize of considerable worth and, as if to confirm that, North America's first oil discovery was made at Oil Springs, Ontario, in 1857.

Further discoveries were made in the second half of the 19th century, adding to Canada's widening range of minerals. In the 1870s phosphate was found on both sides of the Quebec/Ontario border near Buckingham, Quebec, to the north east of Ottawa, and near Kingston on Lake Ontario. Mines were developed in both locations with product being exported to England for a while. Canadian demand in the later part of the 19th century was good and over half a dozen mines operated at one time. However, phosphate prices collapsed and the mines became uneconomic, with phosphate mining ending in 1892.

Mineral discoveries continued to be made in both Quebec and Ontario in the latter part of the 19th century. Asbestos was found on the southern side of the St Lawrence in the Eastern Townships to the east of Montreal and WH Jeffrey, a wealthy local farmer, opened the first mine there in 1881 and became a major producer for over a hundred years. Despite the collapse of asbestos usage in recent decades, small-scale mining at what was called the Jeffrey Mine continued until 2011.

Of even more significance was the first discovery of nickel made at Sudbury, Ontario, in 1883 by Thomas Frood, with mining commencing in 1886. The first mine at Sudbury was the Stobie which, along with the adjacent Frood deposit, discovered at the same time, formed the basis of International Nickel (Inco). Although war and economic recessions meant that mining was not continuous throughout the 20th century, the Frood mine was only exhausted in 2000, and therefore finally closed. The Stobie mine continues to operate to this day. Over its life Frood mined 5 million tonnes of nickel and copper and 55 tonnes of gold, making it the first of Canada's giant mines.

THOMAS FROOD (1837-1916)

Thomas Frood was one of a large number of amateur mineral prospectors who opened up the great Canadian north for mining development. Frood, however, was one of an exclusive group who could claim a major discovery, which in his case was the Sudbury copper and nickel district of central Ontario, the home of Inco and Falconbridge, Canada's two largest nickel miners.

Frood was born in McNab in Renfrew County, eastern Ontario, in 1837. His parents Thomas and Barbara Frood were immigrants from Scotland and his father farmed in McNab. He was educated, as was the practice then, at home and as a youth also worked on the family farm. He took up teaching himself and for many years taught in a variety of Ontario townships as Canada developed its public school system. He also worked in the army medical corps between 1866 and 1871 during the Fenian raids over the border made by dissident Irishmen resident in the US. Following this, Frood became a chemist and opened his own chemist's shop in Southampton, a seaside town on Lake Huron. He also married his first wife, Mary, in 1865 and they had two daughters.

In 1883 Frood, who was teaching in Kincardine – another Lake Huron township – decided to leave teaching and head north to work for first the Canadian Pacific Railway and then the Crown Lands department as a ranger. It was at this time that Frood became interested in prospecting and taught himself the basic tenets of the craft. This led him to identify the Murray copper deposit when supervising the laying of rail lines near Sudbury. This happened in 1883 and others pegged the lease. However, the next year Frood identified mineralisation at the prospect that became Inco's great Frood mine which only closed after more than a century of operation in 2000. He introduced two other prospectors to the Frood lease. None of them had the required capital to develop a mine there after they had completed their prospecting work, so they had to sell the Frood prospect to Canadian Copper for which Frood received $12,500; worth over a quarter of a million dollars in today's money. Frood continued to prospect in the Sudbury area, discovering amongst others the Lady MacDonald and Copper Cliff deposits, the latter of which remains an operating mine in the 21st century. However, as the years went by he was unable to recreate his huge discoveries made in the mid-1880s.

Although Frood's Sudbury discoveries were impressive, they did not make him a rich man. Apart from the Frood mine, any properties he pegged he eventually sold on for often just a few hundred dollars. It was almost as if the chase was as important as the winning; having said that, he had little good to say about the mine owners who he felt had cheated him and other prospectors in the Sudbury area, out of their just rewards. Frood's first wife died in 1886 and he married again three years later, moving to the town of Wallace Mine on the north channel of Lake Huron overlooking Georgian Bay. Here he worked as a farmer but continued to invest in small scale mining and timber projects and devoted an increasing amount of time to boosting the attractions of the northern Lake Huron area for settlement and investment, particularly in agriculture. Thomas Frood died in 1916 and was buried in Kincardine. Whilst he does not figure prominently in the conventional annals of Canadian mining history Frood's legacy, the mine to which he gave his name, makes him a true giant amongst Canada's minerals prospectors.

British Columbia continued to throw up new mineral finds as the century came to a close. In 1893 the Sullivan lead/zinc/silver deposit was discovered at Kimberley in the south of the province and mining continued throughout the 20th century, with the mine only closing in 2001 having produced 10 million tonnes of both lead and zinc over its long life, as well as over 9,000 tonnes of silver. In 1896 the great Klondike gold placers were discovered, with the rush, which we looked at earlier, really putting Canada on the world map. Although the Klondike Rush was white hot for only a brief few years, mining continued for over 100 years and historic production from all Yukon mines adds up to well over 400 tonnes since gold mining started in the territory.

One more great mineral find was made before Canada achieved de facto independent status after the end of the First World War – the silver deposits of Cobalt in Ontario on the Quebec border north west of Ottawa. These rich deposits, which eventually yielded more than 18,000 tonnes of silver, were literally unearthed during the construction of the Temiskaming and Northern Ontario Railway line from North Bay. Mining started in 1903 and lasted until 1989 when the last mine ceased production. However, hope springs eternal in mining, and bullish expectations in some places for a sustained leap in the silver price could yet see silver mining again in the future at Cobalt.

7. LATIN AMERICA

The first major political event of the 19th century in South America was the recognition by the great powers of independence for the states which made up the Spanish Empire in Latin America. This happened at the end of 1824 and it ushered in the first major economic event of the century for Latin America – the 1825 mining boom and bust. Up to his ears in this latter event was none other than Benjamin Disraeli, destined in due course to become one of Britain's greatest Prime Ministers when the country was at the peak of its powers.

In the opening years of the 19th century there was a rapid build up in European banking loans to the nascent republics of South America. There is some argument whether the loans fuelled the mining boom or merely reflected the belief amongst European financiers, particularly those in London, that with independence from the broken Spanish Empire looming growth prospects in Latin America looked promising, and with Madrid out of the picture foreign finance would be very welcome. In fact foreign money had been flowing into the region for much of the first two decades of the 19th century and although mining was an attractive investment target it was not the only one as British railway builders, for example, later discovered.

DISRAELI, THE SPECULATOR

It is here that Benjamin Disraeli enters the picture. He was anxious for quick wealth and was excited by the potential for making money in an independent Latin America that eagerly sought foreign investment, particularly in mining. However Disraeli's first instinct, as he pondered with friends what action to take in the market, was to short Latin American mining stocks. Many of them had done very well as international recognition of the new republics loomed – Anglo Mexican Mining had risen from £33 to £158 in a month at the end of 1824 and Colombian Mining rose from £19 to £82 in the same period.

Disraeli was dazzled by such performance and fell in with J.D. Powles, who headed a leading South American merchant trader heavily involved in mining and mining promotion. Disraeli then turned buyer, using borrowed money, but the boom was ending and Disraeli's investments tanked. In order to try and salvage something from this disaster Disraeli was persuaded by Powles to write a research pamphlet on Latin American mining prospects which was called *An Enquiry into the Plans Progress and Policy of the American Mining Companies*. Much of the statistical work was drawn from a review of Latin American mining prospectuses by the London stockbroker Henry English. The pamphlet contained much useful and interesting background information, but despite that it really constituted little more than a puff for some of Disraeli's ailing share holdings.

The pamphlet, nonetheless, was quite widely read and could be described as an early example of financial mining research. Disraeli followed it up with two other pamphlets aimed at mining – *Notes on the American Mining Companies* and *The Present State of Mexico*. Unfortunately for him none of these efforts were able to revive his portfolio. Amongst the companies that Disraeli described were Anglo Mexican and Colombian, as mentioned above, United Mexican Mining Association, the Company of Adventurers in the Mines of Real Del Monte, the Association for Working the Mines of Tlalpuxahua and Others in Mexico, Imperial Brazilian Mining, The Provinces of Rio de la Plata Mining, Pasco Peruvian Mining, Chilian Mining, Anglo Chilian, Chilian and Peruvian Mining and General South American Mining. The companies all had large boards of London directors (often more than ten) which included the great and the good in the form of members of parliament, landed gentry and City professionals. The capital raised could be as much as one million pounds but many of the companies had a short life. Some, like Imperial Brazilian, lasted rather longer, as we will see below.

MEXICO

Although this London-based speculation might have been described as reckless, Latin America did have interesting mineral potential with silver to the forefront, particularly in Mexico. Mexico was a country with a long history of silver mining and the Valenciana mine near the old colonial city of Guanajuato in Central Mexico was one of the biggest, with shafts reaching down beyond 1,500 feet. The new investment coming from Britain was expected to be repaid within two years with distributable profits thereafter, and to encourage this investment local duty on the value of mine gate metal had been reduced from 30% to 6%.

The Anglo Mexican Association was formed to renovate Valenciana and other Mexican mines, including Concepcion near the town of Catorce, also in Central Mexico. Other historically worked mines in the same region, including the very rich Rayas silver mine, were to be renovated by the United Mexican Mining Association following its London incorporation. Others of the newly formed associations targeted regions such as Real del Monte with its rich silver mines, which had been the subject of an historic labour dispute in the previous century.

One of the major problems that dogged silver mines in Mexico was the very high working costs. Mines tended to use old, inefficient technology and the manner of the development of the mines was also wasteful. Sizeable galleries were often dug out to access ore instead of using more conventional tunnels, shafts tended to be poorly located, and ore and waste had to be removed from the mine by hand which meant that the mines had large and unproductive workforces. There was no hydraulic pumping of water in most of the mines; water was being raised to the surface by hand, using bags which wore easily. The technology of using a stamp mill to crush ore was also ignored, and instead a laborious system was followed of hand crushing of ore and then further crushing by mules to produce a final paste-like concentrate.

Much of this under-investment was a result of the relationship between Spain's Latin American colonies and Madrid, where the colonial power lacked dynamism, seeing the colonies simply as providers of wealth. This situation was also exacerbated by the early 19th century fight for independence, which led to many mines being abandoned and others being operated on a shoestring budget during the struggles. Unfortunately, as independence rolled over Latin America in the first quarter of the 19th century, the newly independent governments did little better, engaging in an international borrowing spree reminiscent of more recent events which have battered the Latin American sovereign debt market. The opportunity then fell to British capital, in particular, to re-finance the ageing mines on the back of excitement about the returns that mine modernisation and restructuring could generate.

Whilst Mexico attracted a substantial share of British capital, the names listed in Disraeli's pamphlet indicate a continent-wide interest from London.

COLOMBIA

Colombian Mining was one company formed to exploit the silver prospects of old government-owned mines near Mariquita in the centre of the country. As well as much needed capital equipment, the company also provided experienced labour from Britain, although the heavy drinking of some British miners upset the local authorities. The man whose task it was to manage Colombian Mining's operations was Robert Stephenson, son of George Stephenson who built the legendary railway engine The Rocket.

Colombian Mining's operations at Mariquita continued until 1875 when they were acquired by the Mariquita Mining Company.

BRAZIL

At least one of the companies that Disraeli was pushing, Imperial Brazilian, established a profitable mining business in the state of Minas Gerais in Brazil at the old and rich Gongo Soco mine, which lasted from 1826 until 1856 when flooding led to the mine's collapse. Over 30 years it produced more than 400,000 ozs of gold, the revenue generated being divided 60% to costs, 20% to shareholder dividends and 20% to the Brazilian government in taxes, making it one of the few profitable companies to come under Disraeli's gaze.

Brazil had been a major gold producer since gold was first discovered in Minas Gerais at the end of the 17th century. Gold production is thought to have been around 1,200 tonnes between 1700 and 1820, at which time British capital was allowed in, following independence from Portugal, to revive the ageing mines of the region. The British had in fact benefited for decades from the growth of gold mining in the pre-independence era as Portugal ran a permanent trade deficit with Britain which was plugged by gold deliveries from Brazil.

Perhaps the most enduring of the early British-promoted 19th century ventures in Brazil was the St John del Rey Mining Company, which operated for over a century from its formation in 1830 until 1960, primarily producing gold from the Morro Vehlo mine near Novo Lima in Minas Gerais. It then fell under the control of US mining group Hanna.

St John del Rey's early operating years were undistinguished, mining some poor deposits in Minas Gerais, but in 1834 after raising new capital it bought the Morro Vehlo mine and its luck turned. Following this St John del Rey, along with Imperial Brazilian, provided their British shareholders with highly satisfactory returns for many years; a lot of other London backed Brazilian gold companies in the 19th century proved poor investments in contrast.

All in all around 20 companies, including Imperial Brazilian and St John del Rey, were floated in London between 1824 and 1898 with Imperial and St John producing well over 80% of the gold mined by these new entities. However, although the overall returns to St John del Rey from Morro Vehlo were ultimately very rewarding, they did not come without serious headaches. The mine was worked for many years as one huge excavation from surface, supported by timbers and pillars. In 1867 a fire destroyed these workings but the mine was re-opened in 1873 and the orebody was accessed by two shafts which went down more than 1,000 feet. In 1886 disaster struck again as the mine workings collapsed following a catastrophic timber burst. When the mine was once more brought back to production in 1892 the mine design had been changed and the workings were now accessed through two new shafts outside the mineralised lode, which was now mined at several levels and reached by drives.

MORRO VEHLO

Morro Vehlo, which is now operated by AngloGold Ashanti, has always been one of the world's deepest mines and with some current workings at just under 10,000 feet it remains so today. When St John del Rey, by then part of Hanna Mining, exited in 1960 to concentrate on its iron ore interests, the mine had produced almost 450 tonnes of gold over its life; today its annual output runs at above 250,000 ozs. The ore is primarily composed of pyrrhotite, quartz, and iron and lime carbonates. It is also refractory, which provided treatment problems for St John del Rey in the early years until technological improvements allowed for a finer grind which ultimately raised recoveries to around 90% and improved the ore grade to above 12 grams/ton. The complexity of the mine's structure meant that in the 19th century working methods were labour intensive – much of this was slave labour until the end of the century. Today, despite the depth of Morro Velho's operations, labour productivity at the mine is significantly higher than AngloGold experiences in South Africa. As for the future of the mine, a further decade of production at least is expected, which would put it within sight of its 200th anniversary of operation.

Although Morro Velho was the star in Minas Gerais's golden crown, by the end of the 19th century there were ten other gold mines, seven of them controlled by British companies, still in operation – Passagem, Faria, Santa Querita, Morro da Santa Anna, Itabira, Carrapato, Conquista-Xicao, Sao Bento and the two locally owned mines Juca Vieira and Florisbella. Although these were considerably smaller operations than Morro Velho they nonetheless underlined Brazil's global position as an important gold producer before the great gold rushes of the second half of the century. Gold in the 20th century, however, was to belong to another country – South Africa – where British capital again was to play a major role. For Brazil, the old saying about Minas Gerais having a heart of gold in a breast of iron proved a prophetic comment about the direction in which its mining industry began to move in the new century.

THE CORNISH INFLUENCE

Another important link between Great Britain and Latin America in the area of mining development was the increasing use of Cornish miners, as investment in modern mining equipment was made. Nowhere was this more evident than in the mines of the Real del Monte in the state of Hidalgo in central-eastern Mexico. These mines, which have a recorded history going back to the arrival of the Spanish in the 16th century and are believed to have produced more than a billion ozs of silver and 6 million ozs of gold over the years, are still in production today.

Cornishmen worked and lived for more than a century here and many of their ancestors remain today.

Establishing the mining community was not without great risks and in the early days many Cornish miners died of disease in transit before reaching the Mexican silver fields. The job of bringing in and transporting the capital machinery needed to renovate and upgrade the mines was also fraught with peril as transport links were rudimentary at best, adding to the death toll. As well as introducing modern mining methods to this region of Mexico, the Cornishmen also brought football to the new nation, something that Mexicans of the 21st century might consider a more important legacy.

The Cornish also emigrated to other South American countries as the 19th century mining boom continued and expanded. They played an important part in the re-birth of the Brazilian gold mining industry and were also to be found in considerable numbers in Peru where major new investment in modernising the silver mines was needed. They were active in Colombia and in Chile as well.

Whilst the bulk of the Cornishmen associated with the renewal of Latin America's mining industry were traditional underground miners, albeit usually skilled operatives, Cornishmen were also very active in setting up some of the London companies established to invest in the process of renewal. They also held many of the key management positions at the mines themselves.

The opening up of mining opportunities in Latin America, and in places like Australia and South Africa, also provided business for the engineering industries specialising in mining equipment which had been built up around the world-class tin and copper mining industries of Cornwall. Unfortunately, Cornish mineral output in the second half of the 19th century was unable to meet UK industrial demand driven by growth engendered by the Industrial Revolution in England. But Cornwall had the men who could criss-cross the world and make good that raw material deficit.

PERU

Mining in Latin America has always been a fraught and uncertain business and Peru, currently reasonably well regarded as a destination for foreign investment, has a history of volatility. In the 19th century, in the early decades of independence from Spain, Peru was keen, like other countries, to attract British capital to fund the modernisation of its historic silver and gold mines. It is interesting to note that there was already a sector of locally-operated silver mines in Peru before the British arrived. These were very large in number but very small in scale, often operating to a depth of no more than 25 feet. By the end of the century there were thousands of mines in Peru and around 80% of them were silver operations.

Peru was also an important producer of copper, with a thriving export trade with Great Britain in the 19th century. The expansion of its copper production came as a result of the onset of the Industrial Revolution and the arrival of British capital and mining expertise in the country. So it was that even though the majority of the mines in Peru were silver operations, it was the copper mines, particularly of the Cerro del Pasco Corporation, that were the dominant revenue producers in the Peruvian mining industry then.

By today's standards the level of Peru's annual metals production at the end of the 19th century was modest with copper totalling around 9,500 tonnes, silver 5,700,000 ozs and gold 32,000 ozs. However, it is worth remembering that in the late 18th century, on the eve of the Industrial Revolution, world copper production was probably only around 10,000 tonnes in total, down even on the estimated rates at the peak of the Roman Empire and the Chinese Sung Dynasty centuries before. During the 19th century silver production remained fairly stable on the basis of start and finish, but did fluctuate materially over the hundred years, with events such as the War of the Pacific with Chile (1879/83) taking its toll.

Copper on the other hand grew steadily over the period, although part of this was due to mines such as Cerro del Pasco altering the mix of the ore they mined to favour copper-rich material. By the end of the century copper had overtaken silver in terms of value of metal mined in Peru. Indeed silver production in Peru was slightly behind Chile, and significantly behind Bolivia and Mexico. The silver mines often provided useful by-product revenues through the extraction of tin; Peru had a busy export trade in tin to Great Britain.

Peru's mines were mainly located in the Andes and stretched the length of the country, with Cerro del Pasco in the province of Junin being the area with the largest concentration of mines and the greatest output. By the 1880s there were almost 700 mines in the Cerro del Pasco area; many of them small-scale, family-owned enterprises that provided an income stream for the owners. When the mines became uneconomic, judged simply by matching outputs and receipts, the owners would close them down.

Foreign capital coming in organised its operations on a more commercial basis, with a more formal workforce than the small owner-run mines where the labour force was usually cheap indigenous Indian labour. Much of the labour force, as in other Latin American countries in the early decades of the century, was forced, even slave, labour, but, as elsewhere, in the later decades the forced labour system faded away to be replaced by more conventional hired labour. Working days were long and wages poor, especially where miners were in a contract arrangement that required them to provide items, such as mining equipment, from their own resources.

Further foreign assistance for the mining industry came in the form of railway construction, where British capital played an important part. This enabled the mining industry to move away from the mule trains of earlier decades and use railways to transport metal to the cities for either local sale or export. Although railway building began in the middle of the century it was really in the 1870s and later that construction began to accelerate, partly as a result of the expansion of copper mining.

CHILE

Although it was eventually to become one of the most important copper mining countries in the world, Chile, for the first part of the 19th century, was primarily a gold producer. In the early decades of the century it was responsible for over 15% of the world's annual gold production, and in the first decade that added up to Chilean output of around 100,000 ozs. Mines were largely located in the north of the country in the Atacama Desert region. Silver mining was also fairly widespread and the Chanarchillo mine near Copiapo, discovered in 1832, became one of the largest in the world at that time, stimulating the building of the first railway in South America. However, the industry was dogged by having to import mercury, needed to separate silver from the ore, using a long and expensive transport route from Peru, due to Spain's enforcement of specific trade routes on its Latin American colonies. Copper mining in the north began to expand following independence in 1810, and between 1820 and the end of the century Codelco, the state-owned copper producer, calculated that Chile's output may have been as high as two million tonnes, modest perhaps by today's standards but a substantial industry at that time which benefitted from the 19th century industrialisation of Great Britain and then of the US.

Mining methods at this time were very basic; shafts and adits were eschewed in favour of much smaller and shallower pits dug to access near-surface, high-grade ores. These mines were worked by owner miners, often in partnerships, who used at most a handful of labourers; in some cases the mines were operated by entrepreneurs leasing the ground from landowners for a royalty. The ore was rich and required very little capital input, and treatment techniques were traditional and relatively basic. It was not until later in the century on the back of rising demand that new technology, often supported by British capital as elsewhere in Latin America, was introduced, increasing working efficiency and better controlling costs.

The mine that was to make Chile into a 20th century copper mining giant was Chuquicamata located in the Atacama Desert, which as we have seen earlier has been the site of copper mining for many centuries. Increasing industrialisation and the close diplomatic ties between Chile and Great Britain led to a rush of activity in the Chuquicamata area when the War of the Pacific ended in 1884 with Chile annexing territory to its north from Peru and Bolivia. The decades spanning the end of the 19th century and the beginning of the 20th century saw a huge inrush of speculative miners who worked high-grade surface veins which ran as high as 15% copper. The camps set up around the Chuquicamata were disorganised, unsanitary and violent, and there were numerous disputes over lease titles. However, a highly prospective copper mining region had been opened up, and when the late century nitrate boom, which we mention below, subsided and copper came back into fashion, US interests were ready to enter the Atacama Desert targetting the lower grade ore that had been ignored by the independent miners.

Chile's other giant copper mine, El Teniente, situated near a dormant volcano some 50 miles from Santiago, also has an old if modest history as a producer going back into the 16th century when Jesuits operated a small mine there, which was then known as Socavon de los Jesuitas. The Jesuits were expelled and their estate, including the mine, was acquired by Spanish aristocrats who operated a small but financially unsuccessful mine on the site in the early 19th century. Eventually the property was bought by an American mining engineer, William Braden, on the advice of Marco Chiapponi, an Italian mining engineer. Braden then approached American Smelting's Barton Sewell and other US mining men and in a joint venture funded the development of a new underground mine in the first decade of the 20th century. This became the El Teniente mine, the story of which we will pick up later.

In the latter years of the 19th century and into the first decade of the 20th century, the mining emphasis changed from copper to nitrates. As with elsewhere in Latin America, British capital was critical, and a Yorkshireman, John North, was the leading source of this capital. The problem for copper was that the high-grade surface deposits had run out and the industry, faced with easy pickings from nitrates, was not ready to spend heavily on looking for copper mineralisation at depth until just before the First World War, when American capital, led by the Guggenheims, entered the scene for the first time sure that the war would lead to a massive increase in demand for all raw materials. The American challenge to British hegemony in Latin America had begun.

JOHN NORTH (1842-1896)

John Thomas North was known as the nitrates king as a result of his dominance of Chile's nitrates industry in the latter part of the 19th century and his web of listed nitrates companies in London. He was also an active promoter and owner of coal mines in Chile and the UK and gold mines in Australia.

The son of a prosperous coal merchant, North was born in Holbeck, near Leeds in Yorkshire, in the north of England, in 1842. He served an apprenticeship in a local engineering firm, Fowler & Co, and then went out to southern Peru in 1869 where he installed machinery to treat nitrates. At the time nitrates – nitrogen rich salts – were beginning to be the fertiliser of choice for farmers, and Peru and Bolivia had the largest reserves of the mineral. In 1879 the War of the Pacific broke out between Chile and Peru and Bolivia and as a result there was concern for the nitrate fields in southern Peru. One of the main problems was that the Peruvian government had issued government nitrate bonds in 1875 in an attempt to nationalise the largely British and Chilean owned nitrate deposits in Peru. With the outbreak of war the value of the bonds plunged and North, who had by then established his own nitrate works and invested in a water company which supplied the desert-located nitrate fields, began to purchase these nitrate bonds at less than 15% of face value. North's speculation paid off when Chile was victorious in the War and annexed the nitrate provinces of Tarapaca and Antofagasta. At the same time Chile confirmed the ownership of the nitrate fields through ownership of the old Peruvian bonds in 1882.

North, through contacts in the old Peruvian nitrates set-up, had inside knowledge that the Chileans would favour the old ownership structure and he encouraged other owners to join him in a cartel which would dominate nitrate production and influence world prices for the vital commodity. With his coup complete North controlled the most profitable of the nitrate operations and he set about exploiting his position by returning to London in 1882 where he fed a developing nitrates boom on the Stock Exchange by floating a number of nitrates companies. Included in the list were Liverpool Nitrate, Colorado Nitrate and Primitiva Nitrate. The flotations were highly successful and substantial dividends were paid almost immediately. North was also investing in enterprises in Chile including the Nitrates Railway, Tarapaca Waterworks, Bank of Tarapaca, and London and Nitrates Provisions Supplies. He also invested heavily in Chilean coal and iron ore mines on the Bio Bio River in the south-central region.

North's almost monopolist position, certainly in terms of control, in the Chilean nitrates industry eventually led him into clashes with the government of President Balmaceda, who felt that North's central position in the nitrates industry was unhealthy in the light of the importance of the industry as a source of government revenue. In the end, a short civil war in 1891 lasting eight months and consisting of just one battle saw Balmaceda overthrown. By now North was looking beyond Chile for investment opportunities and he became involved with King Leopold of the Belgians in the Congo, where he invested in Anglo-Belgian India Rubber. He also had developed North's Navigation Collieries Ltd in South Wales which was immensely profitable and helped support a lavish lifestyle of grand parties, horse racing and travel. He also built a mansion near Greenwich, Avery Hill, where he lived with his wife and three children.

North became intrigued by the Australian gold boom of the 1890s and decided to get involved, putting his name to the promotion of Londonderry Gold and Wealth of Nations. Both stocks were popular counters but soon crashed and burned. Around this time he also made significant investments in Belgian cement and Cairo trams. He was not finished with new challenges though and went into politics in the UK, narrowly losing a by-election in West Leeds to William Gladstone's son Herbert; a further indication of the popularity North enjoyed despite his sometimes dubious business contacts and methods. Although he had largely, without comment, exited from Chilean nitrates he retained some executive positions and it was while chairing the board meeting of Buena Ventura Nitrates in London in 1896 that he had a fit and died. His funeral in Kent drew huge crowds of mourners, matching the numbers who attended the funerals of the great Randlords like Rhodes and Barnato.

Now little remembered in the UK, North was a substantial figure in the development of the Chilean mining industry after the War of the Pacific. He was lionised by some in Chile and hated by others and still remains a controversial figure in that country's history. In the UK he died relatively poor, much of his wealth having been simply frittered away, but he was able to make endowments to Leeds Infirmary and Leeds University. Unfortunately Avery Hill could not be maintained and his widow was forced to sell the mansion and estate. The park is now council owned and remains one of the few memorials to an extraordinary if faulted man.

8. COPPER AND OTHER MINERAL RICHES IN THE USA

Central to the rise of the United States to industrial superpower status in the 19th century was its huge store of indigenous raw materials. However, for many years after gaining independence from Great Britain, the US had been an importer of manufactured goods from Britain and Europe, its exports being raw cotton and other agricultural products – hardly the stuff of an advancing economy. As the century progressed the US began to experience growing immigration, and the economy began the process of industrialisation, particularly in the northern states. It was after the savage Civil War, which lasted from 1861 until 1865 and stimulated a huge weapons programme in the north, that the US began to flex its muscles, and challenge the economic and political supremacy of Europe – in particular that of its former colonial ruler, Great Britain.

TECHNOLOGY DRIVES THE COPPER SEARCH

The first copper discovery in the US was in the east at Simsbury, Connecticut, in 1709 during the colonial period. After independence, the expansion of the USA westwards – as settlers moved away from the relatively heavily populated original 13 colonies – led to copper discoveries; first in Michigan's Upper Peninsular in the 1850s, and then in the following decade the famous Butte copper deposits were discovered in Montana. The copper deposits of Arizona were found a decade later and then in the 1890s copper was discovered at Bingham Canyon, originally a gold mine, in Utah. This particular mine is still in operation today and is also still a world class producer.

These discoveries enabled the US to exploit the technological advances that the 19th century was throwing up by using its own raw materials. Of particular note was the development of electricity and associated scientific advances – the telegraph, the telephone and the power generator. Here Sir Michael Faraday's discovery of electromagnetic induction in 1831 and Sir Joseph Swan's invention of the electric lamp in 1860 were but two critical advances. The US's leading position in utilising these technological advances added further weight to its push for world power status, as well as underlining its need for the copper that was the essential element in these new technologies, particularly due to its role as the most effective element for conducting electrical current.

As we have already seen, the use of copper in the manufacture of weapons remained the key to demand for centuries. In the end, the invention of firearms and the accelerating use of iron and steel in the replacement of items formerly made with bronze shifted demand for copper to coins, as well as larger weapons such as cannons, and as a material for artworks, such as sculptures. As science inched its way forward, copper became widely known as one of the key elements in its advance and, though patchily, techniques for mining copper ores evolved as new deposits were discovered and new treatment techniques were developed and applied to different types of ore. Also, new uses in weapons and defence were found, such as in the copper cladding of ship hulls as a method of resisting attacks by invertebrates in tropical waters. It also allowed faster sailing, important in battle, as seaweed and other growths did not become attached to the hull.

As the 19th century advanced the main copper mining areas of the world were Spain and Chile. The former's mines were under British control and supplied the smelters of Swansea in south Wales, and Chile's mines supplied the smelters of the eastern US seaboard in New York, Baltimore and Boston. The Americans were anxious to develop their own raw material resources for both strategic and economic reasons, but the supply routes from Chile to the eastern seaboard were well established. There was also the issue of transporting mid-western concentrates east to the smelters. The routes taken by the settlers moving west provided some guidance, but large quantities of ore were better suited to carriage by water and the Michigan Upper Peninsular mines, developed by Calumet and Hecla, sought to find a water route using the Great Lakes.

GETTING THE FIRST COPPER TO EASTERN MARKETS

One major snag in the search for a route for the transportation of copper through the Great Lakes was the rapids at Sault Sainte Marie, which formed the boundary between Lake Superior and Lake Huron. The copper laden boats crossed Lake Superior to the edge of Lake Huron but there was no navigable channel between the two and they had to break their passage at the Lake Huron rapids and transport their ore overland to re-load onto boats on the other side of the torrents. After that they crossed to Lake Erie and then went via the Welland Canal around the Niagara Falls into Lake Ontario and then through a variety of locks to the St Lawrence and then on to the northern Atlantic. This made the Michigan mines marginal due to the very high transport costs. It was therefore decided in the early 1850s that a canal should be built to by-pass the rapids, opening up Lake Superior and the affected states to increased trade. This canal was also needed to carry iron ore from the Michigan Peninsular to the iron and steel mills of the Great Lakes in Michigan, Ohio and Pennsylvania.

However, first there was the requirement to get the US Congress to pass a bill enabling the construction of the canal and this was no straightforward matter with a large number of southern states opposed to anything that might increase the economic power of the north. After much angst in Congress the canal appropriations bill was duly passed and a canal was constructed just before the outbreak of the Civil War. After the Civil War, US economic activity plunged and copper mining on the Michigan Peninsular was sharply reduced.

Prospecting continued though and so did politicking, with the result that in pursuit of protection for Michigan copper mining the state persuaded Congress to apply import tariffs to copper as it did to iron and steel. Although resisted by the government, tariffs were introduced for copper and the long-term result was a collapse of Chilean copper imports and the closure of smelters in the east as new plants were established near the Great Lakes. At the same time railroads were being built across the US further enhancing the viability of mineral deposits that would formerly have been too remote to be developed (with the exception of gold and silver, with their high value-to-weight ratio). Coal discoveries also made viable the building of smelters in these remote areas.

THE GREAT ANACONDA

The next great discovery was Butte in Montana where Anaconda started production in 1884. The Michigan mines were based on the mining of very high-grade but scattered boulders of copper ore, both on the surface and underground. Butte, in contrast, was a vein system which started as silver near the surface and became copper rich at depth. As we mentioned earlier, Marcus Daly, who was an associate of George Hearst, discovered the copper potential and sought to greatly expand the Anaconda mine which he owned – Hearst was one of Daly's partners in this enterprise which turned a silver mine into a copper mine.

For a short period at the end of the century the Rothschilds became involved as investors, but just before his death in 1900 Daly involved the Rockefellers in Anaconda and the company name became Amalgamated Copper until 1915 when it reverted to Anaconda again. The Rockefellers were involved in building monopolies, or trusts as they were called then, and Anaconda became their copper trust, rivalling their oil trust, Standard Oil. For a few years in the 1920s Anaconda was one of the largest companies in the world but the Great Depression of the following decade cut it down to size as the copper price, along with Anaconda's share price, collapsed. Recovery for both Butte and Anaconda had to await the outbreak of war in 1939.

By 1983, almost 100 years after Anaconda started producing copper at Butte, mining ceased; over this period it is believed that the Butte operations of Anaconda produced around $300 billion worth of copper.

SOUTH TO ARIZONA

The search for minerals in the US, and in particular the search for copper, was not confined to the states and territories in the Midwest and central US near the Canadian border – it began to push south and west to Arizona, which had become a US territory in 1863, and to New Mexico. The first discovery in Arizona was the New Cornelia copper deposit near Ajo. This led to problems with Mexico, which thought the deposit might lie its side of the newly drawn border with the US. Although in time a substantial operation was established at Ajo, in the beginning the ore was shipped through San Francisco to Swansea in Wales to be treated.

The first major copper mine developed in Arizona was Magma Copper's Silver Queen mine in 1871. In 1877 copper was found in the Warren district at Bisbee near Tombstone and after that the United Verde mine was developed near the town of Jerome, named after the Jerome family, one of whose number was the mother of Winston Churchill. The United Verde mine was highly profitable and over its life, which finally ended in the 1950s, produced over $1 billion worth of copper and by-product precious metals.

Although Arizona has remained the US's prime copper producing state to the present day, Utah's Bingham Canyon mine, now in the ownership of Rio Tinto of the UK, became a major copper producer in 1896 and Utah has remained a rival to Arizona ever since. For 30 years before that Bingham Canyon had been first an important gold producing area, and then a silver and lead producing area. Copper had been observed at Bingham Canyon ever since the area had been worked for gold, but the problem was that the copper was generally rather low in grade and only the odd high-grade section, as was found at the Highland Boy mine within the Canyon, was considered economic.

Enos Wall, a miner with 30 years experience, noticed in the ground above the Canyon that there were potentially large quantities of lower grade copper and he acquired land to develop an operation using huge surface mining and transport equipment to mine ore in sufficient amounts to make the operation viable. The company he formed to develop the operation was the Utah Copper Company and it was he who named the mine Bingham Canyon after the district. In due course, the mine became one of the largest open pit operations in the world. The geological structure that Bingham Canyon ore was found in was porphyry, fine grained granite with embedded crystals and copper mineralisation spread or disseminated evenly throughout the rock. This porphyry setting was not uncommon but economic working depended on large-scale mining and therefore on strong demand, something that at the end of the 19th century was beginning to be seen as the US economy expanded.

In due course prospectors and mining companies found more porphyry copper deposits in Arizona, New Mexico and in Nevada, traditionally a gold mining state. These discoveries, which were mined for decades, played an important part in the rise of the US to sole economic superpower status in the 20th century. The search for porphyries also spread beyond the continental US into Alaska and ultimately to Latin America and the Far East.

THE DREAM OF DIAMONDS

Whilst in recent years Canada has emerged as a major force in world diamond mining, the US has yet to find any major diamond deposits. The Crater of Diamonds in Arkansas and the now closed Kelsey Lake mine in Colorado were two of the better known operations, but the former is really a tourist attraction where visitors are allowed to dig on payment of a daily fee and the latter a very small mine intermittently operated in the late 20th century. However, for a few months in 1872 it was believed that a significant diamond find had been made in America's Midwest. It had not, but a major scam was nonetheless perpetrated, beginning in San Francisco and stretching as far as London, the inevitable first call for speculative mining exploration funds.

It had started with the deposit of some uncut diamonds and rubies at the Bank of California by two prospectors, Philip Arnold and John Slack, in March 1872. William Ralston, the President of the Bank contacted Asbury Harpending, a business entrepreneur with whom Ralston and the bank had had extensive and successful past dealings. He told Harpending that diamonds worth $125,000 had been deposited in the bank and that these had come from a new diamond discovery somewhere in the Midwest. Harpending, who was in London at the time, repeated the story to an intrigued Baron Rothschild before heading back to San Francisco.

In the meantime, Ralston had told the story to George Roberts, an experienced mining engineer, who met with Arnold and Slack and arranged to send two geologists to examine the deposit which was thought at that stage to lie somewhere in Arizona. The deal was that during the final stage of the journey the two geologists would be blindfolded to protect the exact location of the deposit. The examination of the deposit was successfully completed with confident reports of its prospectivity. Harpending was by now back in San Francisco and Arnold and Slack suggested that they return to the deposit and bring out stones to the value of two million dollars, a plan which Harpending endorsed. In the end they brought back an alleged one million dollars' worth. Some of the stones, around 10% of the sample, were sent to New York where Tiffany the jewellers valued the parcel at $150,000.

Whilst this was going on Harpending and a group of associates was putting together a financing proposal for the diamond deposit. Henry Janin, a highly respected mining engineer, was engaged to make a trip to the deposit, which was finally revealed to be somewhere in Wyoming, to give it a thorough examination and he took a team of helpers with him. Janin, though pressed for time to complete his examination, was enthusiastic, diamonds and other precious stones were found, Janin thought it all looked very promising, but in his enthusiasm Janin had not done proper due diligence on the deposit which was to prove fatal for investors, but surprisingly not for his long-term reputation.

Harpending and his associates were delighted with Janin's news and the financing slipped into gear. A company with $10 million capital was formed and subscriptions were sought from Harpending's associates and from other interested professional investors. Around $2 million was raised from 25 substantial San Francisco investors, the funds being held at the Bank of California. As part of the formation of the company the last interests of Arnold and Slack, who had already received substantial payments for some of their interest in the deposit, were bought out for $300,000. Samples from Janin's visit were forwarded to Tiffany again and also to Baron Rothschild, who became the group's agent in Europe. Tiffany's valuation was less heady this time but this was ignored.

Further visits were made to Wyoming but due to the winter lockdown at the deposit, the sale of shares to the public was delayed until the next spring. This was fortunate as out of the blue in November 1872 a telegram arrived from Wyoming from a Clarence King, head of the US government's 40th Parallel Survey, who claimed that the diamond deposit had been salted with precious stones from outside – the whole thing was a hoax. Janin was despatched to meet up with King and examine the area of the deposit once again; shamefaced Janin reported back to Harpending that King was correct. The stones unearthed had been skilfully but clearly dug into the area. In addition, the fact that diamonds and other precious stones including rubies were found together was geologically flawed.

It is believed that the diamonds originally came from South Africa, where new mines were opening up at the time. The uncut stones were also eventually judged to be worth $35,000 and of industrial quality which would explain why Tiffany the second time around placed a low value on the new parcel of stones sent to New York by Harpending. Nevertheless, it is extraordinary that so many experienced miners and financiers were taken in by Arnold and Slack. Interestingly both Arnold and Slack escaped prosecution – Slack simply disappeared and Arnold denied complicity saying that any salting was done by "the Californians". He also pointed out the earlier high valuations of the stones by Tiffany, but of course Tiffany had no experience in valuing uncut rough stones, their expertise was in polished goods.

For the US the search for commercial diamond deposits continued following the Wyoming scam. Over 130 years later the country has yet to open any substantial diamond mines, in some contrast to its neighbour to the north.

LEAD EXPANSION

We have already examined the Mississippi Valley lead mines developed by the French and the arrival of eastern adventurers following the retreat of the French from America. The original owners of the land, a group of American Indian tribes, were dispossessed of the economic value of the lead and by the middle of the 19th century lead mining was expanding, rapidly bringing into the picture one of the US's most powerful miners, St Joseph Lead, which eventually became St Joe Minerals and was taken over by Fluor in 1981. The market for the lead also began to change, with the route down the Mississippi and then to Europe being replaced by a route via Lake Michigan to the eastern states where demand for lead had begun to grow.

By the middle of the 19th century the Galena district in Illinois was producing around 25,000 tonnes of lead a year, the market price then was $66 per tonne. The Missouri lead mines had been in decline for some years but in 1867, following the end of the American Civil War, St Joseph Lead established a mine at Bonne Terre in the southeast corner of the state where earlier the French had mined surface lead ore. The Bonne Terre mine contained flat sheets of galena very close to the surface which made mining straightforward, but the relatively low grade lead (around 6%) and difficult weather conditions meant that financial returns were often poor as mining could not continue unbroken, and the smelting technology had not progressed much beyond the old stone hearth system.

In time the mine probed deeper where ore was not only richer but mining at depth could continue uninterrupted. After a mill fire in 1883 St Joseph was able to build modern treatment facilities and reorganise the mine's underground operations. Returns improved and the company began to acquire additional ground in the area for exploration and exploitation. This became of particular importance during the First World War, which the US entered in 1917, as St Joseph was a key supplier of lead to the American war effort.

BUNKER HILL, IDAHO

Another major 19th century lead discovery was the Bunker Hill at Kellogg, Idaho, in the Coeur d'Alene district, originally made famous as a destination for gold prospectors. In 1885 Noah Kellogg, a gold prospector, was grubstaked by two local storeowners, John Cooper and Origin Peck, and set off to explore around the south fork of the Coeur d'Alene River. Kellogg and his backers were interested in gold, but Kellogg instead found galena which did not interest Cooper, so Kellogg approached Phil O'Rourke who was something of a lead and silver expert and he accompanied Kellogg back to the river.

There a number of outcrops were identified, including the famous Bunker Hill outcrop. Legend has it Kellogg's pack donkey, mesmerised by the shining ore (not that galena actually shines after weathering) drew Kellogg's attention to it. Whatever the truth, the fact that an economic lead mine at Bunker Hill was in prospect attracted Cooper's attention, and, believing that whatever his original reaction to the galena find he still had a legally enforceable arrangement with Kellogg, he went to court and was awarded a share in the prospective lead mine.

Working of the surface ore began almost immediately in 1885, the discovery year, but the Bunker Hill mine's early days were difficult ones. The ore was initially rich and shipped to San Francisco for treatment but the rich ore soon ran out, and then began a long and expensive process of raising capital for the building of a mill on site to concentrate the lower grade ore that had quickly become the mine's standard offering. The process included the incorporation of the mill as Bunker Hill and Sullivan Mining and Concentrating in 1887. After that, further capital was raised to build a new, larger mill and other treatment facilities, but the raising of the capital took time, and in due course the share of the original discoverers and grubstakers, including Noah Kellogg, had been diluted down to almost nothing.

Now with the investment and re-financing complete, in 1892 severe labour problems hit Bunker Hill, and indeed other Coeur d'Alene district mines, and lasted five years. Despite that, the mine paid its first dividend in 1894 with its ore grade around 11% lead and 5 ozs of silver, although by the standards of the time these grades were on the low side. However, the labour disputes, which started with one that was about the structure of medical insurance, unsurprisingly undermined efficient working, and the hiring of contract miners by the district mine owners, which included Bunker Hill, further exacerbated things. In the end, violence between miners and the mine owners escalated and troops had to be called. Mine property was damaged and two men were killed when fighting broke out. These skirmishes continued on and off for several years and the district's reputation plummeted as it acquired an image of lawlessness.

METALS AND THE ADVANCE OF CIVILSATION

We alluded earlier to the link between metals and the advance of civilisation. Much of this advance, certainly in more recent centuries, related to the development of new technology, and this became a particular feature of the economic landscape with the Industrial Revolution. We have already described the impact of technological advances on the demand for copper; another dramatic demonstration of the way metals changed the shape of whole industries was the construction of the first iron-hulled warships in the second half of the 19th century.

The most striking example of the huge significance of this change was the clash between Confederate iron warships and Union wooden warships in 1861 on the Mississippi River during the US Civil War. The complete obliteration of the Union wooden ships spelt the end of centuries of wooden warships and the accompanying end of the wholesale destruction of oak forests; in their place came the iron ships which led to a massive increase in warship firepower, which was necessary to destroy the new ships. Of course the great British and French navies, and indeed others, had been developing ironclad warships for some years before the US Civil War clash, but the Mississippi battle was the first practical demonstration of the power of the iron warship and the inevitable and immediate end of the wooden warship. It was also yet another example of the way that access to metals stimulated research and development, and led to the implementation of new technology that supported economic and technical advance.

9. MINING IN THE EAST

Whilst the economies of the East have become the powerhouses of the modern world, the 19th century belonged to the West, and industrial development in the East lagged far behind the economies of Western Europe, particularly Britain and Germany, and the fast growing United States.

JAPAN AS A MINING NATION

Today we recognise Japan as one of the great modern industrial and technologically advanced economies, but in the latter stages of the 19th century, as an essentially rural economy, it trailed behind the West in terms of the Industrial Revolution. It was determined though to accelerate the pace of industrialisation and try to catch up with the West, and the country did have an historically important mining industry, one that ironically – in the light of the Japanese economy today – was an exporter of raw materials, particularly copper.

COPPER

The largest copper mine in the country was the Ashio mine in central Japan on Honshu Island. It had operated for centuries as an effectively publicly owned mine, producing around 1,500 tonnes of copper annually, but by the beginning of the 19th century it was closed due to low reserves. In 1877 the Furukawa company bought the Ashio mine, re-opened it and expanded output after an exploration programme disclosed new ore sources. By the end of the century Ashio was producing over 6,000 tonnes of copper per year, a quarter of Japan's output, and Furukawa, which had bought the Ani copper mine from the government in 1885, with Ashio and its other copper mines, was responsible for over 60% of the country's production. Copper exports, 82% of the country's copper production, were an essential part of Japan's economic strategy to modernise and build an industrial infrastructure. To do this, foreign currency had to be earned and copper represented almost 10% of Japan's export earnings by value at the time.

With the Meiji Restoration in 1867, Japan instituted a supposedly enlightened form of government under the Emperor, which led to the emergence of a new ruling class with far-reaching economic and imperial ambitions. It was they who drove the industrialisation strategy and Furukawa joined their ranks, its foreign currency earning copper mines being central to fulfilling Japan's strategic ambitions.

MODERNISATION AND ACCOMPANYING PROBLEMS

Significant modernisation was also taking place at Ashio to enable production to be increased – new machinery was bought and foreign consultants were brought in to upgrade mining methods. This led to the introduction of horizontal mining, increasing productivity and efficiency. In 1888 the British South East Asian trading giant, Jardine Matheson, signed an exclusive copper marketing agreement with Furukawa. The rather primitive smelting set up at Ashio was also upgraded in 1893, which materially reduced the time taken to produce finished copper. Transportation links were also improved. As hoped, the growth in the Furukawa company's copper output provided the needed foreign currency for the resurgent Japanese economy and its imperial ambitions.

However, as production levels increased a major problem arose with a very modern ring to it – environmental damage. One of the main problems that those affected by this environmental damage had to face was that the government's commitment to economic growth and regional expansion meant that it was far more interested in building a modern industrial infrastructure, and in particular encouraging growth in the mining industry, than it was in assuaging the concerns of the rural population about the consequences of this strategy.

The primary problems related to the pollution of nearby rivers by mine and smelter waste run-off, often caused by flooding, which also led to farmland and villages being swamped with water containing toxic mine waste. Woodland and grassland near to the mines were also contaminated by mine and smelter pollution and gradually became barren and dead. Locals were vociferous in their complaints and demands were made to the government in Tokyo that the Ashio mine be closed, but that was never likely with Japan waging war against China and being reliant upon copper earnings. In the end, compensation was paid but it was ungenerous. In 1907, the mine itself was the centre of some of the worst labour unrest in Japan's history, but despite that it had a long life, closing in 1973.

Other copper mines operating at the end of the 19th century, in addition to the Furukawa mines, included the Kosaka mine of Fujita Gumi (now Dowa Holdings), the Besshi mine of Sumitomo and the Akasawa mine, which was re-named the Hitachi mine in 1905 when acquired by Nippon Mining. Many of these mines had been operating since the early 17th century. In the late 1800s there was an extensive programme of privatisation in the Japanese economy and many government-owned mines were sold to growing industrial groups such as Sumitomo.

The concept of growth when used in company with Japan's mining industry sounds strange to modern ears, used as they are to hearing about the country's huge appetite for mineral imports. However, in the last 20 years of the 19th century copper output rose by 520%, coal by 850%, silver by 570% and gold by 680%, as Japan's own industrial revolution gathered pace. The position of these minerals in terms of Japan's exports was also impressive, with copper providing 5% of Japan's exports by value in 1895 and coal 6%. At that time over 50% of Japan's exports were in the form of raw materials, with raw silk being the largest earner.

The Kosaka mine was originally a loss-making silver mine, but was re-invigorated when in 1877 the introduction of German technology led to a major improvement in the recovery of both silver and copper.

GOLD

Gold was also mined in Japan. A notable example was the Sado gold mine on Sado Island, which lies off the coast of the Honshu province of Chubu. It was first found in the second half of the 16th century and was only exhausted and closed in 1989. Although it was one of the largest gold mines in Japan, its lifetime output of 2.5 million ozs of gold, with silver as a by product, was modest. Sado, which early in the 20th century came under the control of Mitsubishi, was a narrow-veined deposit, which made mining expensive but high grades of 20gms plus were a helpful mitigation.

The Toi gold mine, perhaps second only to Sado in size, and also situated in Chubu, had a similarly long history, having started up in the 16th century. There were also other smaller gold mines in the region, such as Yugashima, which was established around the same time. Although closed in the 1960s, both Sado and Toi have museums open to the public depicting the history of their operations. Today Japan has one gold mine, Sumitomo's Hishikari, which opened in 1985. Hishikari, which is very high grade (40 gms per tonne), is the country's biggest ever producer of the yellow metal, having mined around 200 tonnes (6.5 million ozs) over its life up to now.

OTHER ASPECTS OF THE INDUSTRIALISATION STRATEGY – IRON, STEEL AND COAL

Central to the industrialisation strategy was the development of an iron and steel industry, but Japan could only provide around half of its iron ore needs. As an inevitable part of the Meija industrial revolution, ironworks based on western technology were built in the late 1800s and in the early stages some of them were sited close to indigenous supplies of iron ore. Two of the main producers were the Kamaishi and Nakakasaka mines. However, it was not long before iron ore was being sourced from other parts of the East and Far East, including India and China, and as Japan's industrial expansion accelerated, its own iron ore mines became increasingly irrelevant; mirroring the experience of Britain, then still the world's leading economy.

Japan also had a significant coal-mining sector in the 19th century. The Miike mine, probably the largest mine, was developed at the start of the Meija Restoration and was nationalised in 1872 before being sold to Mitsui in 1899. Until well into the 20th century the majority of Miike's labour force consisted of state prisoners, and working conditions were harsh and dangerous. This slave element in the labour force did not disappear until after the Second World War, although Mitsui employed more free miners after it assumed control. In the early days Japanese mines also employed women, particularly miners' wives, to work in the mines, something that continued well into the 20th century.

In terms of size the Horonai and Takashima mines, along with Miike, were Japan's largest coal producers at this time and working practices were unsophisticated, with coal seams being blasted using gunpowder and then the material being carried to a coal bin in baskets before being loaded into hand pulled trucks to be hauled out of the workings. Once on surface it was hand-sorted to get rid of stones and other waste. The first mine to mechanise was the Hokutan Yubari mine on Hokkaido Island, the most northerly of Japan's islands.

Hokkaido, along with Kyushu, Japan's most southerly island, became the centre of the coal mining industry and other key coalfields on the island included Rumoi, Tenpoku and Kushiro. Although these fields had access to the coast they were still a long way from the country's main ports, and transportation remained a problem until investment in roads and rail had been completed. Other larger coal mines operating at the end of the 19th century included Kaijima Ohnoura, Mitsubishi Sinnyu, Mitsui Tagawa, Kaijima Ohtsuji, Mitsui Hondo, Futase Coal and Mitsubishi Ouchi.

Important though coal mining was to Japan, the level of the country's production by the end of the 19th century – at 7.3 million tonnes – was dwarfed by the UK at 250 million tonnes and the US at 200 million tonnes, which underlines the fact that Japan's industrialisation lagged decades behind the West. However, by the time of the Second World War Japan had become a large coal producer in world terms, driven by its imperial expansion which had led it to invade China, and was to lead it into war with the US. The peak year for output was 1960, when around 55 million tonnes of coal were produced. By the new millennium all domestic production had ceased.

The conditions in many of these mines were dangerous and countless accidents occurred in the 19th century; a trend that continued throughout the following century and cumulated in the horrific coal dust explosion in the Mitsui Miike mine in 1963 when 458 miners died and hundreds more suffered from carbon monoxide poisoning. Since its mines had proved so dangerous over many decades it is perhaps not surprising that Japan has increasingly sought to lay off the mining risk elsewhere, thus becoming an importer of raw materials for its hungry industrial machine. Its attitude to what is probably the most important raw material, food, is rather different, with self-sufficiency in its staple foodstuff – rice – being an article of national faith.

MALAYAN TIN

One of the most sought after metals towards the end of the 19th century was tin from the Far East. This was the heyday of the British Empire and also of Britain's Industrial Revolution and its requirements for raw materials had outrun its own indigenous supplies. Tin had become increasingly important due to the advent of tin plate cans in the early part of the 19th century. This had a profound impact on preservation times for food, important in an age before universal refrigeration, and it also made, during this golden age for exploration, the supply of expeditions easier and therefore more certain. Its properties as a metal used in the early soldering of electric components also became increasingly important as the Industrial Revolution became more sophisticated.

Britain had been increasing its influence in the various sultanates that made up the Malayan Peninsular and outlying states such as Sarawak during the 19th century. The whole region was prospective for raw materials such as timber, gold and most importantly tin, and its climate and the fertility of the land led to the expansion of rubber, palm oil and tea growing.

Tin had been mined in Malaya for centuries but the discovery of rich new deposits in the 1820s led to a substantial influx of Chinese immigrants who provided both capital and labour to drive the industry's revival. They settled in the western state of Perak where they uncovered rich alluvial tin deposits in the Kinta Valley, and the wealth generated from tin mining played an important part in the development and expansion of cities and towns in the region, such as Kuala Lumpur.

Most of Malaya's mines were alluvial, latterly using big industrial dredges. In the early days alluvial tin was primarily mined using a gravel pump where tin bearing ore was hosed under high pressure, washed down into what might be described as a bin and then pumped out along a long, sloped and rutted wooden trestle which captured the tin in the ruts. Small-scale artisan miners made do with a simple panning process much like that used by alluvial gold miners. Malaya also had a number of small open cast mines which used mechanical diggers to extract the ore, and some of these survived the larger alluvial operations which were progressively closed down in the later years of the 20th century.

In the last quarter of the 19th century tin mining began to expand quickly, but internal political disputes led to an increase in British control of the tin mines and of Malaya itself. To meet rising British demand for Malayan tin British capital began to flood in and substantial investment, particularly in the form of giant dredges, was made in order to expand output. Chinese operations, which were often small scale, became increasingly marginalised and their previous control of the industry slipped away to the British. As often happened during British colonial rule, substantial infrastructure projects, rail and road, were implemented in order to provide rapid access to the tin fields and to Malaya's ports. Almost from a standing start, by the end of the 19th century Malaya produced over 50% of the world's tin.

One of the earlier figures in the revival of tin mining in Malaya in the second half of the 19th century was Loke Yew. Other Chinese entrepreneurs, such as the Khaw family, also acquired substantial interests in the growing tin mining sector.

LOKE YEW (1845-1917)

Loke Yew was born in China in the Heshen district of Guandong province in 1845. He was one of four children in a poor farming family and worked in the fields from an early age. He had relatives living in Malaya, which was then a British colony, and in 1858 at the age of 13 he left China to join them, hoping to better himself there.

He was found a job in a general store in Singapore where he worked for four years and saved almost $100, which he used to start his own store. When the business was established, Loke, intrigued by talk in the Singapore business community about the new tin areas being uncovered in the north of the country, left his staff to run the Singapore store and in 1867 headed to Perak where he joined the exploration rush. His first four years were difficult and he had limited success, with his costs at around $60,000 over the period. But he was certain that his luck would change so he persisted and he eventually found a rich tin deposit in Kelian Bahru.

This was the breakthrough and in due course Loke acquired or uncovered a number of other tin deposits and he also branched out into rubber plantations. The profits from the tin were sufficient to encourage Loke to expand his business interests beyond natural resources into pawnbroking, alcohol distribution and gambling. He was very close to the British administration in Malaya and this was a material support for his growing business interests. He expanded into motor vehicles, trading, shipping, engineering and rice. He also acquired property and land in Malaya and back in China.

During the 1870s the Lurat Wars led to the shutdown of the tin mining industry. The unrest was put down by the British, but Loke's tin mines were badly damaged. In due course, tin mining recovered and Loke continued to build his tin interests until he was the largest force in Malayan tin, and therefore in world tin. His fortunes were particularly boosted when he expanded his tin operations into the Kinta Valley and Selangor. He also teamed up with another Chinese tin baron, Thamboosamy Pillai, to establish the New Tin Company. The two were also founders of the prestigious Victoria Institution, which is still one of Malaysia's leading secondary schools today.

Although Loke's business activities clearly took up a lot of his time he still managed to find space for a family life. He had four wives, the last of whom he married in 1914 at the age of 69 and with whom he had three children. In all he and his four wives had seven sons and four daughters. Although extremely rich, Loke was a simple, even humble, man and was not given to great shows of ostentation, but he did build a very fine house, Loke Mansion, in Kuala Lumpur which was finished in 1904, and though it is no longer in the Loke family it is still standing. In his later years he was very active in the Kuala Lumpur community, endowing schools and hospitals, but he never forgot China, and Hong Kong University was one recipient of his generosity.

He was much admired by the British in Malaya and he was appointed a Selangor State Council Member in the 1870s. In recognition of his role in the development of the city of Kuala Lumpur and his support of the government of British Malaya he became a Companion of the Order of St Michael and St George. One of his last business projects was the refinancing of the Kwong Yik Bank in 1915. Loke died of malaria in 1917 and his fortune was measured at $20 million ($350 million in today's money).

Although he was the most powerful force in world tin in the late 19th century with over 12,000 miners working his alluvial deposits, Loke was an outstandingly successful businessman because he was flexible and because he could therefore take advantage of opportunities even if they were outside his then present interests. Although his family retains status in modern Malayasia, Loke has become a rather forgotten figure perhaps as a result of his support for the British, or perhaps because tin no longer drives the Malaysian economy and tin barons have been supplanted by other business giants.

The rise of British incorporated tin mining companies really got underway in the early decades of the 20th century with, among others, Pahang Consolidated in 1906, Kamunting Tin Dredging in 1913 and Petaling Tin in 1920. However, two earlier non-Malayan companies to incorporate were Gopeng Tin in the 1890s and a French group, Kinta Societe des Etains, in 1886. Following Malayan independence in 1960 when the country became Malaysia, many tin companies, having moved their operating offices to Malaya many years before, began to consider re-incorporation in the newly independent state. This process was put in train in the 1970s when other UK incorporated mining companies were also moving out of the UK due to high taxes and exchange controls.

TIN IN THAILAND AND INDONESIA

Around the same time as Malaya's tin industry was being developed, in the later 19th century, there was a material increase in interest in tin mining in Thailand, known as Siam until 1939, on Malaya's northern border. The mining of tin had, since the 16th century, been centred on what is now a major international tourist destination, the island of Phuket. Phuket first attracted the attention of the Dutch who for a short time in the 17th century had a tin trading monopoly there, and the French obtained a similar trading deal in the 18th century. The British pushed into the country in the 19th century and in 1825 the Siamese government signed a trade treaty with Britain, which included tin. With the establishment of British interest in Siam, the Chinese, who were flooding to the new tin fields of Malaya, also went north to participate in the expansion of Phuket's tin production and try to establish themselves there.

The first of the new age Siamese tin mines was developed in 1809 following the discovery of economic tin deposits at what is now called Baan Thungkha. One of the larger mines was the Mhuang Harb tin producer. These operations were locally run with Chinese expertise, but by the turn of the century foreign miners were allowed to own and operate mines in Siam, and particularly on Phuket. The Australians also began to take an interest and were responsible for the establishment of the famous Tongkah Harbour company in 1907 which operated the first offshore dredging operation. Siam/Thailand maintained a key position in world tin mining until the latter years of the 20th century when reserves began to be exhausted, as eventually happened in Malaysia, and tin prices collapsed.

As in neighbouring Malaysia and Thailand, tin mining in Indonesia, formerly known as the Dutch East Indies, has been going on for centuries, but it was not until the 19th century that tin began to be exploited on a commercial scale. One of the biggest sources of Indonesian tin was found on Bangka Island in the early 18th century, which remains one of the country's largest sources of tin today. Another major source of tin was Belitung Island where the Dutch began tin mining in the first half of the 19th century; the island gave its name to the main Dutch tin miner there, Billiton, now one of the world's largest mining groups.

The Chinese, as elsewhere, played an important part in the expansion of the tin mining industry and it was their labour that primarily worked the tin deposits. Mining methods were usually gravel pumps or open pit working, conditions in the mines were very tough and some indentured Chinese miners even disfigured themselves in order to be dismissed and sent home. Another kind of disfiguration caused by tin mining was of the jungle environment of the region, where trees and vegetation disappeared as the tin ore was blasted and scraped from the surface and shallow deposits, leaving in many cases a growing wasteland of polluted lakes formed as a result of the mining.

The marketing of Bangka tin in the 18th century was carried out by the Dutch East Indies Company, the VOC, which had a close relationship with the Sultan of Palembang who controlled Bangka Island. The main markets for tin were Canton in China and India, although in the following century as British influence in the region grew, demand shifted towards Europe. The Sultan encouraged small scale tin mining by locals even though the tin had to be sold to the VOC; unsurprisingly both the VOC and the Sultan did well out of this as local miners did not get the market price for their tin. By the end of the 18th century tin smuggling had become widespread as local miners sought higher free market returns and the city of Singapore, which provided a market for smuggled tin, was close by.

One of the leading smugglers on Bangka was Bong A. Siong, a Chinese mining entrepreneur. Since many officials of both the Sultan and the VOC were involved with smuggling, Siong operated unfettered for many years. Eventually caught and charged when the Sultan's tin returns slipped to unacceptable levels, he was sentenced to death but this was commuted to imprisonment, and he was eventually released because without his supervision tin output from his mines had crashed, making the Sultan's financial position even worse.

When the VOC collapsed in the early years of the 19th century the British entered the picture with a short period of administration between 1812 and 1816, and eased the Sultan out of the picture, organising a tightly controlled tin mining and marketing monopoly which the Dutch inherited when they resumed control of Indonesia in 1816. The mining of tin during the 19th century saw local private activity wither away to nothing as large-scale mining increasingly predominated.

BORNEO DIAMONDS

Whilst the 19th century was, as far as diamonds were concerned, the age of Africa, particularly in the latter decades when Brazilian production fell away sharply, the end of diamond mining in India meant that there were limited local supply sources for the Asian and Far East regions. One relatively small diamond miner was the Dutch colony of Borneo, then part of the Dutch East Indies. There were two main prospective diamond locations, one on the west side of the island the other on the southeast side. As in Brazil, the diamonds were primarily found in an alluvial setting, often buried in the gravel beds of local rivers where they had been deposited from sources upriver, most likely the mountains that pepper the huge island. The mining method used was to dig pits in the riverbeds; these pits went down to the underlying rock where the diamonds had been deposited. It was both crude and dangerous work as the pits usually had to be reinforced. Sometimes the gravels were found above the water course, which made things easier for the miners.

The value of Borneo's diamonds is thought to have been between $3 million and $4 million in the early years of the 19th century (50,000 plus carats), although this may have been a tenth of the value of the output 50 or so years previously. By the end of the century output had melted away to just a few thousand carats. The problem with Borneo's diamonds was that they were no better in quality terms than those of Brazil and very inferior in comparison to the diamonds beginning to flow in large quantities from southern Africa in the second half of the 19th century.

There were some much prized blue/white stones mined but, on the whole, 5 carat plus stones were very rare and very large stones almost unknown. Perhaps no more than half a dozen 100 carat plus stones were found in Borneo, with just one outstanding 367 carat stone, the Danau Rajah or Mattan, the authenticity of which is disputed as it is generally thought to be simply rock crystal. By the end of the 19th century, output had also fallen right away and the Dutch were unable to put into practice a plan to mine Borneo's diamonds on a large scale to boost production and extract resulting mining cost economies. As the 20th century progressed Africa rendered Borneo's diamonds irrelevant.

10. KING COAL

It is no exaggeration to say that without the development of a coal mining industry the Industrial Revolution would have stalled before it had gathered any momentum. Just to contemplate powering machinery using wood burning boilers alone on any but the most modest of scales nullifies the idea of an industrial age. And the arrival of electricity, which transformed industry and society, would not have been possible without coal. This may seem strange to us now as the post Revolution age that we are living in, turned its back on coal as a resource for anything but power stations, as other fuels – cleaner and more efficient – were utilised, and now we face a new age where we may be forced to ultimately reduce the use of fossil fuels to generate power as these fuels gradually run out. But without coal, none of these advances would have been possible in the first place.

Coal was also at the centre of the first efficient public lighting system through the use of gas manufactured from coal – often known as town gas. The main countries at the forefront of coal gasification were the UK, the US, Germany and France and the era of development stretched from the late 18th century to the middle of the 19th century. Whilst gas was originally used for lighting, and later for heating, cooking and refrigeration, by the later years of the 19th century with the invention of the incandescent light bulb, electricity began to make itself felt and in the 20th century, with the arrival of large scale coal fired power stations, the age of technology was born.

COAL IN THE UK

The first nation to fully embrace the Industrial Revolution was Great Britain, the leading military power in the world at the start of the 19th century, and also a leader in industrial innovation. However, coal had been mined in Britain since, and possibly even before, the coming of the Romans in the 1st century. Originally it had been used in fires and forges for working metal, a role that it had held for many centuries in other parts of the world. Mining in these earlier times was quite crude, favouring surface accumulations of coal, and when these were exhausted shallow drifts would be driven into the coal to allow mineworkers to mine it at shallow depth. As long as the British economy remained primarily agrarian, the use of coal was not widespread. Gradually, a number of significant engineering advances stimulated interest in coal mining and provided the Industrial Revolution with the means to materially quicken the pace of development.

Coal and the invention of machines, to both improve coal mining and to utilise the power that coal could generate, drove the early decades of the Industrial Revolution. Thomas Savery and Thomas Newcomen were credited with inventing the steam engine in the 18th century, which allowed pumping to take place in the mines to remove water as coal mining moved to new depths. It was, however, James Watt who refined the steam engine so it became more efficient and reliable and importantly more powerful, so it could drive production in the new factories that Richard Arkwright had introduced to the textile industry.

As the Industrial Revolution gathered pace, other activities were swept up in the advance of technology. One of the most significant beneficiaries of the invention of the steam engine was transport, which up until then had largely used barges, coaches and farm carts. George Stephenson's Rocket changed everything, ushering in the era of steam engines, which led to the building of railways, first in Britain and then around the world. It also revolutionised travel and commerce by sea, providing reliable and fast passage for ships and thereby slashing sailing times and stimulating trade. The fuel all these inventions used for power was, of course, coal and amongst the earlier British coal barons was Thomas Powell.

THOMAS POWELL (1779-1863)

Britain's position as the birthplace of the Industrial Revolution was in part due to the development of its coal mining industry. One of the most foresighted of the coal owners was Thomas Powell, who was born in Chepstow, in southeast Wales, in 1779. His father, a timber merchant, died when he was 14 and Powell took over the business, despite his youth, in order to support his mother.

The timber business did not provide a secure living for Powell and his mother so he began to explore other possibilities and he alighted on coal in South Wales, finding it to be an abundant, but hardly developed, source of fuel. Powell decided that the invention of steam-driven machinery in the latter part of the 18th century was a revolutionary event and that coal was the fuel that would drive the machinery of industrialisation. Therefore, in 1810 he bought his first coal lease at Llanhilleth and with two others worked the coal seam on the property, gaining first hand experience of coal mining techniques and conditions. Over the next 20 years Powell bought other leases and developed mines but coal's advance, though perceptible, was slow and Powell almost went under. He persisted though and bought more coal seams around Newport and in the nearby valleys. A ruthless operator, he once had to pay damages for undermining a neighbouring coal lease without agreement.

By 1830 Powell was through the woods and as demand for coal, spurred by accelerating industrialisation, rose sharply he began to be seen in South Wales as a shrewd and successful businessman whose vision for the black fuel had begun to pay off handsomely. Powell's Welsh coal was of the highest quality and his mines were conveniently located for the canal system and shipment to markets in the UK and France. Already strong in the south, the lifting of internal shipment restrictions in 1834 opened up the fast-growing industrial regions of the north for Powell, where he undercut northern coal owners; a sign of his increasingly aggressive business tactics, which started with a failed attempt to create a coal price cartel, the Newport Coal Association in 1833.

He also began to invest in transport in order to control the shipment routes to his UK markets and also was a major investor in the Taft Valley Railway, which linked Newport with England. Entering his sixties, Powell opened his two biggest mines in Aberdare, Tir-founder and Mountain Ash in 1842, thereby making him the largest coal producer in the world at the time. He now had business interests covering banking as well as coal and transport and in 1863, appreciating that the day of the larger-than-life coal owner had past, he decided to secure the future of his Welsh coal empire by merging his companies with those of another coal owner, Sir George Elliot. The company, Powell Duffryn Steam Coal, became a public company in 1864 just after Powell's death; the eight collieries in the group produced 400,000 tonnes of coal annually. In the following 50 years output soared in line with industrialisation and economic growth.

Whilst there can be no arguments about Powell's achievements in building his coal empire and providing the fuel for the new industries, his record as an employer was poor and his ruthless approach to operating his collieries and unsafe operating methods made him no friends amongst his workforce. He faced a number of strikes in the 1850s as workforce unrest grew; one followed a dispute over a 15% wage cut Powell was demanding. He brought in English 'scab' labour, broke the strike and then cut his Welsh miners' wages by a further 5% when they returned to work.

Despite his ruthless drive, Powell was a family man; he had three wives and it was his third wife, Anne, whom he married in 1833, who bore his five children – two daughters and three sons. In 1860 Powell built the Manor House at Coldra Woods outside Newport, which he gave to his son Thomas when he got married. The house is now part of the Celtic Manor Resort where golf's Ryder Cup was held in 2010.

Thomas and Powell's other two sons, Henry and Walter, took over the running of Powell Duffryn on their father's death in 1864. Tragedy, however, was just round the corner. In 1869 Thomas Powell and his wife Julia were on safari in Abyssinia (Ethiopia) when they were murdered. In 1881 Walter Powell, a Conservative MP, went missing, presumed drowned, when hot air ballooning over southern England. It is likely that many a grieving Welsh miner's family, suffering the loss of a breadwinner due to Powell's indifference to proper safety, will have viewed these two tragedies as some sort of rough justice for their own loss.

COAL USE EXPANDS

Coal's greatest moment still lay ahead. As the 19th century rolled on the use of town gas, which was made from coal, had led to the development of gas lighting for both homes and city streets, which had a favourable impact on reducing crime rates. Gas lighting was also installed in many factories, increasing industrial output. London led the world in implementing gas street lighting during the first decade of the century. As gas lighting spread, gas mains were laid and in the early 20th century gas began to be used for cooking and heating.

But the revolution that had been wrought by coal had not run its course, for in the later stages of the 19th century electric power began to make itself felt as the future source of lighting. Electrically-powered trams were also introduced on to city streets around the world. As electricity distribution systems were expanded across the industrial world the power stations constructed to deliver the electricity used coal. Further, in those days of industrial expansion the key industry was iron and steel, and furnaces in the plants that produced these metals also mainly burnt coal.

THE SOURCE OF BRITAIN'S COAL

The source of Britain's coal was its own mines, which were primarily to be found from the Midlands north into Scotland and west into Wales. Fields were established in South Wales, Lancashire, Nottinghamshire, Yorkshire, Warwickshire, Northumberland, Ayrshire and Lanarkshire, and individual collieries numbered in the thousands, both big and small. There was virtually no coal in the south of England – Kent was the only mining area and was small and hardly profitable. Whilst coal provided the UK with the wherewithal to pioneer the Industrial Revolution, mining areas did not really benefit economically from this. For all their life they remained relatively poor and were politically radical, with often appalling labour relations; both when the industry was privately owned and after nationalisation in the late 1940s. Productivity was also patchy, particularly following nationalisation.

It is ironic that despite what it did for the UK in terms of wealth and industrial growth, coal brought limited economic advantages to the mining areas themselves compared with other industrial activities in other parts of the country. Whilst steel drove South Wales into a new dawn of prosperity in the 1960s, this wealth barely touched the South Wales coal mining areas. Their prosperity, compared with that of booming Coventry in the 1960s, when the British car industry was in its prime, is stark. Perhaps coal mining in the UK was always too heavily politicised, a product of the ruthless practices of 19th century mine owners like Thomas Powell, amongst which were the use of women and very young children at extraordinarily low rates of pay, which led to the Mines Act of 1842. Miners were therefore usually socially isolated, had very strong internal loyalties and were often opposed to social change where it affected their personal lives despite their political radicalism and their relentless push for better working conditions.

HISTORIC COAL MINING IN BRITAIN

As we mentioned earlier, coal had been mined at least from the time of the Romans. In these early times, the use of coal was not widespread; transport links were fairly rudimentary and the only way that coal could be moved long distances was by boat. Whilst there is evidence that coal was used for heating in basic furnaces in the medieval period, as indeed it had been used in the Bronze and Iron Ages, wood was far more common as a fuel, as it could be found close to communities whose members in those days never strayed far from the villages and towns in which they had been born. Often small tithes had to be paid for the right to lift wood, but these were much lower than the cost of coal away from mining areas at a time when transport was still difficult and costs could raise the price of north-eastern coal tenfold when delivered to inland areas in the south.

In the northeast of England around Newcastle, coal mining and the shipping of coal became widespread in the 13th and 14th centuries. The trade led to considerable disputes over shipping rights between the crown, the church and the elders of Newcastle itself. The coal was described as sea coal and it was the case that coal clearly washed on to the beaches on the northeast coast, but some believe that the coal from the Newcastle area was called sea coal because it was shipped by sea to buyers in the south of England and on the continent. Whilst coal was a significant industry in the northeast in those days, it is interesting to note that records show 15,000 tonnes of coal being shipped from Newcastle in 1378; hardly a significant amount in today's terms, but clearly big business in the medieval age.

18TH CENTURY EXPANSION

Between 1700 and 1800 coal mined annually in the UK increased from 2.7 to 10 million tonnes. However, with the onset of the Industrial Revolution production really took off, reaching 250 million tonnes by 1900. This underlines the fact of Britain's economic power. The US, with its ambition to overhaul the UK as an economic force, produced 200 million tonnes in 1900.

As the industrial age dawned, the use of coal increased as its superiority to wood and charcoal, in terms of the heat it could give off, recommended it as the fuel to be used in many of the new processes that were developed during the Industrial

Revolution. One of the earliest industrial uses of coal was in smelting pig iron, whose rising output mirrored that of coal. Another of the earlier customers for coal was the textile industry, which, armed with cheap raw materials from the empire, was expanding rapidly in the late 18th century. With the advent of suitable metal products, the textile looms that had been made of wood and largely powered by water – limiting where the factories could be built – could now be powered by steam made from coal burning. The number of power looms in Britain rose by almost a hundred times between 1813 and 1850, stimulating an enormous increase in demand for coal.

At the same time the issue of poor transport links in the UK was being addressed through a large-scale building programme of waterways, canals and then railways, which enabled coal to be quickly and cheaply transported from coalfield to customer, and in the case of the railways, also provided a new customer. In time, the arrival of town gas and gas lighting, and then electricity at the end of the 19th century, increased the market for coal in the UK, as did the growth in steel usage.

MINING CONDITIONS

Coal mining accidents and deaths were frequent and although larger mines could afford to bring in safety improvements and in many cases did, some much smaller mines tended to be lax in this area, often with disastrous results. Whilst there were many cases of deaths from rockfalls and other problems relating to mining activity, very often in the 19th century accidents were caused by gas and coal dust explosions. The British coal mining industry was no exception and over 12,000 miners were killed in accidents between 1855 and 1867.

This poor safety record inevitably led to the rise of unions within the mining industry, although in the early days they were not always very influential. In time their power increased and by 1912 the movement was able to organise the first national coal strike. It is well recognised that during the 20th century coal mining union power in the UK increased and battles with the owners, whether private or state (following the nationalisations of the late 1940s), regularly ensued. Some argue that miners unions abused their power in the latter part of the 20th century and this directly led to the massive contraction in coal mining in the UK and its denationalisation. Certainly, today's industry is a shadow of the mighty beast that powered the Industrial Revolution in the 19th century, as we will discover later on.

At the time of the Industrial Revolution, UK coal mines were primarily financed by private owners, and often the scale of the development was very modest, the capital being invested simply to provide owners with a flow of income. In the 19th century, large flows of capital were being directed towards gold mining across the empire and in Latin America. The development of railways also attracted huge amounts of capital from London investors, which was employed in the UK and also around the world. Unglamorous coal had to fight for every penny, but nevertheless a huge coal industry was built up on the back of modest capital inflows and an often-exploited workforce. This shortage of capital was to become an increasing problem in the 20th century, distorting the split between ownership and control.

Coal, as we have seen, was a key element in the Industrial Revolution and any country in the 18th and 19th centuries with geopolitical ambitions had to have ready access to the fuel, preferably from within its own borders. As the nation that led the world into the industrial age Britain was fortunate in having its own considerable coal resources to exploit. Other European countries keen to exploit the new industrial technology, and also anxious to make their mark on the outside world, particularly with regard to acquiring overseas possessions, both geographical and economic, followed suit. One of Britain's greatest political rivals during this period was France.

FRENCH COAL MINING

Organised coal mining had been carried out on a small scale in France since the 12th century but it was at the start of the 19th century that the development of new mines began to accelerate. The main coal mining areas in France were in the northeast in the Nord-Pas de Calais region, in Lorraine, in the Loire, and lignite was mined in Provence in the southeast. Early in the 18th century French coal production was a modest 50,000 tonnes but by the end of the century, notwithstanding the French Revolution, coal output had risen to 600,000 tonnes, of which half came from the mines at Anzin in the Nord.

In time smaller mining areas in the south were developed, including the Carmaux mines in the Tarn and the Cevenol mines near Ales. The Cevennes became the heartland of French Protestantism following the massive persecutions of Protestants in the 16th and 17th centuries. It is interesting, therefore, from a social point of view, that later in the 19th century and in the 20th century, as coal mining lost its attraction as a job for the locals, Catholics from Italy, Spain and Poland were employed to fill vacancies and work side by side with Cevenol Protestants in the mines.

For much of the 19th century France imported substantial amounts of coal, mainly from England, and in the middle of the century imports ran at 40% of coal used by French industry in the north. In the first half of the century the Loire, at the confluence of the Loire and Rhone rivers, was the largest producing area but ultimately the Nord-Pas de Calais took over. One of the problems that French coal mining faced was slowing productivity improvement combined with slowing production growth over several decades into the early 1900s. This meant that the industry always suffered by comparison with England, Belgium and Germany, particularly as mining conditions were far more difficult than those in the huge, rich fields of Belgium. Distribution of coal was primarily by canal, and that worked efficiently in getting product to the main industrial and iron-working areas of the north and the centre. However, there was widespread price fixing, which benefited the profits of the coal mines at the expense of their customers; something that continued well into the 20th century.

BELGIUM

Squeezed between France and Germany, the basin which hosted the coal mines of Belgium, now silent, as they are throughout Europe, in fact ran from northern France through Belgium into Holland. The largest of the early 19th century mining areas was Hainault in the west. The coal basin here was over 40 miles long and up to 8 miles wide and consisted of multiple seams and qualities of coal going down as deep as 900 feet. The Belgian mines tended to be poorly ventilated, and for that reason many more shafts had to be sunk to access the economic seams than was the case in Britain, and the galleries leading from the shaft were limited by this issue. In due course technology – in the form of more powerful steam driven pumps – helped to mitigate the problem.

Output levels by 1840 across the region saw Belgium producing 3.2 million tonnes annually, France 2.9 million tonnes and Germany (Prussia) 2.4 million tons; these figures, however, look modest when compared with Britain's 24 million tons.

Financing the growth of Belgium's mines in the 19th century was largely achieved through private equity, with the establishment of limited companies to own and run the mines. The earlier small-scale artisan industry had largely consisted of small operations owned and worked by small groups or associations of miners. When Belgium separated from the United Kingdom of the Netherlands (Holland) in 1830 this led to closer government regulation of all aspects of mining, but it was unable to control the over-investment in coal that led by 1840 to substantial overproduction.

However, aided by a growing transport system in the form of rail and waterways, and industrial growth in France and Germany, Belgian coal production rose to 13.7 million tonnes in 1870. This growth went hand in hand with rising iron ore production and the rise of Belgian iron and steel makers. However, as the 19th century closed, Belgian coal production was rising more slowly – its peak of around

35 million tonnes being reached in the mid-20th century. It then faced a long, steady decline until the the 1990s, when the mines were largely gone.

GERMANY

Inevitably the largest producer of coal in continental Europe was, in the end, Germany. In the first half of the 19th century Germany, as we have seen, lagged behind producers such as France and Belgium, and way behind Britain, but the rapid expansion of its railway network in the second half of the century, in response to economic growth powered by the Industrial Revolution, changed that. Its reserves were massive and it was able to expand output until by the end of the century, only Britain produced more coal in Europe. From 1850 Germany, on the back of the expanding rail network as well as the introduction of modern mining methods to the backward local industry, increased coal production to almost 30 million tonnes by 1870; output then accelerated reaching almost 200 million tonnes by the outbreak of the First World War.

Hand in hand with coal output, iron and steel production rose rapidly and by the early 20th century had surpassed that of the UK, far outrunning the ability of local deposits of iron ore discovered in Lorraine, then under German control, to supply the furnaces of the Ruhr, which led to imports from Sweden and Spain.

The main coal fields were to be found in the Ruhr Valley in the west of the country and were a continuation of the rich coal seam that ran from France through Belgium. In time areas to the north were developed and by the early 20th century the whole Ruhr produced 60% of Germany's coal. The Saar Valley, in Bavaria on the border with Lorraine, also became a substantial coal producer. Another big coalfield was established in Upper Silesia in Prussia, which is now in the western portion of Poland, and this area provided Germany with around 20% of its coal output. There were also very large reserves of lignite, which were located in Saxony in central Germany, and by the end of the 19th century output had reached over 80 million tonnes annually. Although the heat output from lignite was low, it was cheaply mineable from large open pit mines and consumed locally. Ownership of the mines was either in the hands of the state, the province, or private interests – often based on old aristocratic landowning structures.

THE REST OF EUROPE

The widespread nature of coal resources means that most advanced or advancing economies in the 19th century developed coal mining capacity. One of the more important in eastern Europe was Poland, which had been mining coal since the mid-18th century. The main coal areas then were in the Kielce Heights in the southeast of the country and in Silesia, which at that time was under the control of Prussia, as mentioned above. The extensive shallow coal deposits in Silesia played a key part in making the region a major industrial centre, with rail, steel and manufacturing being developed in tandem with the coal mines. Zinc and lead mines were also developed as the area's industrial expansion continued in response to Europe's industrialisation.

Above we touched on the importance of Asturias in Spain as an iron ore and iron making province, but the smelting side of the industry meant there was a demand for coal. Originally Spain was a large importer of British coal through its northern ports, but in due course it began to exploit its own coal resources near Asturias as the iron ore smelting industry expanded. The eastern Pyrenees near Bergueda in Catalonia was also a coal mining area, albeit a rather remote location where mining started in the mid-19th century. The quality of coal mined, lignite, was poor and the work as so often in coal mines was dangerous, the miners working in galleries with no significant investment in mechanisation until the 1960s. The mines then closed in the 1980s.

Hungary was another important coal producer in Europe in the 19th century, although small in relation to Great Britain and Germany. The industry's financial structure was also interesting in that around 10% of mine operating income was paid to the king, although in due course a more conventional mining tax was introduced. Austria also had a significant coal-mining sector, and countries like Sweden, Ireland and Italy also had a number of small coalmines. Italy was perhaps better known for its export of coal miners to large producing countries like Britain and the US rather than the mineral itself.

THE UNITED STATES

As the American colonies/United States developed and expanded in the 18th and 19th centuries the issue of energy became increasingly important. The abundance of wood in rural areas where the early settlers lived had meant that wood had for many years been the prime source of energy for heating and cooking. It had also been the prime source of fuel for industrial machinery. However, the dawning of the industrial age led to rising interest in the use of coal, which was far more compact for transport and storage in terms of measured energy per unit. Also, as the urban areas of the eastern US expanded following the War of Independence, the use of coal increased, although it was still largely imported from the UK; a trade that had been going on for a hundred years.

After the War of 1812 when UK coal imports were disrupted, local sources of bituminous coal from Virginia and then anthracite from Pennsylvania became more popular, and the building of the Delaware and Hudson Canal supported the growth of that trade. With the coming of the canal, it was far cheaper to transport large quantities of coal to New York than to bring wood from rural areas to the eastern cities.

As an independent nation the US had a number of political aims of which the main two were the expansion of its boundaries within the North American continent and the other, as the 19th century progressed, was to make sure that it remained in touch with Europe (particularly the UK) in terms of the developing Industrial Revolution. As the US expanded geographically it found that many of the newly formed states had significant coal resources, as well as other metals, which meant that the necessary resources were available to pursue the industrial advance it desired.

At the same time, the widespread coal resources aided the expansion of the US as it allowed new centres of population to be established great distances from the east coast. This led to tragic clashes with the indigenous Indians as free land was enclosed by the government and then sold or leased to farmers, ranchers, mining companies and later urban developers. The expansion to the west stimulated the building of the railroads and, once routes had been established, centres of population began to grow up alongside the routes.

One of the most important coal mining areas was Ohio, where coal had first been found in the mid-18th century. Ohio's coal resources and the proximity of the Great Lakes for transport made it a natural place for urban growth, and of course ultimately for the railroads, where the Baltimore and Ohio railroad gained its charter in 1825, to be followed by many other companies throughout the 19th century.

Coal was first discovered in Belmont County on the border with that other great coal mining state, West Virginia. The early mines were all underground operations and primarily mined by immigrants from Great Britain. Mining methods were fairly crude, with all the work being done laboriously and dangerously by hand, but as time went by the operations slowly became more efficient, with distribution in particular moving from crude wagons and carts, which traversed difficult terrain in delivering the product to customers, to canal boats, which, though slow, were able to establish timely delivery schedules. Canals also enabled coal to be shipped long distances within Ohio and, along with the railroad, helped the development of industrial hubs such as Cincinnati, Cleveland, Columbus and Toledo.

As coal began to replace wood in providing heating and energy for Ohio, as elsewhere, production slowly increased. Then in the second half of the 19th century coal production, driven by technological advances, began to soar. Home coal oil and street gas lighting became increasingly widespread and coal began to power electricity-generating plants and other industrial processes and forms of transport, such as riverboats and railroad engines.

The second half of the 19th century saw the acceleration of productivity in the mines as mechanisation was introduced. However, full mechanisation was not achieved until the 1930s and Ohio's coal mining remained primarily underground until the 1950s when opencast mining, a particularly labour efficient and capital intensive form of mining, became the main mining method used. Even so, opencast mining did have a small corner of the Ohio industry in the early 19th century when coal seams came to surface. However, if the seam persisted at depth, mining then went underground; full scale opencast mining had to await the development of specialised machinery in the 20th century able to strip overburden and then dig the coal out in large quantities at a low unit cost. The state has had two production peaks over the last century and a half, in the 1920s and then again in the 1970s; currently output at around 26 million short tonnes is at a similar level to that of the mid-1990s, as shown below.

The chart shows Ohio coal production from 1850 to 1996.

OHIO COAL PRODUCTION (1850-1996)

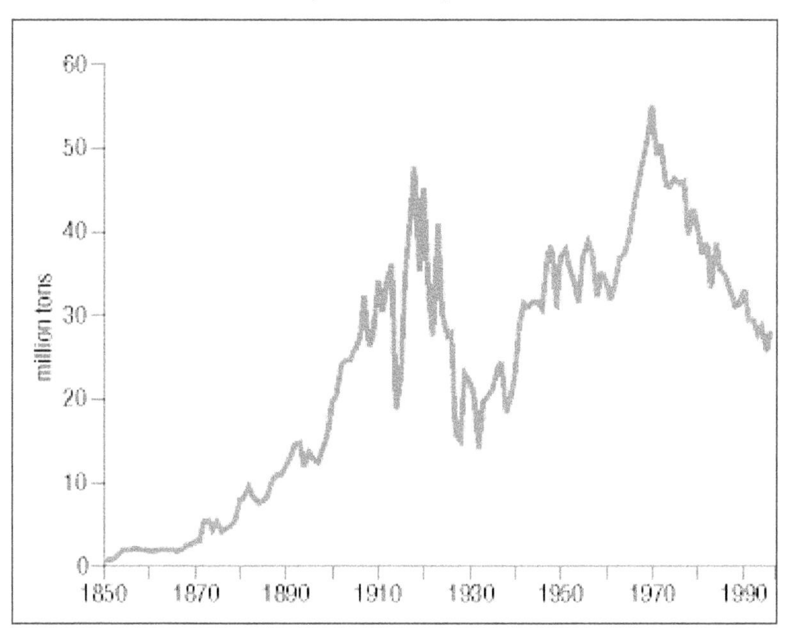

Source: Ohio Department of Natural Resources,
www.dnr.state.oh.us/Portals/10/images/geofacts/no14a.gif

COAL IN WYOMING

Wyoming today is by far the largest coal producer in the US, its output of low sulphur coal coming almost entirely from huge opencast deposits with which the state is very well endowed. Historically coal was first mined in Wyoming in the mid-19th century, having been discovered by the Fremont Expedition in 1843, but for years output was tiny, providing remote single operations, such as forges, with fuel for their furnaces. Interest in Wyoming's coal increased later in the 1860s as the Union Pacific Railroad pushed its way west, and extensive coal deposits in the south of the state determined the railroad's route across Wyoming. Whilst the coming of the Union Pacific and its coal requirements led to a substantial increase in the number of people settling in Wyoming, the state's economic activity remained predominantly ranching and farming. When the railroad came this gave Wyoming mining companies the ability to transport coal east to more populous markets.

Wyoming's coal production, which today exceeds 400 million tonnes per year, climbed over the second half of the 19th century from 1 to 3 million tonnes. Virtually all of this output was from underground mines and in the last years of the century the industry was hit by a number of serious accidents, mainly caused by explosions in the pits. Not surprisingly labour relations in the pits were strained by these incidents and it is no accident that Wyoming's coal future became one of open pit mines, high productivity and high wages.

SOUTH AFRICA

Although coal was first used in southern Africa many centuries ago by indigenous tribes to supplement traditional wood sources for fire – used in basic metalworking, heating and cooking – it was in the mid-19th century that coal began to be used on a regular, if limited, basis by European settlers. Even then the mines were small in scale and largely on the surface.

In the earlier part of the 19th century, the search for coal was primarily in the Western Cape as a result of the arrival of the British and the depletion of forests in the area. The search was largely unsuccessful with uneconomic shows of poor quality coal being found around Cape Town and east towards Franschoek. In the end coal had to be imported from Britain, but the search went on and moved to the Eastern Cape where a small mine was established near Molteno in the northern part of the province. By the middle of the century steam ships calling at Durban encouraged coal exploration in Natal. In 1865 a small surface coal deposit – discovered by Peter Smith, a farmer – near Ladysmith began production and 7,000 tonnes of coal was mined over the next seven years. From this modest beginning the Natal coal mining industry sprung. Later on in the 19th century coal was discovered by the Voortrekkers in eastern Transvaal. The foundations for the huge modern South African coal mining industry had been laid.

Whilst the link between minerals, economic growth and technological advance over the centuries is of particular interest to us, sometimes minerals have an important role to play for more mundane reasons. Coal's role as the key to electricity generation and therefore the development of power for the modern economy is well understood. However, as we see in South Africa, the search for coal became urgent as forest depletion meant that alternatives were needed to meet the basic requirements of heating and cooking, not to mention transport, as steamers calling at South African ports replaced sailing ships.

Another major breakthrough for South African coal came on the back of a geological survey commissioned by the Orange Free State government of President Brand in 1876 and undertaken by George Stow, an English doctor with a passion for geology. The government hoped that Stow would find the Free State prospective for diamonds and gold, but instead he found coal near what is now the town of Vereeniging along the Vaal River on the border with the Transvaal (now Gauteng). The government, disappointed that coal, for which there was no local market at the time, was all Stow could find, abandoned the survey. But in fact a market for coal was not very far away in time, with diamonds being discovered at Hopetown in 1867 and then in 1871 the fabulous New Rush discoveries around Kimberley. Galvanised by these new discoveries and by the later gold discoveries of the Witwatersrand in the mid-1880s, governments in the Cape Colony and Natal decided to expand the size and scope of their very modest railway systems.

The essential infrastructure to enable coal to be transported was slowly being put in place. Equally importantly, the expansion of mining in South Africa led to a rapid denuding of traditional wood resources used by the diamond and gold diggers for fuel, hastening the need for another base fuel. The first coal baron was Sammy Marks who, with his cousin and partner, Isaac Lewis, established Lewis & Marks; initially a provider of goods and services to the Kimberley diamond diggers, on the back of the opening up of the Kimberley diamond fields.

It was a meeting between Marks and George Stow that led to Marks appreciating the potential of Stow's Vereeniging coal discovery. Stow was sent to Vereeniging to buy prospective coal-bearing land. The first operating mine, which Stow established in 1881, was the Leeuwkuil pit which became the foundation of the giant Vereeniging Estates coal company. Later Lewis and Marks founded the Union Steel Corporation (USCO) to build a steel works in 1912 in Vereeniging, which eventually was integrated into the South African para-statal ISCOR in the 1920s.

SAMMY MARKS (1844-1920)

Sammy Marks was born in Neustadt-Sugind in Lithuania in 1844 in an area of Imperial Russia known then as the Pale of Settlement – the only place Russian Jews were allowed to reside. The son of a strict but poor Orthodox Jewish tailor, he was offered a trip to England to deliver some horses to Sheffield when he was 17 and decided to stay in England to escape poverty and persecution back in Lithuania. In Sheffield he met the Guttmans whose daughter, Bertha, he was to marry years later.

Learning of the diamond rush in South Africa, Marks set sail for Cape Town, arriving in 1869 and headed for Kimberley. He had brought with him a set of fine silver cutlery from Sheffield and this he sold to buy a cart and stores to take up to the booming diamond fields, a shrewd decision bearing in mind that suppliers of goods tend to do better than most prospectors and miners in the mining camps. His cousin Isaac Lewis left Lithuania to join him and the two set up store in Kimberley under the name of Lewis & Marks, which was to become one of South Africa's biggest industrial groups. As was the practice in those days, the Lewis & Marks company as it expanded was incorporated in the UK where finance was readily available.

In Kimberley, Marks did not just run the store, but he also became involved in diamond trading. As well as this, he was an investor and it is thought that he made a lot of money out of the sale of the French Company to Barney Barnato. After a few successful years in Kimberley Marks headed for the Highveld where stories of gold finds were rife. He and Lewis were drawn to the eastern Transvaal and the Barberton area where gold interests, including the Great Sheba mine, were added to Lewis and Marks's growing portfolio of operations. Marks also acquired leases on the Witwatersrand when gold was discovered there in 1885. Prominent amongst those were East Rand Mining Estates, which included the Grootvlei and Palmietkuil farms. It was at this stage that Marks happened upon two things that were to be critical in the further expansion and prosperity of the group – coal and President Kruger.

In the Transvaal the main attention was being paid to the Witwatersrand gold rush, but the pace of growth in gold discoveries and the growth of Johannesburg from a tent township to a permanent city was so fast that energy needs became critical. This led to the discovery that the Transvaal was as rich in coal as in gold. Marks very quickly moved into coal, making a number of acquisitions in the area and placed them in Lewis & Marks, where eventually these coal interests were amalgamated into Vereeniging Estates and Transvaal Estates and Development. Marks directed his business from Pretoria, deep in Boer territory, and became very friendly with President Kruger; so friendly indeed that in 1898 Kruger gave Marks free run of the South African Mint to produce 300 gold 3d pieces (known as *Tickeys*, which were usually silver) to distribute to friends. The Boer War was a difficult period for Marks as he had friends on both sides of the conflict. This, however, allowed him some influence in the difficult negotiations that led to the ending of the conflict and in 1910 he was appointed a senator in the first Union Parliament, a position he held until his death.

In 1884 Marks returned to Sheffield where he married Bertha, the daughter of his old friends the Guttmans. She was 19 years younger than Marks. They had nine children – five boys and four girls – who were educated in England. Marks built a family mansion, Zwartkoppies Hall, near Pretoria which still stands today having been restored by the South African government in 1984 as a museum illustrating Marks's life. The mansion acted as more than just a family home, even at the time, as Marks was a generous and expansive entertainer in keeping with his substantial wealth. In the latter years of his life he made a number of philanthropic donations to Jewish charities in South Africa, particularly in the education field. He also continued to build Lewis & Marks, which though predominantly a mining house had also acquired businesses in liquor, building materials, agriculture, glass and flour milling. In 1912 he founded Union Steel.

Marks is perhaps less well known than Randlords like Rhodes, Barnato and the Oppenheimers, but he was truly South Africa's King Coal and he remained faithful to his adopted home, South Africa, by building his mansion in Pretoria rather than Park Lane in the old country.

RISING MINE OUTPUT FUELS SA COAL DEMAND

The last two decades of the 19th century saw huge developments for diamonds in Kimberley and gold on the Witwatersrand. These were accompanied by a surge in railway building as the Boer Republics and the British colonies of the Cape and Natal all realised that these great mining enterprises had to have proper connections with the coastal ports. At the same time, this surge in industrial and mining activity needed increasing amounts of fuel and power.

Initially, coal for the Witwatersrand gold fields and for the fast growing town of Johannesburg was brought from Vereeniging, but quickly the growth of Kimberley absorbed all that output and the gold mines had to look nearer to home. Two sources were opened up: what is now the Witbank-Middleberg field in the Eastern Transvaal, 60 miles from Johannesburg, and the much closer Boksburg coal field in the East Rand, which overlay some of the gold reefs. As coal output rose and the railways were completed in the last decade of the century disputes broke out between the new railway companies and the coal miners over carriage rates. These disputes have continued off and on to this day.

The mining methods used depended on the size of the mine and the capital it employed. Having accessed the coal seam, miners would use hand auger drills to make four feet deep holes in the seams and then break the coal for hauling by blasting. The smaller mines would then haul the coal to surface using labour and mules, the larger ones would have steam engines to power mechanical haulage. In those mines with no methane gas problem, lighting was often by candle but some mines had already began to install electric plants for lighting – Brakpan in the East Rand did this in 1891.

The corporate bones of the industry to come were also laid down around this time. In the Orange Free State what was to become Marks & Lewis's flagship mining interest, Vereeniging Estates, had been a development pioneer with Bedworth Colliery, Central Mine and Cornelia Colliery. The Witbank/Middleberg field followed with mines such as The Douglas, Witbank and Transvaal and Delagoa Bay, with Witbank being arguably the industry leader with its high-quality steam coal, relative shallowness and comparatively accessible working faces. In Natal, Dundee Coal was incorporated in 1888 and two years later the Newcastle, Natal Collieries and Estate Co was formed and financed in London.

South African coal, in terms of its quality, was at the premium end of the scale in Natal, slightly more mixed in Transvaal and less good in the Cape and Orange Free State. It was generally low in sulphur and had good energy output, but the best coal, particularly in Natal, lay at depth and here methane was a problem. This contributed to a number of fatal accidents in South African coal mines over the years as the shafts were extended to access deeper coal seams; the worst being the Glencoe colliery disaster in Natal in 1908 when 75 miners were killed. These seams though were by no means as deep as those in the UK, from where the South African industry got a lot of its design and operational management. In time shafts – which had originally been vertically sunk on the British model – were sunk as declines, as management adjusted to the shallower seams being worked. This enabled many miners to access the working seams on foot.

The width of the coal seams varied quite markedly from field to field and in the East Rand, which was conveniently positioned to supply the nearby gold mines, some coal deposits overlay the gold reefs. Indeed, the Apex mine produced both gold and coal, as did the Brakpan mine, although the latter quickly became solely a gold producer. Seams, for example, in the Witbank/Middleberg mines went down no more than 300 feet where mineable coal was found in seams of between 6 and 12 feet.

The Anglo-Boer War between 1899 and 1902 temporarily disrupted the South African coal industry but also marked the industry's graduation from development to growth status as the new century evolved. We will pick up this story later on.

ASIA AND THE FAR EAST

Asian and Far Eastern coal mining, which long ago overtook all but the US amongst the developed world in terms of both reserves and production, was of very modest size during the colonial period of the 19th century. This is not surprising when one considers that relatively advanced countries such as Australia were economically small and the export markets that thrive today in the Far East did not exist then. The Industrial Revolution, a major consumer of coal, was also slow to take off in the Far East, with former economic giant China wracked by political instability, India still a peasant society and Japan, whose 19th century coal industry we have already touched on, only beginning to industrialise in pursuit of geopolitical aims that were to have such an effect on the region and the world as the 20th century progressed.

INDIA

India is now the third largest coal producer in the world, after China and the US. In the age of the Industrial Revolution India was an agrarian, aristocratic society largely under the control of the British, for whom it was a supplier of raw materials and also a market for finished British goods.

The first commercial Indian coal pit opened was the Ranigunj mine in 1775 in West Bengal, a region which today abuts Bangladesh. Although the development of the pit was supported by Warren Hastings, Governor of Bengal, the product, when shipped for testing by the British military in Calcutta, did not pass muster, giving off 50% less heat than coal imported from England. Despite this unpromising start, further small pits were opened and by 1840 output from Raniguji mines reached 50,000 tonnes. The early pits were small-scale, open cut operations, but in 1815 the first underground mine was commissioned.

The first pits were British controlled but in 1835 the Carr Tagore Company, headed by Prince Dwara Kanath, was established. Eventually, by 1843, the Bengal Coal Company Ltd was incorporated as an amalgamation of many of the West Bengal coal operations. Calcutta was the main market for Bengal coal, and transportation slowly upgraded from bullock driven carts and barges to rail. Output was also on the rise and total Indian production reached around 6 million tonnes by the end of the century – much of it coming from mines owned by the railways. At the same time development of the Jharia coalfield in the north east of India started, encouraged by the East Indian Railway extension of the Grand Chord line; today Jharia is the source of most of India's coal. An important factor in the expansion of Indian coal production was the establishment of the Geological Survey of India in 1851.

CHINA

Like India, the Industrial Revolution came to China far later than in the West, which meant that during the 18th and 19th centuries it was primarily an agrarian society, despite its history of technological advances over millennia – something we have mentioned before. It had, however, an ancient history of coal mining and coal was used primarily in the smelting of metals, which the Chinese had been doing since the 1st millennium BC. It was also, due to the relatively large size of its population and therefore of its economic activity, the largest producer and consumer of coal in the pre-industrial world, until in the mid-18th century the Industrial Revolution began to spread across Europe.

China remained a desirable trading partner, however, and it was demand from visiting foreign traders for coal to fuel their ships on the return journeys that led to foreign and local investment in an expansion of the Chinese coal mining industry by the middle of the 19th century. Chinese coalmines were relatively shallow compared to mines in Europe, the deepest shafts going down no further than 500 feet – by contrast some British pits went down over 1,000 feet. This did not mean that Chinese pits were particularly safe though, and the industry has been dogged for centuries by poor safety. Mechanisation was also very rudimentary as labour was plentiful and cheap, but unfortunately it was not very well trained as the majority of workers spent only limited time in the mines; many of them had agricultural jobs and responsibilities to fulfil as well. This remained a problem until the end of the 19th century, when foreign groups were allowed to invest in Chinese mining.

Local demand from the Chinese navy and from the railway also led to a quickening of interest in coal, as did the arrival, if late, of industrialisation, mirroring trends in Japan. The first foreign-backed mine was the Jilong on the island of Taiwan. This was followed by the development of new foreign-controlled mines on the mainland in Hebei (north east China), Inner Mongolia (north China), Liaoning (north east China), Shandong (central eastern China) and Shanxi (central China). The prime initial movers were the British, as usual, the Germans and the Russians; later the Japanese, geopolitically aggressive towards both the Chinese and the Russians, began to take an interest in Chinese coal resources as well.

During this period there was also substantial investment put into railways to haul coal to markets. Before the intervention of foreign capital Chinese coal mines usually tended, because of problems of distance, to confine sales to regional customers. At the end of the century and in the early years of the 20th century further major deals were sealed between the Chinese and foreign interests to further invest in Chinese coal. However, internal political unrest and the increasing involvement of Japan in China's affairs, which was to lead to war and Japan's eventual invasion of China in 1931, saw many of these arrangements unravel.

AUSTRALIA AND NEW ZEALAND

Today Australia is the largest coal exporter in the world, but in the beginning things were rather more modest. The early development of this great industry in the 19th century followed a similar pattern to that experienced in other parts of the world. The technological developments of the Industrial Revolution required power for their full potential to be realised, and wood, for centuries the prime fuel for heating and cooking, was not equal to the challenge. In Australia, with its close connection to the powerhouse of the Industrial Revolution, Great Britain, the arrival of the railway, essential in binding this huge country together, led to a ready and growing market for coal.

The first coal was discovered in New South Wales's Hunter Valley, still a large producer today. A convict, William Bryant, is credited with the discovery in 1791, and ten years later a mine was established there. To begin with, coal steadily replaced wood in heating and cooking, but by the middle of the 19th century the establishment of a railway network in the east and the increasingly frequent arrival of steamships from Europe and the US led to a rapid growth in coal demand. Discoveries of coal were also made in Queensland on the Brisbane River and at nearby Ipswich in the 1820s, but it was to be another 20 years before the first mines were established there. By the end of the century coal output in NSW and Queensland amounted to around 2.5 million tons, far behind the output of the *mother country*, which is of course a contrast to today's situation. As the 19th century came to a close the development of town gas for lighting, heating and cooking promised a new era for coal, as did the discovery of electricity.

Across the Tasman Sea the search for coal was also on in New Zealand. The first producing mine was opened in 1849 near Dunedin, South Island, and over the next few years several other small-scale pits were worked. Coal mining also spread to other prospective areas of South Island – Kaitangata in 1858, Green Island in 1861, Shag Point and Malvern Hills in 1862, and Collingwood in 1868. Coal mining also started in 1875 on the North Island at Kawakawa. The two major producers on the South Island were the Brunner and Buller mines, which eventually merged and became responsible for producing half of New Zealand's coal. As in Australia, and indeed elsewhere, the expansion of New Zealand's coal industry was built on the impact of the Industrial Revolution and the rise of railways and steamships. By the turn of the 19th century New Zealand's coal production had reached 1 million tons.

But King Coal's key role powering the Industrial Revolution did not end there and we will return to coal when we arrive at mining in the 20th century; a century that saw great changes in the sources and production of energy.

11. CONCLUSION

The century when the Industrial Revolution flowered was an amazing period of technological advance and of scientific discovery. The 20th century was to prove no less extraordinary with economic growth powering ahead on the back of quite unforeseeable developments and inventions in industries, such as transport and communications. This progress was in spite of two devastating World Wars and once again almost permanent political and military conflict. As the natural resources industry had been more than equal to the challenges of the Industrial Revolution, so in the 20th century we will see it rising to the challenges of that century with equal success.

Source: De Re Metallica

Twig divining and trenching for minerals in Germany in the Middle Ages

Source: Mining

German iron ore miners in the Middle Ages in the Harz Mountains using the overhand stoping method

Hydraulic mining with pressure hoses in the Californian Sierras. Circa 1850s

The first railway built in 1873 to access the Bingham Canyon copper mine in Utah, now operated by Rio Tinto

Source: Consolidated Gold Fields

Magma Copper's Silver Queen mine in Arizona in 1871 which latterly became the expanded Superior copper mine

Alluvial diamond mining in 19th century Brazil

Source: Consolidated Gold Fields

The Big Pit at Kimberley in 1875 before Rhodes consolidated the workings

Cecil Rhodes painted by Sir Luke Fildes, a portrait which Rhodes hated

Source: Consolidated Gold Fields

Source: Mining

The Kimberley diamond pipe in South Africa with its hundreds of individual claims and extensive network of haulage ropes, 1877

The tree which today marks the site of Paddy Hannon's 1893 gold discovery, Kalgoorlie, Western Australia (inset, plaque detailing the discovery)

Source: Hilary Coulson

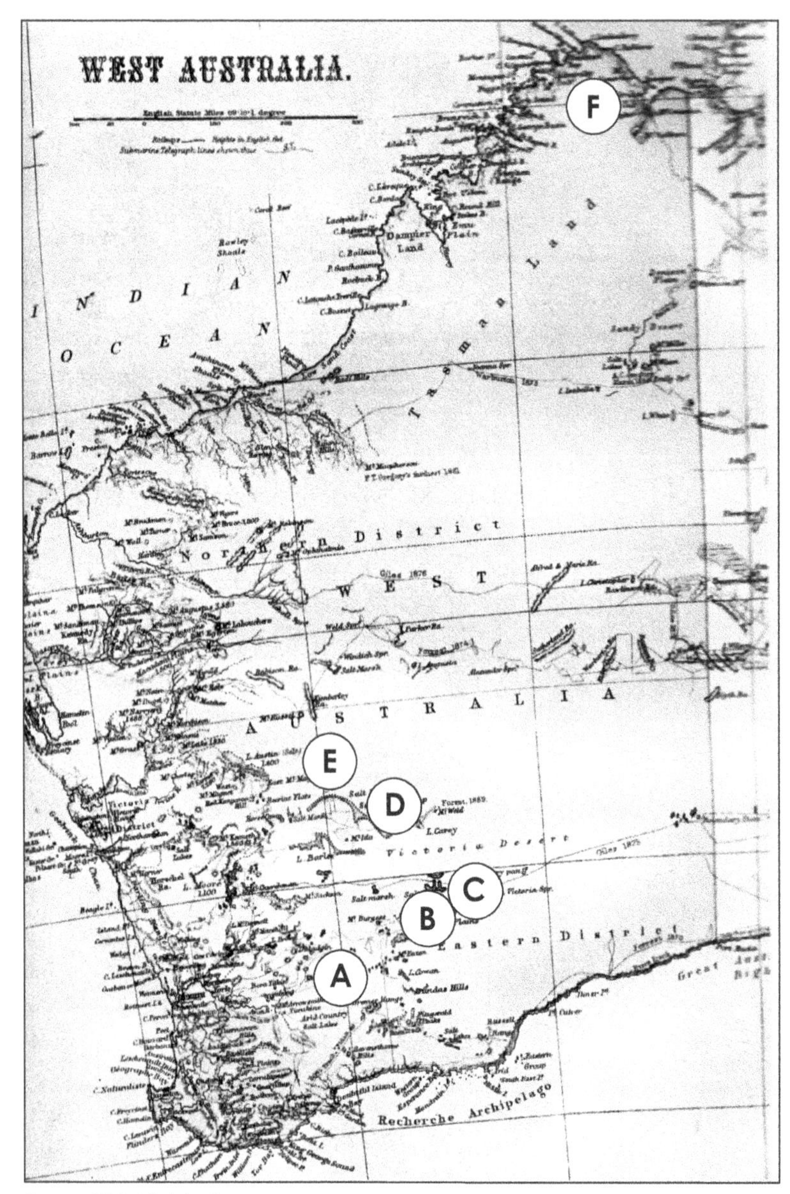

Source: W & AK Johnston

The main gold fields of Western Australia, 1878.
A) Southern Cross.
B) Coolgardie.
C) Kalgoorlie.
D) Leinster.
E) Lawlers.
F) The Kimberleys.

Source: Consolidated Gold Fields

Bettington's Horse gathers outside the headquarters of Gold Fields in Johannesburg before their aborted mission to rescue Dr Jameson, 1898

Source: EA Hegg

Prospectors climb the Chilkoot Pass on their way to the Klondike, 1898

Source: Meet me at the Carlton

A throbbing Eloff Street, Johannesburg in 1905, just thirty years after the first Witwatersrand gold discovery

Source: Diamonds in the Desert

August Stauch (mounted second from left) leads the expedition which found the Pomona diamond field in South West Africa (Namibia) in 1908

THE MODERN AGE (FROM 1900)

1. BROKEN HILL AND OTHER AUSTRALIAN GIANTS

The development of Australia's gold mining industry was a story of romance and drama, and like all the gold rushes of the 19th century, attracted prospectors and speculators from around the world. However, it was arguably less important to the emergence of Australia as a significant economy than the discovery and development of the Broken Hill mines in the far west of New South Wales (NSW) was, in the 1880s.

BROKEN HILL

SILVER IS DISCOVERED

The story of Broken Hill began on the NSW state border with South Australia when silver was found at Thackaringa in 1876. Prospectors, excited by the news, pushed west from the copper discoveries at Cobar. The Thackaringa silver deposits, however, did not have a long life and soon petered out. In due course prospectors found gold in modest amounts at Mount Browne, 200 miles north of Thackaringa, and this started a mini-rush to the new goldfield. Conditions were appalling with temperatures holding steady for weeks at an end at over 100 degrees Fahrenheit. Disease was rife and the cost of supplying food and water so high that Mount Browne never really had a chance of measuring up to other gold fields in Australia and around the world. Interest soon switched back south towards Thackaringa, where prospectors had begun to search for silver again.

The first of these new discoveries was the Umberumberka silver and lead mine and this spawned the town of Silverton, which today – though pretty much a ghost town – is a magnet for tourists and film makers. At the time it was the centre for some highly profitable silver mines and a busy stock exchange. The grades found at three of the mines were astonishing, even for those times – rock from the Chanticleer mine ran 10,000 ozs to the ton, the Marine ran 10,724 ozs and the Hen and Chickens mine assayed a less juicy but still mouth watering 3,400 ozs to the ton.

Intermittently, droughts created problems for miners and prospectors, and the mid-1880s witnessed a particularly serious one, but Silverton survived and the town's population based on silver mining was in the thousands by 1885. The Silverton stock exchange that year also saw it start trading the shares of Broken Hill Proprietary, then merely an exploration company with leases a few miles to the east of the town. But as the silver began to run out again, interest in Broken Hill, despite problems of geological interpretation, was on the rise.

BROKEN HILL IS BORN

Broken Hill had attracted prospectors for many years, but surface samples taken had yielded only small shows of lead and silver mineralisation. What fascinated those people who had stopped, looked and wondered about Broken Hill, named after its jagged top, was the size of the area stained by the oxidisation of the surface iron and manganese. Those oxides were much larger than others in the area, but those samples that had been chipped from the surface were unexciting, giving no hint of what actually lay beneath this stained cap.

In September 1883 sheep boundary rider, Charles Rasp, a German who had become interested in prospecting in the wake of the Silverton silver rush, was working close to the Broken Hill and decided to take a few samples of his own. Two contractors, David James and Jim Poole, excavating a dam nearby, were shown the samples by Rasp and were intrigued by their weight. With Rasp they put a little money up to secure a mining lease over part of the Hill. They then took further samples and sent these off to Adelaide for assaying; the results showed lead and some silver present.

Whilst the paucity of the silver was a disappointment, the three decided to persist and hired a miner to drill a single hole to see what might be found at depth. The three partners also took on four more pastoralists, including the station manager, George McCulloch, and with the expanded syndicate having contributed £500 the partners were able to secure almost 300 acres of ground. Additional drilling was done on behalf of the syndicate and it further confirmed the presence of lead and silver in the Broken Hill mineralised lode near the surface. The silver assayed around ten ozs per ton, a respectable 300 plus grammes in today's silver mining world but disappointing then, particularly when compared to the riches of nearby Silverton.

Perhaps if the syndicate had been made up of geologists the drilling programme would have been abandoned there and then, for high grade silver not lead and silver ores was what the group was hoping to find. The syndicate pressed on believing that higher-grade silver might well lie deeper. Further drilling confirmed the presence of substantial quantities of lead, but high-grade silver still eluded the drillers. This was a difficult time for the project with money very tight and so the syndicate had to pass over the opportunity to buy the freehold of the Broken Hill leases.

When the lode finally showed its potential this decision proved to be an expensive if unfortunately inevitable one as Broken Hill had to pay millions in royalties to maintain the lease when it came up for renewal in the early 1900s. Shares in the syndicate also occasionally changed hands, often at prices that would eventually be seen as bargain basement. Also for the first time in 1884 a mining man, W.R. Wilson, bought a stake. Some of the new members of the syndicate were only in for the short-term ride but others stayed the course and reaped huge rewards in the form of capital growth and dividends.

DEVELOPMENT STARTS

In the same year – 1884 – the project finally began to come together. A small-scale shaft was sunk and larger quantities of ore were mined for analysis. Some of the samples, which were now being drawn from deeper parts of the lode and thought to be chloride of silver, were sent to nearby Silverton for assaying and they came back showing up to 700 ozs per ton. The project was beginning to reward the faith its largely pastoral owners had shown in funding such a speculative venture.

The syndicate was now headed by W.R. Wilson and George McCulloch and two newer members, William Jamieson, a government surveyor, and Bowes Kelly, pastoralist and investor. They decided that the time was ripe for Broken Hill to be incorporated and floated on the stock exchange, with a view to raising sufficient funds to develop a mine. In today's heavily regulated market it is interesting to note that the prospectus issued for the Broken Hill Proprietary Company (BHP) offered very little information on the mine and its prospects; perhaps the owners thought any detail superfluous, instinctively knowing that Broken Hill would become one of Australia's largest mines paying out many millions of pounds in dividends over the years. Also that it would, in the end, become one of the largest mining companies in the world, with a global reach both in terms of shareholders and investments.

The issue price was £9 a share and when the first offer was made to investors in nearby Silverton in 1885 they wolfed down the lot. The next offering was in Adelaide where the prospectus had to be re-written but after that the shares sold quickly. Further sales were made to Sydney and Melbourne investors where the offering was less well received. The more sophisticated eastern investors thought the shares extremely over-priced, but in contrast the owners, who had retained control of the company, holding 14,000 of the issued capital of 16,000 shares, thought the listing price had been too low. The new capital enabled a more thorough exploration of the Broken Hill lode and soon the first ore, just under 50 tons, was dispatched to Melbourne for smelting; this high-grade batch yielded around 800 ozs of silver to the ton. BHP decided to build its own smelters at Broken Hill; they had a capacity of 100 tonnes a day and opened in early 1886.

OVERSEAS CAPITAL BEGINS TO FLOW

Although BHP and other mines along the Broken Hill lode were beginning to do very well, these were economically difficult times for the region; labour was cheap and readily available, which swelled mining profits. The mines' product was generally sold overseas and to help this process a railway linking Broken Hill with the South Australian coast was completed in 1888.

The growth of the mines at Broken Hill began to attract overseas capital, particularly from Britain, and the price of Broken Hill shares led by BHP began to soar. Originally offered at £9 to Silverton investors in 1885, BHP shares rose to £409 on the Melbourne exchange in early 1888. Speculation spilled over into other Broken Hill shares including two stocks well known to more recent Australian mining investors, Broken Hill South and North Broken Hill (now part of the Rio Tinto group). Price rises amongst other stocks were often meteoric and huge short-term profits were made by investors in producers and explorers alike.

The level and intensity of activity in Broken Hill stocks was very similar to that which was to rage 120 years later over the nickel discoveries in Western Australia. Some of the claims made by stock promoters about unproven ground in the wider Broken Hill area would have even embarrassed Aussie Boom brokers and entrepreneurs in the 1960s, and would have caused life-threatening palpitations amongst today's securities compliance officers. Such was the appetite amongst investors for Broken Hill shares that BHP split off some of its unutilised ground into three new companies to feed the frenzy, making even more money for itself and its own investors.

Behind this speculation there was, however, real growth amongst the producers and BHP was able to start paying dividends from an early stage. The town of Broken Hill also grew exponentially, reaching a population of 20,000 by 1891, although like most mining rush towns the growth in services to support this rate of expansion was wholly inadequate. Broken Hill was filthy and disease ridden but its development provided dynamism and economic prosperity at a time when Australia's economy was depressed.

The product from Broken Hill was primarily silver and lead, although there was zinc in the ore which became more important in later decades. In the 19th century silver was still considered a monetary metal and lead was beginning to benefit from the development of automatic weapons firing large quantities of ammunition, as well as scaled up wars like the Boer War in South Africa and the looming First World War.

Developments in medicine and public health initiatives were helping to increase longevity and the Industrial Revolution had already had a huge impact on moving people from the land to the cities where mass housing was built. Lead was an increasingly important component of the construction equation, both in terms of housing and commercial building, where the metal was extensively used in piping and roofing. The coming of electricity and motor vehicles also boosted demand for lead as battery technology was developed, and its use as an additive in petroleum products further increased demand. Huge discoveries like Broken Hill and Mount Isa (which we will come to later), were a critical part of the coming of the modern economic age and provided the size of resource that could support the establishment of the wider consumer economy that developed as the 20th century progressed.

BUILDING BHP

But before all that could happen, the full potential of Broken Hill needed to be realised and that meant the establishment of a professional mining structure. BHP was essentially the creation of pastoralists and they managed the company from Melbourne. This was all right when the company was in its infancy but inadequate as it became a listed company with ambitions to increase output and add value at the smelting and refining stage. So it was that the BHP board in 1886, following flotation, realised that a new approach was required if the company was to expand its Broken Hill project to exploit rising demand for lead and silver.

A search was therefore instigated for a mining engineer experienced in running large operations and a top metallurgist. The search took the company to the US. There BHP hired William Patton from the giant Comstock mine in Nevada and Herman Schlapp, a metallurgist from Colorado. The hirings were well timed because the Broken Hill lode, though exceptionally rich, was not easy to mine, and as operations on the huge lode were expanded, difficulties concerning ground and orebody stability arose. Mining was dangerous and though techniques were developed to try and improve safety and productivity they were not always successful; fatalities, shifting in-situ ore and fires all dogged mines along the lode. The movement of the ore beneath the surface meant that many of the buildings on the surface were subject to movement also, which was particularly unfortunate for the operating precision of many of the treatment plants.

BHP TURNS TO STEEL AND THE COLLINS HOUSE GROUP TURNS TO BROKEN HILL

Huge change was in the air in Broken Hill towards the end of the 1890s. The monetary role of silver was on the wane and the very success of Broken Hill threatening oversupply had led to a weakening lead price as the century came to a close. The complexion of Broken Hill also began to change as companies, including BHP, decided to relocate their smelters to the South Australian coast around Port Pirie and Port Adelaide. Although Broken Hill would rise again, the excitement generated in the early years had faded away as the new century dawned. For BHP itself, the issue arose as to its long-term future, for though it owned the richest part of the Broken Hill lode, it also owned the shallowest and it had mined it extensively.

Close to its relocated smelters at Port Pirie BHP also owned the iron ore deposit of Iron Knob, whose product was used in the lead smelting process, not for the making of steel. In due course we will come to the great modern industry of iron ore exports and Australia's role in it, but it was at Iron Knob that BHP decided in 1911 to explore the feasibility of establishing a steel-making plant at Port Pirie. However, when the decision was taken to enter the iron and steel-making industry BHP located the works in New South Wales at Newcastle near the coalfields of the Hunter Valley, shipping the ore from Iron Knob to the Newcastle works.

Whilst BHP's days at Broken Hill were numbered the great lode itself, dominated by producers like Broken Hill South, North Broken Hill and the Zinc Corporation, continued to pour out its wealth throughout the 20th century. During its 120 years of operation over 200 million tonnes of ore have been mined from Broken Hill and mining continues to this day. After BHP's heyday on the central structure of the lode, the targets became the deep but rich ores of the lode at its south and north ends. Mining at Broken Hill was broadly profitable but there were periods when the going became very difficult. One such period was in 1904 when William Baillieu, a Melbourne financier, and brothers William and Lionel Robinson, Australians with very strong London connections who had made a fortune in gold mining in Western Australia, moved in on Broken Hill. They formed an alliance that was the foundation for the Collins House Group which, from its headquarters in Melbourne, dominated Australian mining for much of the 20th century.

WILLIAM BAILLIEU (1859-1963)

William Baillieu spans the Industrial Revolution and the Modern Age but his impact on Australia's mining sector only really began in the 20th century with the founding of the Collins House Group in Melbourne in 1904. He was born in Queenscliff, Victoria, in 1859 as one of fourteen children (ten boys). His father was James Baillieu from Wales who married Emma Pow from Somerset in 1853.

James Baillieu became a merchant navy man and on a voyage to Australia jumped ship and settled in Queenscliff in 1853, where he met Emma when working for the government health service. From modest beginnings he saved sufficient money to build a hotel in Queenscliff. William Baillieu, along with all his siblings, was educated at the Queenscliff Common School until the age of 14 and he then joined the Bank of Victoria in Queenscliff. After progressing in the bank and, as was his natural wont, making influential friendships along the way, Baillieu turned to estate agency and auctioneering in Melbourne, where he was highly successful for a time.

However, in the 1890s a huge boom in Victorian land and property, which Baillieu and his partners had done well from, collapsed and he had to start again. He formed a new estate agency, WL Baillieu, with his brother Arthur and also linked up with another brother, Prince, in a new stockbroking firm, EL & C Baillieu, in 1892. It was at this stage that Baillieu began to take a close interest in mining and in 1897 he spent some time in London building contacts that were to prove invaluable to the Collins House Group, which was the market name given to the family business interests which were loosely known as Mutual Trust.

In 1887, at the age of 28, he married Bertha, the daughter of a close friend and business associate, Edward Latham. They had four sons and four daughters. Whilst the Baillieus lived comfortably and Baillieu built a fine home, Sefton at Mount Macedon, he was not a great entertainer, preferring to concentrate on family and close friends. His wife died in 1925 and his daughter, Vere, took over the running of Sefton. In 1902 Ballieu went into politics in Victoria and served as a liberal minded legislator, minister and eventually honorary leader of the Legislative Council until 1922. Baillieu's early exposure to mining had been coal in Gippsland and gold in Victoria – the Duke United and Jubilee mines. In 1904 he pulled back from estate agency to pursue his growing business interests – he was on the boards of Dunlop Tyre, Carlton

Brewery, Wunderlich Roofing, Hampden Cloncurry Copper and Mount Morgan Gold. Baillieu also had become heavily involved in the future direction of the lead and silver mines at Broken Hill and he led the reorganisation of the Hill, following the retreat of BHP, incorporating North Broken Hill in 1905, and then founded the Zinc Corporation which was to metamorphose into Consolidated Zinc and later became a part of Rio Tinto in 1962.

Baillieu's main focus at this time was finding a treatment process for the tailings dumps at Broken Hill. This followed a meeting with one of the Carlton Brewery chemists, Belgian Auguste de Bavay, who was researching flotation as a way of separating metal from ore. The successful development of the process in 1904 revolutionised the technology of ore treatment and Baillieu made a fortune. He also joined the board of Melbourne Electricity in 1908 and this led him to form Great Morwell Coal to supply the growing utility. In 1910 he established his group in new premises in Collins Street; the street name was used thereafter to describe the Baillieu empire. With the coming of the First World War, he was in the thick of reorganising Australia's smelting industry to meet war demands, forming Broken Hill Associated Smelters and Metal Manufacturers and taking control of Electrolytic Refining. Towards the end of the War, Baillieu formed and supported what was to become EZ Industries, building a zinc plant in Tasmania.

After the war he also took an active interest in the welfare of returning veterans and their families, as well as the families of Australia's fallen. His sons served with distinction in France and all returned safely. Baillieu also had to deal with post-war labour troubles, which he did with some understanding, but this period, with all its socialist rhetoric, did worry him as a staunch liberal free enterpriser. He remained active throughout the 1920s, expanding his metals interests and always on the look out for new avenues, such as paper and cotton.

The 1930s were altogether more difficult and Baillieu faced a formidable headwind in the shape of the Great Depression. He fought hard against the idea of deflation as a tool to right world financial problems and pursued ideas of writing down debt to reduce interest payments. The strain proved too great though and his health collapsed in 1932. He was sent to England by his sons to recover, but died from pneumonia in 1936. His estate was worth just £60,000 suggesting possibly the effects of the Great Depression but more likely the wise disbursement of assets during his lifetime. Although William Baillieu does not have the glamorous reputation of some of the Randlords of South Africa, his achievement in building up the Collins House Group into Australia's biggest mining house was considerable and Collins House continued to grow for decades after Baillieu's death. Many of the companies it controlled, like Western Mining and Consolidated Zinc, became latter-day mining giants.

SMELTING EXPANSION AND COLLINS HOUSE

Broken Hill provided the raw materials for smelters both in Australia and Europe, but this meant that much value added fell to the processors, and ambitious plans were formulated by Collins House to build or acquire smelters for both lead and zinc; the latter metal having been boosted from annoying by-product to key strategic metal by the First World War. Eventually the group gained control of BHP's smelters at Port Pirie in 1915 and the next year planning started for a zinc smelter in Tasmania. By the outbreak of the Second World War all of Broken Hill was owned by Collins House companies and this coincided with the loosening of the group as the Zinc Corporation increased its London contacts. In due course this was to lead to the birth of Rio Tinto, which eventually took over North Broken Hill; more recently Rio was the corporate target of BHP, the original Broken Hill miner.

By the final decade of the 20th century Broken Hill was owned and operated by one company, Pasminco, but many years of low metal prices led to that company's collapse. When Pasminco was eventually restructured and re-floated as Zinifex, the Broken Hill mine was not retained.

Although Broken Hill's role as the largest producer of lead, zinc and silver in the world has ended, it continues to this day to produce metal – now under the control of Perilya, which bought it from Pasminco's administrators in 2002.

MOUNT ISA

Whilst BHP was the earliest of Australia's great mining groups and Broken Hill itself was the first significant base metal discovery in the country, there were more to follow as the 20th century progressed. Although many, such as Mount Lyell Copper and Renison Tin in Tasmania, did not match Broken Hill for size, some, like Mount Isa in Queensland and Olympic Dam in South Australia, did.

Later in a separate section we will look in some detail at the legendary Australian nickel boom in the late 1960s which saw the discovery of Western Mining's Kambalda nickel mines in Western Australia, and at the discovery around the same time of the fabulous iron ore deposits in the Pilbara in the north of Western Australia. There were also the large energy discoveries of coal, uranium and natural gas, all as a result of the revival of interest in mining in Australia following the extended period of worldwide economic growth in the 1950s and 1960s.

PROSPECTING AT ISA

Mount Isa, which is still in operation today, grew out of some casual prospecting by John Miles, a man who had fashioned a modest living from seasonal farm work in eastern Australia and a bit of mining at Broken Hill. Camping overnight at Leichhardt River in Queensland in 1923 on his way to prospect for gold in the Northern Territory he stumbled upon some galena, heavy with lead. After having its lead richness confirmed by the government assayer for the state, he shipped some ore to market and received a good cheque in return.

This attracted attention from a few prospectors who travelled to what was to become Mount Isa and did the same thing. However, still in economic recession after the First World War, there was no rush to peg leases. In 1924 a particularly visionary ex-miner, William Courbould, decided to try and put together a decent parcel of land in the area and float it on the stock exchange. He found the early prospectors, including Miles, keen to sell to him and so Mount Isa Mines was listed in Sydney in that year.

Although the modern Mount Isa is a multi-metal operation with copper being its leading income producer, the mine started off as a lead and silver mine, with zinc being extracted later. In the 1920s the motor car industry was beginning to expand worldwide and lead-based batteries were being introduced into the manufacturing process, leading to a sharp increase in the demand for that metal. The discovery and development of Mount Isa seemed very well timed to meet the demand flowing from the birth of a new mass-market industry. Despite that, the early years at Mount Isa were difficult; losses were frequent and the company was constantly seeking capital injections for the mine. This was primarily as a result of having a treatment plant in such a remote spot – 700 miles from the Queensland coast and the main population centres – and the heavy infrastructure spending required to service the mine.

DEVELOPMENT, LISTING AND THE 1930S DEPRESSION

Following the 1924 listing of the company, its fortunes passed into the hands of Leslie John Urquhart, a Scotsman, who had made his money in Russian mining but had lost his investments there when the Bolsheviks seized power in 1917. His Russo-Asiatic Consolidated company provided the essential capital needed to take the Isa mining complex forward, with projects not only to expand the mine's resources and its treatment plants, which were truly cathedrals in the desert, but also to build a modern town with advanced facilities. In due course Urquhart split his interests into two companies and Mount Isa was placed in Mining Trust, which also owned the Britannia Lead Smelter in England, and became one of Mount Isa's key customers.

With the 1930s Great Depression looming closer, the mine unsurprisingly fell on hard times and Urquhart's resources were no longer adequate to support it. At this stage the Guggenheims and their American Smelting and Refining Company (ASARCO) came to Urquhart's rescue, committing substantial new investments to keep Isa afloat. Eventually ASARCO's stake in Mount Isa reached 53% – an interest that was eventually sold off in two tranches in 1987 and 1996. The mine, however, which had a history of disappointing metal recoveries, also had water problems, which became more serious as mining pushed deeper. ASARCO quickly regretted its investment as Isa seemed a bottomless and unprofitable pit, but no one would take it off its hands.

COPPER

At this stage geology began to turn in Mount Isa'a favour. The mine's exploration programme in the early 1930s began to find copper adjacent to Isa's main lead, zinc and silver orebody, the Black Star. Thus two very large, but separate, orebodies fortuitously lay next to each other. Initially the assays from the drilling were low grade but in due course, and conveniently in time to support the war effort in the 1940s, Isa began to produce large quantities of copper. It was not plain sailing, however, and the large copper treatment facilities were constructed from bits and pieces brought in from defunct copper mines in the region to save money. One of the problems that Mount Isa faced at its lead, zinc, silver and copper operations were metallurgical difficulties that led to poor recoveries. When metal prices were strong this was tolerable, but it hurt when metal prices fell. A huge surge in the lead price in the late 1940s, from £21 to £91 a tonne, led to Mount Isa's first dividend payment in 1947.

In the 1950s profits remained strong and £21 million was made in ten years, providing resources for further dividends, bonus payments to the miners and, most importantly, funds for exploration. As the drill programme matured, the reserves and resources at Isa for both lead and copper soared as new orebodies were uncovered. Plans were hatched to more than double production, and huge infrastructure projects to build a power station, expand the 600-mile railway and provide adequate water were put into action.

As a remote but large operation, Mount Isa experienced the problems that other similar mines in Australia did – virtually all economic activity in the town depended on the mine and on its prosperity. The people of Mount Isa were therefore fortunate that despite its problems in the 1920s and 1930s the mine did not close, but the era after the Second World War saw a more militant approach by workers and in 1964, a few years before the Aussie nickel boom, an eight-month strike broke out at the mine, perhaps the most bitter in Australian labour history. The strike unsurprisingly concerned wages and working practices, but the bitterness was exacerbated by lack of support from the main Australian labour leaders, as well as some high-pressure legal manoeuvrings by the Queensland government, which was right wing and firmly anti-union. It was finally settled in 1965 and no major disputes have followed since.

EXPANSION, COAL AND A MERGER

Growing production and successful exploration had provided Mount Isa with a boost in the late 1960s when the Australian mining boom was at its height, and the period of strong metal prices in the late 1970s saw keen interest in the company's shares – particularly from US private investors – with the shares doubling in 1979. At the same time, Mount Isa began to expand its coal mining division, which it had developed at Collinsville in the 1950s, initially to provide fuel for its own needs. In due course, further coalmines were brought into production taking the company into the coal export trade with Japan. At times this diversification proved very well judged as during the 1980s and 1990s metal prices performed erratically, creating problems for base metal producers in particular.

Despite that Mount Isa, which by then had been renamed MIM Holdings, pressed on with its growth plan. Nearby Hilton, comprising the George Fisher underground and Handlebar Hill open cast lead and zinc operations, and the rather more distant MacArthur River lead and zinc project, another metallurgically difficult orebody which had had a 40 years gestation period, opened in the 1990s. Later, in 1997, Mount Isa opened the nearby Ernest Henry copper and gold mine, and in the following year opened its first significant overseas mine, the 50% owned Bajo de la Alumbrera gold mine in Argentina. The deep level Enterprise copper orebody at Mount Isa followed and was first mined in 2000. Despite this rapid expansion MIM's days as an independent company were shortening and in 2003 it was taken over by Anglo-Swiss hybrid, Xstrata.

OLYMPIC DAM

Perhaps the greatest Australian mineral discovery of the second half of the 20th century was what is now called Olympic Dam in South Australia. First discovered in 1975 by Western Mining, the project was originally known as Roxby Downs. Most of Australia's great mineral discoveries have come about as a result of observation of surface indicators by prospectors and pastoralists. Olympic Dam was different – its copper orebody lay almost 1,000 feet below the surface and was targeted by Western Mining geologists, following an extensive modelling exercise based on geological mapping and an intensive aero-magnetic programme. Over the next four years a large drilling programme was carried out to establish the size of the deposit, which proved huge, and to assess the economics of developing a mine. In 1979 Western Mining approached BP, which was then building up a substantial mining division (long since sold off), to come in as a partner to develop the deposit. BP acquired a 49% stake and Western Mining retained 51% and operating control.

THE LONG GESTATION

Between 1979 and 1985 when the final decision was taken to proceed to the mine development stage, the joint venture spent around $A150 million in accessing the orebody to extract bulk samples for metallurgical testing in a pilot processing plant. Due to the fact that there was also uranium present in the ore, a small concentrator had to be built so that sample concentrates could be produced and tested to evaluate the possibility of separating the uranium; this was successfully achieved. It is unlikely that any previous mine consumed as much capital as did Olympic Dam to reach a development decision. The mine opened in 1988, 13 years after the discovery was made, a very long gestation period – even in an era of huge mega-mine developments. For Western Mining the wait must have been very frustrating as its famous company-making Kambalda nickel mine in 1966 was in production a year after the first ore shoot was discovered.

Although during feasibility study work the orebody was initially accessed using a shaft, the mine was developed with access being via a decline and an extensive network of drives. Ore is raised to the surface using the old shaft. The remoteness of the mine, over 350 miles north of Adelaide, meant that a town had to be built, plus facilities for family living, and this was located at the old Roxby Downs sheep station, a few miles from Olympic Dam. Water was piped 60 miles to the mine and purified in a desalination plant, and electric power lines were also run in; on-site power generation being impractical at the time. As well as copper and uranium, Olympic Dam also produces gold and some silver, with a copper grade of just under 2%, uranium (around 0.8 lbs per tonne) and gold a rather skinny 0.27 gms per tonne, although in the future Olympic Dam may well be able to mine higher-grade gold pockets.

A STORY OF EXPANSION

The mine has been expanded considerably over its life, having started in 1988 with a production capacity of 45,000 tonnes of copper and 1,000 tonnes of uranium oxide, and currently produces annually around 235,000 tonnes of copper, 4,500 tonnes of uranium oxide and 100,000 ozs of gold. A massive open pit is in the planning stage, which would increase output by six times, with annual production reaching 750,000 tonnes of copper (as much as the whole Zambian copperbelt), 19,000 tonnes of uranium oxide and 800,000 ozs of gold. This expansion, however, is on ice until the economic outlook improves.

When planning permission is granted for the open pit development, the construction period will be staged over 11 years and when finished Olympic Dam will be one of the largest base and precious metal mining operations in the world, although its annual planned output in terms of ore mined (72 million tonnes) will still be behind the levels of Western Australia's Pilbara iron ore fields.

One of the early problems Western Mining faced at Olympic Dam was the contained uranium oxide. Apart from the need to separate the uranium from the copper ore to enable a treatable copper concentrate to be produced, there was the need to sell the uranium and in the 1980s Australia had a ban on the export of uranium from new deposits for environmental reasons. However, such was the importance of Olympic Dam to South Australia that the federal government exempted its uranium output from the ban. As the years have passed the significance of Olympic Dam's uranium resources has grown to such a point that now it is thought to be the largest uranium deposit in the world.

An important factor in the mine's expansion plans has been the arrival of BHP Billiton, which took over Western Mining (or WMC Resources as it had become) in 2005. Western Mining itself had acquired joint venturer BP's stake in 1993 under a pre-emption agreement dating from BP's entry into the JV. Whilst a significant company in size, WMC did not have the financial muscle of BHP, the sort of muscle that was needed to expand output from a deposit that had grown so materially over time that resources should last for at least another 100 years.

LESSER LIGHTS

Above we have looked at three of Australia's greatest mineral provinces in the modern age. The Australian mining industry, however, was considerably more than just three giant mines, and we have already mentioned two of the lesser lights – Renison and Mount Lyell on Tasmania. Queensland, home of Mount Isa, was an important base metal producer long before the mighty Isa came into existence and many of Australia's 19th century copper producers were to be found in the northwest of the state, in what became known as the Cloncurry copperbelt. Many of these eventually closed in the metal price slump which followed the end of the First World War and so we will look at them in the context of the 20th century. Indeed, some of them have been revisited in recent years to evaluate reopening possibilities as metal prices have boomed.

COPPER, GOLD AND URANIUM IN QUEENSLAND

The town of Cloncurry was founded in the 1870s by Ernest Henry, after whom one of Mount Isa's most recently developed mines is named. Amongst the mines established in the region were copper mines at Mount Cuthbert and Mount Elliott, and a number of gold mines. Among these gold mines, Mount Morgan is perhaps the best known due to it being discovered by Englishman William Knox D'Arcy, who later founded what today has become British Petroleum (BP).

Mount Cuthbert's genesis owed much to the state railway being extended by the Queensland government towards Cloncurry in 1906, which made deposits in the area accessible and enabled product to be more economically shipped to market. It built a small smelter, as did nearby Mount Elliott, and with feed from other group mines such as Kalkadoon and Mighty Atom, the Mount Cuthbert Company had an annual production capacity of around 2,000 tonnes of copper.

Found in the 19th century, the Mount Morgan gold mine operated from 1883 for almost 100 years, before closing after the gold price peaked in 1980 and then dived. During its long life it produced around 8 million ozs of gold, of which 170,000 ozs was produced in 1899 alone – a very large amount of annual output in those days, although production levels could fluctuate quite widely. In the ten years after its closure in 1981, it re-treated substantial quantities of tailings to extract further gold but the operation could not be sustained when the gold price once more weakened in the 1990s. Today the treatment of the old tailings has re-started, proving once again that the best place to develop a new mine is where an old one existed previously.

Mount Morgan also produced large amounts of copper over its life; in some earlier years over 7,000 tons. In the latter decades of the 20th century its smelter, under the ownership of Peko Wallsend, substantially increased copper and gold output, being fed by a number of Peko's relatively small operations in the region, such as the Warrego mine.

Perhaps the most famous mineral discovery in Queensland, after Mount Isa, was the Mary Kathleen uranium deposit that was found in 1954 and which we will look at later. A substantial proportion of the mines discovered in Australia's outback over the years have tended to be gold, as small-scale mines generating attractive returns for successful prospectors were able to be established without the need for major transport links and infrastructure spending. The presence of nugget or free gold made the economic mining of gold possible in such conditions. Having said that, many of Australia's non-precious metal discoveries have been quite modest at times, from the 19th century discoveries in South Australia through to some of the discoveries made in Queensland and New South Wales.

MISTY TASMANIA

At the end of the 19th century some prospecting had moved offshore to the cool and misty island of Tasmania and over the years a number of important mineral discoveries were made. Perhaps the best known was the copper deposit at Mount Lyell in the northwest part of the island, although in terms of geological significance the tin discoveries of Mount Bischoff and then Zeehan were arguably more outstanding.

Mount Lyell Gold Mining (Mount Lyell), formed in 1888, was primarily a gold operation when mining started. In due course, with the mine making no money, the owners, led by James Crotty, decided to shift emphasis to the copper that the alluvial gold operation was allowing to run away from the sluice boxes that the ore was washed in. During this early period a number of other small gold operations were to be found in the area; the richest (indeed richer than Mount Lyell itself) was North Mount Lyell, and at this time a fierce corporate battle raged which in the end led to James Crotty leaving Mount Lyell and gaining control of North Mount Lyell.

North in reality had the better grade ore, but Mount Lyell had a more efficient treatment process, and in 1903 North and Mount Lyell merged after years of, at times, bitter wrangling and in doing so they became the dominant mine in the area. Mount Lyell, being remote, was more than just a mine, having built a direct railway to the port of Strahan on the west coast and also a hydroelectric scheme for the mine. A new open pit mine, the West Lyell, was established in 1934 and produced over 1 million tonnes of copper and almost 1.5 million ozs of gold before closing in 1972. For many years Mount Lyell was controlled by the UK gold giant

Consolidated Gold Fields after it purchased a 60% stake in the company in 1964. Mining at Mount Lyell finally ended in 1993/94 and, after a century of river pollution from mine waste, a major remediation programme was undertaken.

Today the mine is open once more under the ownership of the large Indian copper group, Sterlite Industries, and supplies around 30,000 tonnes of copper in concentrate for smelting and refining in India. Whilst in modern times Mount Lyell, which retained its stock market quote after Cons Gold's acquisition, was an unreliable market performer, those who were in at the start in the 1890s made many times their money. It is also interesting to observe that for many years in the 1890s and early 1900s Mount Lyell and shares of smaller companies in the Tasmanian copper play dominated trading on the Melbourne Stock Exchange.

Perhaps it is Tasmania's remoteness that, therefore, made it attractive to difficult people seeking something different, but just as Mount Lyell and North Mount Lyell squabbled for years over the ownership of the Mount Bischoff tin mine to the north of Lyell, so did German and Cornish mining men confront each other, although their dispute was based on whether German mining methods were more efficient than Cornish techniques, which the Germans thought were old fashioned and wasteful.

Mining at Mount Bischoff started in 1875 using the waterfall at nearby Waratah, but in the early 20th century the mine operated both open cut and then underground before closing in 1929. It re-opened in 1941 to provide tin for the war effort but eventually closed in 1947. During its life it produced over 60,000 tonnes of tin and is now controlled by Metals X Ltd, which also owns the old Renison Bell tin mine, once part of the Consolidated Gold Fields group, about 50 miles from Mount Bischoff at Zeehan. Production started at Renison in the 1890s but was often only marginally profitable until sulphide flotation technology improved recoveries and subsequently a major investment was made to create the largest underground tin mine in the world in the 1950s. The mine closed in 2003 but re-opened in 2008 with a long-term aim to produce around 8,000 tonnes of tin annually.

Another of Tasmania's great mines, the Rosebery zinc mine located on the island's west coast, was discovered in 1893, at much the same time as Lyell and Renison. However, it was not until the Collins House Group established EZ Industries and built a smelter at Risdon outside Tasmania's capital, Hobart, that Rosebery, which EZ purchased in 1920, came into its own. Having said that, it took a great deal of painstaking research before an economic process was developed in 1921 to smelt Rosebery's zinc concentrate.

By the 1960s EZ's Risdon plant was the third largest zinc complex in the world and was Tasmania's largest consumer of electricity. EZ was eventually taken over by North Broken Hill Peko and the Rosebery mine is now owned by Chinese controlled Minmetals, having previously been part of Pasminco and then Zinifex before their collapse, and Oz Minerals. Rosebery continues to supply concentrates to Risdon, now part of the Nyrstar group, and to other zinc smelters in Australia and overseas.

2. CANADIAN CENTURY

The 19th century ended with Canada firmly in the world's consciousness thanks to the fabulous Klondike gold rush. By the middle of the 20th century Canada would be established as one of the most powerful economies in the world and an important diplomatic player following its key roll on the Allied side in both world wars. The economic underpinning, which enabled Canada to advance to the edge of major power status, was mining. In 1900 the country produced minerals to the value of US$64 million – by the beginning of the Second World War that figure had risen to $567 million and today it is nearer to $45 billion.

Today Canada's population is only around 35 million, making it very much a mid-range country in those terms, but it is a long-standing member of the Group of 7 (or G7), the meeting of the largest economies in the world. Its standard of living is amongst the highest in the world and its proximity to the world's largest economy, the USA, is of major benefit as Canada is an exporter of high quality, high value, advanced products to its rich neighbour.

Canada's economic power is due to its position as a supplier of key raw materials to the world; it is a major producer of energy, gold, diamonds, base metals and industrial minerals such as potash, and has huge reserves of both heavy oil and uranium. It also possesses one of the world's largest sources of clean water, contained in its extensive and vast inland lakes and in its frozen northern territories. It is raw materials that have made the modern Canada.

THE EARLY DECADES OF GOLD

From the time of the Klondike rush, the metal that has most caught the eye of prospectors and investors in Canada has been gold. After South Africa's Witwatersrand, Canada has in the Abitibi region of Ontario one of the most prolific sources of gold in the world. For this reason the 20th century saw a continuing stream of activity in the area surrounding Timmins in the central region of the province. That interest continues up to the present day.

KIRKLAND LAKE

As we have already seen, the Canadian mining industry entered the 20th century in expansionary mode. We remarked on the first great find of the new century earlier, the silver discoveries at Cobalt south of Timmins in Ontario, near to the Quebec border. However, close to Cobalt lies the rich gold deposits of Kirkland Lake that were first prospected in 1911 – around the time the Hollinger gold mine was being developed relatively close by in the Porcupine district, which was to later become the town of Timmins. The first gold in the area, which was served by the curiously named town of Swastika, was found at Larder Lake and by the 1920s there were four substantial gold mines operating at Kirkland Lake – the Teck-Hughes, Lake Shore, Wright-Hargreaves and Macassa mines. Over the years they produced around 22 million ozs of gold. Although the mines mainly shut down in the 1960s (Macassa struggled on until 1999) they are now being brought back into production thanks to a strong gold price and a successful drill programme to prove up more high-grade ore in the camp.

Among the key figures in the opening up of Kirkland Lake were Harry Oakes, Bill Wright, Ed Hargreaves and the Hughes brothers. American born Harry Oakes, who staked and made a fortune from the Lake Shore mine, was knighted by George VI for services to the island and people of the Bahamas, where he had emigrated in the 1930s. However, both his mining fame and charitable works are arguably overshadowed by his being murdered in 1943 in his house in Nassau, a case that was never solved, and which had echoes of that other great unsolved British colonial mystery of the 1940s, the Happy Valley murder in Kenya of Earl Errol. Both murders also curiously spawned major films.

Kirkland Lake's mines were developed through a shaft and drive system and eventually reached considerable depths; the Macassa shaft went down over 7,000 feet and until about 15 years ago was the deepest single shaft in the Americas. The gold was often found in its native state – i.e. almost pure gold – in the porphyry ore typical of the area and this accounts for the high grades mined, which are also a feature of the drilling being undertaken today. The Lake Shore mine was Harry Oakes's sole mine, but such was its richness that it made Oakes, who was the largest shareholder, one of the wealthiest of all the Canadian mining entrepreneurs, as it paid out dividends of over $100 million. Whilst Oakes cannot match Noah Timmins – whose achievements we will come to below – in terms of mines developed, the Lake Shore nonetheless provided him with an enormous and steady flow of dividends and allowed him to pursue a new lifestyle (an ultimately fatal decision).

LARDER LAKE

Larder Lake itself eventually yielded, in the form of the Kerr Addison mine, one of the largest mines in the Americas, let alone Canada, but from the discovery of gold-bearing ore in 1906 it took 30 years for an economic mine to be established. In the intervening years a large amount of speculative capital was sank in shafts and basic treatment plants, such as stamp mills, but returns were very poor. By the start of the First World War the leases had been amalgamated under the control of Canadian Associated Goldfields.

One prospector in particular, Jack Costello, kept faith with Larder Lake and after the war he played a central role in amalgamating some of the claims into an integrated project sufficiently robust to interest McIntyre Mines, one of the major players in the Porcupine camp. In 1936 the big break was finally made at Larder Lake with the formation of Kerr Addison Gold Mines, whose shares initially were valued at 15c. The old Kerr Addison shaft was de-watered and, after deepening the old workings, several million tonnes of high-grade ore was outlined; in due course this figure expanded to over 50 million tonnes. Faith in Larder Lake had finally paid off and Kerr Addison shares eventually traded as high as CAD$15 by the 1940s.

PORCUPINE

Porcupine's first mine, the Hollinger, grew out of the Acme and Millerton leases, the Hollinger amalgamation being a deal brokered by the area's legendary mentor, Noah Timmins. The Hollinger became one of the largest gold mines in the world but it was not the only mine in that developing camp. The Dome, which is still working today, was another of Porcupine's major mines and the McIntyre was the third; all three were almost smothered at birth by the fire of 1911 which destroyed most of the mining camp and town. However, Noah Timmins's leadership led to the rapid re-building of the town and the mining facilities there. The three mines produced gold to the value of $US660 million, a production volume around 20 million ozs. All three were underground mines but the Dome metamorphosed over the decades into the huge Dome Pit, which although past its peak, is part of today's Porcupine Joint Venture open-cut operation.

NOAH TIMMINS (1867-1936)

At the centre of the historic Canadian mining province of Abitibi is the city of Timmins, founded in 1911 and named after Noah Timmins who developed the giant Hollinger gold mine, which today, almost 100 years later, stands on the edge of revival.

Noah Timmins was born in Mattawa, Ontario, in 1867. His parents operated a general store in the town which serviced the mining community in the area and which he and his brother, Henry, inherited. Both brothers were enthusiastic, if initially unsuccessful, backers of prospectors (grubstaking), and Noah in particular had made himself very knowledgeable about mining and mineralogy.

In September 1903 one of Noah's customers was Fred La Rose, whose story of silver riches in Long Lake near the town of Cobalt so intrigued Noah that, with his brother, he purchased a quarter share in La Rose's leases. With other partners, and not without a few legal and mining problems, the La Rose silver mine was developed and then sold, as were other silver leases acquired by Noah. After that, with his silver profits banked, Noah began to raise his game, with spectacular results.

In 1909 he, with Henry and other associates, purchased the leases under which lay the Hollinger gold mine in what became the town of Timmins, from prospector Benny Hollinger for $300,000; a huge sum at the time for undeveloped ground. The Hollinger mine became Canada's largest gold mine and one of the largest in the world. In 1912 the town of Timmins was incorporated and named after Noah Timmins who remained its generous benefactor for the rest of his life. By then, the Timmins brothers had moved to Montreal and had married the Pare sisters, whose brother Noah's sister Josephine had married some years previously. They had a son, Alphonse, who eventually became involved in Noah and Henry's mining activities as the first manager of the Hollinger.

Noah's mining interests expanded into Quebec where he rescued the ailing Siscoe gold mine and also financed Noranda's Horne smelter. By the 1920s Noah's reach had become Canada-wide as he revived the San Antonio mine in Manitoba, financed the Outpost Island mine near Yellowknife and got involved in developing placer gold mines in the Yukon; he did not, however, forget Ontario, where he developed the Ross mine and Young Davidson mine near Kirkland Lake, close to the Quebec border.

Much of this activity was done through the Hollinger company, of which Noah was president, and which long after his death fell into the hands of the now disgraced Conrad Black, who stripped all the mining interests out of the company and ran Hollinger, fatally as it turned out, as a US-orientated industrial conglomerate. In the 1990s the Hollinger mine, which had closed in 1968, passed into the hands of Royal Oak which collapsed in 1999.

Noah Timmins died in 1936. He is widely acknowledged to be the mentor of the modern Canadian mining industry, known affectionately as the 'Grand Old Man of Canadian Mining' and in 1985 he was posthumously admitted to the Canadian Business Hall of Fame in recognition of his services to mining.

RED LAKE

In the period before the Second World War other major new gold camps were established in Ontario, including Red Lake in the west of the province, where mining continues today, and where production began in 1935 (see later). A couple of years before that, the Little Long Lac mine to the east of Red Lake began producing. The early Red Lake mines included Howey, Madsen, McKenzie and Cochenor-Willans.

Today Red Lake, now owned by Goldcorp, has consolidated into one of the major gold mines in the world, but the birth of the camp in the 1930s was difficult and involved many of the people who had been active in other Ontario gold camps, such as Noah Timmins and Jack Hammell. One of the problems related to the relatively basic methods used in assessing orebody size and the geology at depth in those days, and at Howey and then again at Cochenor-Willans shaft sinking revealed unwelcome surprises.

Fortunately for the camp, the gold price eventually helped to bail out some operations. Additionally, techniques such as sorting were applied, enabling higher-grade material to be put through the mill, thus boosting profitabilty. However, in the early days both Dome Mines with the Howey and Hollinger with the Cochenor-Willans, having got involved, pulled out due to doubts about the viability of both mines. In recent decades exploration at depth at the Red Lake mines has revealed large tonnages of high-grade gold ore, guaranteeing the Camp's long-term survival and prosperity.

The two great mines at Red Lake brought into production after the Second World War in 1948 were Campbell and Dickenson. The development of the

Campbell mine owed a lot to the participation of Dome Mines. The Campbell and Dickensen mines, which are next door to each other, in fact accessed the same orebody, but the Campbell historically had higher grades than the Dickensen, and indeed by the mid-1990s the latter appeared ripe for closure. At that stage Goldcorp, which had acquired Dickensen, instituted a deep-level drilling programme on the property beneath the old workings and found a new high-grade deposit which went into production in 2001. The Campbell and Dickensen properties are now worked as one and recently the Goldcorp group acquired Gold Eagle, which has a high-grade deposit thought to be an extension of the old Cochenor-Willans mine, which was closed down in 1971 because of a lack of ore.

MORE THAN GOLD

The romance of Canada's mining industry had been built on gold, but the vast wilderness in the country's north contained much more than just the yellow metal and prospectors uncovered numerous and significant deposits of base metals and energy minerals, including coal and, particularly, oil and gas.

SULLIVAN

So, though gold was a key component in the rise of the Canadian mining industry in the 20th century, as the decades passed, industrial metals became increasingly important. One of the earliest discoveries, which we have already mentioned, was the Sullivan lead, zinc and silver mine at Kootenay, British Columbia, discovered in 1892 but only brought to full production in 1922, when complex metallurgical problems were finally overcome.

Interestingly a substantial investment was made in treatment facilities at Sullivan, in the shape of the famous Trail smelter, long before a viable operation was fully proved; an extravagance that would be impossible today in our age of feasibility studies and due diligence, but which was not uncommon in those earlier days. Once Sullivan had overcome its problems it went on to produce millions of tonnes of lead and zinc and almost 290 million ozs of silver before it was finally shut down in 2001; indeed after the Second World War it was for a while the largest lead and zinc mine in the world.

THE HORNE

In 1911 legendary prospector Ed Horne, after a disappointingly barren period prospecting for gold in the Kirkland Lake area, crossed the border into Quebec, where a parallel gold hunt was having some success in the Rouyn district. Horne, however, found copper and gold rather than gold alone, which often meant metallurgical problems. The discovery came about as a result of several years of intermittent work by Horne between 1911and 1921 in the area around Osisko Lake. In 1920 Horne established the Tremoy Lake Syndicate to fund a concentrated exploration programme in Quebec. To begin with interesting but not sensational gold indications were uncovered with values around 0.25 ozs per tonne; further work revealed higher values but it took Horne several months to bring in essential outside capital to bolster the exploration programme.

The new participating syndicate, Thomson-Chadbourne from New York, metamorphosed into Noranda Mines, which in due course became one of Canada's leading mining groups. Thomson-Chadbourne paid $350,000 for a 90% stake in Horne's claims, but it was not until the summer of 1923 that it began serious work on them. The first drill hole entered ore at 11 feet and ran for 131 feet, grading over 8% copper and around 0.20 ozs of gold. There was one major snag – Osisko Lake was remote and the ore would have to be processed at the mine, but a smelter required heavy capital expenditure and could not be justified without a significant orebody.

Noranda, looking for both finance and smelter operating experience, approached American Metals who investigated and declined to participate, believing that the orebody would never amount to an economic size being less than 1 millions tonnes at the time. In the end it was Noah Timmins who stepped in with Hollinger Mines providing finance for a smelter; Timmins, with his unsurpassed mining intuition, believed that the orebody would grow. Ultimately the mine produced over 1 million tonnes of copper and 8 million ozs of gold from more than 50 million tonnes of ore, before closing in 1989 after 60 years of operation.

FLIN FLON

In 1928 after many years of investigation of the Flin Flon copper, zinc, gold and silver orebody in the central part of Manitoba, right on the Saskatchewan border, shares in Hudson Bay Mining and Smelting were listed on the New York Stock Exchange. Although the old mine has long closed and the historic copper smelter was decommissioned in 2010, the area is still a substantial producer of copper and zinc, based on new orebodies and treatment facilities. The initial problem with the Flin Flon – named after the hero of a book found in a refuge cabin in the area in 1914 by the discovery prospectors led by Thomas Creighton – was that it was a large, but remote, deposit. Unfortunately, the complexity of the ore required state-of-the-art treatment techniques, available only in Vancouver and Belgium – thousands of miles from Flin Flon.

The man who was at the heart of financing the Flin Flon play was Jack Hammell, another of Canada's legendary mining promoters, and it was he who interested New York brokers Hayden Stone in Flin Flon. They, however, were unsure that the problems relating to its remoteness and metallurgy could be economically solved and they pulled out; Hammell replaced them with Dave Fasken, a wealthy businessman from Haileybury, near Cobalt in Ontario. Further drilling was undertaken but Fasken decided not to advance further funds and so Hammell had to find yet another backer.

This he did in the form of Mining Corporation of Canada, whose Cobalt silver mine was in the process of closing down. Mining Corp, however, wanted additional backing in the form of the US copper magnate, William Thompson; Hammell duly reeled him in. The new proposed structure for Flin Flon was Thompson with 75% and Mining Corp with 25%, but before that could happen the two groups had much work to do at Flin Flon. Exploration shafts were sunk and plant acquired, and Hammell got to work on the Manitoba legislature to obtain its backing for a rail spur to Flin Flon.

The results from the extended work, though promising, were not good enough for William Thompson; it was 1919, post-war deflation bore down on his other mining interests and Flin Flon was still too difficult. He pulled out, leaving Flin Flon's new champion, Scott Turner of Mining Corp, to look elsewhere. Turner tidied up the ownership structure by buying out Hammell and the other original prospectors, including Creighton, and he then spent the next few years reinforcing the Flin Flon database. Then it was back to New York to talk to the wealthy Whitney family who were interested in mining but had no big investment in the industry. By now it was 1925 and though a large 16 million tonne orebody had been outlined, the ore's complexity remained a problem. However, the Whitneys fortuitously had an aptly named company called Complex Ores Recoveries, which set to work to find a solution.

It was in 1927, after much work by Complex Ores on the Flin Flon ore, that Whitney exercised its option and the mine finally went into development under the eye of Whitney's chief mining engineer, R.E. Phelan. The following year it was listed in New York as Hudson Bay Mining – the controlling shareholders were the Whitneys, Newmont Mining and Mining Corp. Initial production was in 1930 in the shadow of the Depression and those early years were a struggle as the treatment process was constantly tweaked to improve recoveries from the mill. The mining plan was also enhanced by finding higher-grade ore at depth and Flin Flon, which had started life as an open pit, steadily metamorphosed into an underground operation. It had been a long wait but with recoveries rising, higher grades being mined, profits being made and dividends being paid, Tom Creighton and Jack Hammell's dream of a mine in the wastes of Manitoba was finally realised.

AFTER THE SECOND WORLD WAR

Although gold remained fixed in price between 1934 and 1971 that did not stop the search for it, certainly in the decade after the Second World War, but steady sales by the US, the ban on owning gold bullion in many countries and the rising interest in base metals to support the post-war economic boom meant that gold exploration activity in Canada did eventually slacken. By 1975 Canadian gold production had fallen to 1.7 million ozs but over the next 25 years, as a rising gold price re-ignited exploration activity, output recovered to around 5 million ozs, a level last achieved in the early 1940s in the wake of the 1934 gold price rise.

KIDD CREEK

One of the biggest base metal discoveries of the 20th century was the Kidd Creek mine near Timmins, in the heart of Canada's gold country. Kidd Creek, which was owned by the Hendrie estate, was drilled in 1964 by Texas Gulf Sulphur after obtaining an option on the ground, and production began in 1966 – initially as an open cut mine. Awareness of a possible orebody in the area dates back to the 1950s when an airborne magnetic survey was flown. The exploration programme that was derived from the survey, and other follow-up work, had a number of targets, most of which were barren when tested, but the fifth target came up trumps with an astonishing 177 metres of core grading 8.37% zinc, 1.24% copper and 3.9 ozs silver in the discovery hole.

An interesting aspect of the Kidd Creek discovery was that it was only 15 miles north of Timmins and lay under some of the most heavily explored ground in the world. However, the search had primarily been for gold and the prospecting techniques used were very much based on finding surface signs of gold mineralisation. One gold prospector's cabin was even built on an outcrop within what became the Kidd Creek lease, but the old timer was not looking for base metals buried beneath clay where the great deposit lay, but for gold. However, once the Kidd Creek discovery had been made there was a huge staking rush in the area which led, as usually happens, to a lot of burnt fingers and just a few very well padded bank accounts belonging to local insiders with claims to sell. Whilst Timmins had thrown up several substantial gold mines, in its time Kidd Creek was the only significant base metal deposit found in the region.

By the end of the first decade of the 21st century Kidd Creek had produced over 9 million tonnes of zinc, over 3 million tonnes of copper and over 300 million ozs of silver in 40 years of full operating. Its deepest shaft had reached almost 10,000 feet, making it as deep as some of the deep-level South African gold mines and with a life stretching beyond 2015. Its ownership had been less stable with Texas Gulf Sulphur being followed by Elf Aquitaine, who then sold Kidd Creek to Canada Development Corp. After that ownership passed to Falconbridge, eventually acquired by Noranda, which in turn was acquired by Xstrata.

HEMLO

To the west of Timmins lies the Hemlo gold camp where promising ground was staked in 1980 by prospectors John Larche and Don McKinnon, and geologist David Bell. The area had been thought to be prospective for gold for over a century but the gold did not outcrop, so it remained undisturbed until science and technology caught up. David Bell's theory was that there was no granite host for the gold at Hemlo, but that gold was still present, although widely disseminated throughout volcanic sediments, as often happened with African gold orebodies. He persuaded Murray Pezim of Vancouver junior Corona to put up exploration funds in exchange for Larche, McKinnon's and his stake in what became the Page Williams mine. Two other experienced promoters, Richard Hughes and Frank Lang, through Goliath Gold and Golden Sceptre, obtained options on ground which became the Golden Giant.

The orebody was relatively narrow and dipped over a strike distance of 2 kms and it reached a depth of 1,500 metres. Initially three mines were established – the Page Williams – owned by Teck Corona after a long legal battle with Lac Minerals – Noranda's Golden Giant and Teck Corona's David Bell mine – all of them mining essentially the same orebody as it dipped and extended. The ownership set up was relatively unusual in its disparity and in due course rationalisation saw Williams and David Bell come under the control of Barrick. Mining began in 1985 at the Golden Giant mine, which was eventually re-named Hemlo Gold, and continued until 2006, during which period it produced over 5 million ozs of gold. The other two mines continue to operate today, with the camp having produced around 20 million ozs in total since 1985.

Whether or not Hemlo has fulfilled its full potential is a moot point and there has been much comment over the years about the original development of the camp as three separate operations and all the expensive duplication that this led to. At their peak the three mines produced 1 million ozs of gold annually and Hemlo was by far the biggest producing gold camp in Canada. However, most of the gold-bearing ore lay below 500 metres, which has inhibited explorers searching for similar orebodies in the area. Over the last 15 years or so only a limited amount of exploration has been carried out at Hemlo, although there are dreamers who believe that, like Red Lake, Hemlo will eventually, at depth, reveal further great gold wealth.

COPPER IN BRITISH COLUMBIA

Canada's largest copper province is British Columbia (BC). A number of low-grade, large-tonnage, porphyry copper and molybdenum mines were discovered in BC in the 1960s, underlining the province's historic record as a major source of copper. Interestingly, the new deposits mostly had molybdenum as the prime by-product metal, whereas traditionally the higher-grade copper deposits had gold as the main by-product metal; geologically, the former came from significantly younger rocks than the latter.

The 1960s was a time of rapid global economic growth, with housing and defence leading to a substantial upturn in demand for copper. At the same time, in the post-colonial era, with nationalisation of mining assets on the rise, mining companies and even customers were looking for new supplies from politically stable areas – Canada was an attractive proposition, although from time to time provincial radicalism did complicate things. Amongst the biggest BC copper producers of this 1960s era were Bethlehem Copper, Highmont, Lornex, Valley Copper and Gibraltar Mines; the first four eventually in 1984 were amalgamated into Highland Valley Copper, controlled by Teck Cominco, which today produces around 150,000 tonnes of copper annually and 4 million lbs of molybdenum.

Bethlehem was originally explored in the early 1900s and the prospect then was called Snowstorm. A number of small exploration shafts were sunk, but no mine was established until in the 1950s large-tonnage, low-grade targets became all the rage and a drilling programme instituted by a small group of prospectors headed by Spud Huestis and backed by Japanese trading house Sumitomo Shoji, outlined such a deposit. Right at the end of 1962 the first shipment of copper concentrate to Japan was made. Three further major low grade mines added to the province's burgeoning copper mining industry. Rio Tinto-controlled Lornex was discovered around this time but did not start production until 1970. Highmont began production in 1979 but shut down five years later, after which its treatment plant was moved to Lornex. Valley Copper was another mid 1960s discovery but did not start production until 1982.

Gibraltar Mines lies to the north west of Highland Valley and was opened in 1972, making it part of the Highland Valley generation of low-grade, high-tonnage BC copper mines. It shut down in 1992 during the extended period of low metal prices in that decade and was acquired by Canadian junior Taseko who reopened the mine in 2004. Its expected mine life takes us well into the current decade and its production rate is around 50,000 tonnes of copper per year and over 1 million lbs of molybdenum.

In the northern part of BC lies the Kemess South copper and gold open pit mine, which was also found in the 1960s, but only started production in 1998 after it was acquired by Northgate Minerals. Its annual output was a modest 24,000 tonnes of copper and 172,000 ozs of gold and the mine recently closed as reserves became exhausted. The original plan was to utilise resources at Kemess North, a few miles away, but Northgate decided to write off the project, following an extended planning application and a frustratingly opaque decision which led to the province refusing development permission in 2008.

Another of the 1960s BC porphyries was Utah's (BHP Billiton) Island Copper mine by the sea on Vancouver Island. The mine operated from 1971 until closure in 1996 and during its life it produced 1.4 million tonnes of copper and 1.1 million ozs of gold as well as a number of other by-product metals, including silver and molybdenum. Although a major mine in its own right, it is perhaps more noted for its environmental arrangements where the tailings from the pit – which consisted of over 300 million tonnes of material – were crushed and laid on the bottom of a nearby lake with minimal environmental impact. When the mine closed, bearing in mind its proximity to the sea, the pit was flooded with seawater to create a lake.

THE CRAZY 90S – LAC DE GRAS TO BRE-X

Although the 1990s were difficult years for the mining industry, good stories can sometimes rise above the crowd and catch the imagination of risk-orientated investors. We later sketch out one of the most significant of such events, the Canadian diamond boom, which started in the early years of the decade with the historic Lac de Gras discovery, at a time when worldwide diamond demand was very fragile. For the moment we will take a brief look at a few less edifying Canadian mining scams.

The 1990s in a sense belonged to Canada because the successful development of a new globally significant mining province, diamonds, opened the eyes of the mining world to the fact that major discoveries were still possible, and in a politically-favourable environment as well. In the early 1990s, at a time when the metals and mining industry was depressed, the Canadian diamond discoveries reminded miners and investors alike that there was still life and excitement left in mining. The discoveries also persuaded the Canadian banking and broking community that there were mining and exploration projects worth financing, and on the back of Lac de Gras Canadian money men spread out across the world with stories to tell. Disaster was not far away!

TIMBUKTU

In early 1996 the Canadian junior Timbuktu Gold astounded the mining world with some extraordinarily high gold assays from exploration drilling on its Sidikali lease in Mali, West Africa. Heavily promoted in London, the Alberta-listed explorer had only been floated in February of that year and in April announced the gold find. News then leaked to Alberta that Timbuktu's President, Oliver Reece, had form with the SEC in the US which he had not revealed to the Alberta Stock Exchange, one of Canada's most relaxed exchanges in those days, when Timbuktu was listed. The shares, which had risen from a couple of dollars to CAD$30 in a matter of days, were suspended and further examination of the Sidikali *discovery* suggested the assays had been massively interfered with.

CARTAWAY

Around the same time the Canadian market was hit by another scam concerning a company called Cartaway Resources, which was engaged in garbage container leasing and removal in Kamloops, British Columbia. A group of First Marathon Securities brokers gained control of Cartaway in 1994, paying 10c a share for a 46% stake in the company. A year later First Marathon placed further shares at 12.5c with a number of their employees and families, with the First Marathon stake rising to 66%.

Following this, in May 1995 Cartaway announced that it was to become involved in mineral exploration and to this end had acquired properties in the Voisey's Bay area of Labrador where a major nickel and copper find had been made by Diamond Fields. Some of these leases, it was later discovered, had been in the possession of the First Marathon group for some months and had not been disclosed when the earlier financings had been undertaken.

In April 1996 Cartaway announced that initial tests on its Voisey's Bay properties strongly indicated that they contained nickel and copper mineralisation. The shares, which had made steady progress over the previous year and had reached CAD$3.76, surged as high as CAD$26 following the announcement. A month later, in May, a further announcement from Cartaway admitted that drill assays did not support the previous optimism and the shares crashed back to CAD$2.78. After this the authorities, led by the Toronto Stock Exchange, descended on First Marathon and over the next two years the TSX, and also Alberta and BC securities regulators, launched a succession of actions which led to several millions of dollars of fines on the brokerage and on the individuals involved. The whole Voisey's Bay story had been a carefully structured ramp which had sent the shares soaring from a few cents to $26 in under two years on less than a wing and a prayer. In 1999 Cartaway was de-listed by the Alberta Stock Exchange, its shares worthless.

THE BRE-X CAPER

The Timbuktu disaster and the Cartaway scam were nothing compared to what followed exactly a year later, when the Bre-X gold discovery in Indonesia was found to be a fraud. Although both Bre-X and Timbuktu were Canadian promotions, the fallout from Timbuktu was in financial terms largely confined to London investors, some of whom had bought the stock from London mining specialists T Hoare and Co. Bre-X was a full-blown Canadian disaster and a huge embarrassment to the North American mining community who had fought each other tooth and nail to buy control of the small explorer, which some thought at the time had made one of the largest gold discoveries of the 20th century.

Bre-X had acquired leases on the Indonesian island of Kalimantan at a place called Busang in 1993. The property had been drilled previously by two small Australian exploration companies who had found gold but thought the deposit, such as it was, too inconsistent to warrant further work. Between 1993 and 1996 Bre-X carried out an extensive drill programme at Busang and over the three years reported a growing gold resource which had somehow eluded the previous Australian leaseholders. Bre-X did not report many individual drill hole results over that period – its practice was rather to issue regular reports of its estimates of tonnage and grade of the whole deposit as the drill programme continued.

In 1993, the first estimate of Busang's size was 20 million tonnes grading 2 gms plus of gold per tonne, with estimated mining costs of $155/oz. The following year the deposit had doubled in size and the grade had increased to 3gms/tonne, a very interesting 3 to 6 million oz prospect, bearing in mind that mining would be via a large and shallow open pit. After that Bre-X started to report the resource simply in terms of millions of ozs and by 1997 the reported resource had reached 71 million ozs. By then Bre-X was being stalked by every major gold company in the world and also by the rapacious Indonesian ruling family headed by President Suharto. The chasing companies were primarily CRA, Barrick, Place Dome and finally and fatally Freeport McMoran, and Placer even proposed a full-scale merger.

One of the things that worried some was the fact that despite the flattering approaches Bre-X could not, perhaps unsurprisingly, conclude a deal because it would not permit any of the suitors to do their own test drilling. In 1997 a deal was finally struck with Freeport which left Bre-X with a 35% free carried interest in the project with Indonesian interests having 50% of Busang, which Freeport would operate. Freeport also was given permission by Bre-X, under pressure from the Indonesians, to check-drill parts of the deposit. Freeport's results showed no gold in the cores – Busang was a scam.

The main personalities in Bre-X were John Felderhof, the Chief Geologist, David Walsh, the president and Michael de Guzman, the senior geologist at Busang. Felderhof, who introduced Busang to Bre-X, was charged with insider share trading but acquitted in 2007, Walsh died of a heart attack in 1998 and de Guzman 'fell' from a helicopter and died in 1997 just before Freeport announced the dud drill holes.

The Bre-X scam was substantially to blame for the collapse of interest as well as confidence in the mining industry in the final years of the 1990s. Already deemed a sunset industry by many commentators, mining also compared poorly in terms of industry prospects and investor rewards with the technology and internet industries which were booming at the time. The collapse of Barings Bank in the mid-1990s did not help sentiment either, with some people believing that the returns on risk capital invested in such venerable industries and institutions could no longer earn the sort of returns which would mitigate the risk element of post-Big Bang banking and broking.

Ten years later this theme was to return to haunt markets horrified by the implosion of the Western banking system, and this after the destruction of billions of dollars of risk capital invested in the internet boom at the start of the new millennium.

FROM FROBISHER TO SOUTHWESTERN

Following the internet securities crash in the early 2000s, mining had begun to stage a major cyclical rally encouraging the incorrigible to return to their bad old ways. Duly in 2007 Southwestern Resources, a Canadian junior, was accused of falsifying drill assays on its Boka gold project in China and the company filed a lawsuit against its former CEO, John Paterson, for fraud in the High Court of British Columbia. None of this should have been surprising for a nation which possibly was the source of the first recorded attempt to tamper with ore samples by adding extraneous gold.

This is believed to have happened on Sir Martin Frobisher's expedition to the Canadian Artic in 1577/78 to find the North West Passage; Frobisher carried back several tonnes of ore from Kodlunarn Island for assaying in London. The ore assayed many times the amount of gold that Kodlunarn ore typically was found to contain four centuries later, suggesting to historians that gold was added to the samples in London before they were delivered to the assayers. The aim no doubt was to tempt the crown into financing another expedition and the distinguished Frobisher may have colluded in the foul deed.

Although often in the shadow of its giant neighbour to the south, it could be argued that Canada, at times a spectacularly beautiful if slightly unassuming country, shows us how America might have developed if Captain Ferguson had shot George Washington at Brandywine Creek in 1777 or Louis XVI, expansive backer of the colonists, had had a crystal ball foretelling the disastrous, for the French crown, events of 1789. However, whatever the case, Canadians have always had an aggressive approach to mining exploration and development, not exactly like, but not overly different from that of their Australian cousins, and rather different from the US where mineral resources are broadly a means to an industrial end, and where financial scams are carried out in different fields and on a different scale altogether. Mining runs deep in Canada's blood and from time to time greed cannot help but bubble to the surface in an industry that is extremely difficult to regulate.

3. THE DEVELOPMENT OF MINING TECHNOLOGY

Mining in recent decades has leaned increasingly towards technology in the all-important field of mineral exploration. However, it would be wrong to assume that this use of technology is a recent development, because science in the form of geophysics and geochemistry has been helping in mining exploration since the late 19th century. Of course many old-time prospectors believed that their local knowledge and *feel* were all that was needed to conduct successful exploration, but as surface ore leads became rarer the search for metals turned towards buried deposits where no outcrops were obvious. In order to identify the most promising targets mining exploration companies have, over the years, increasingly utilised technology and science in their search.

Sometimes the technology is used not in actual exploration and mining but in more prosaic ways, such as the introduction of aircraft to carry prospectors and equipment into the remote Red Lake gold camp in Ontario in the mid-1920s. However, this transport innovation had a huge effect on expenses, both direct and indirect, and on the productivity of exploration.

EDISON'S MAGNETOMETER

One of the earliest examples of technology helping in the uncovering of a major orebody was the identification by Thomas Edison of Falconbridge's original Sudbury nickel deposit in 1901 using an electrical dip needle, a crude early magnetometer, which measures the level of magnetic intensity of buried rock strata. The basic idea is to identify anomalies lying at depth; such anomalies indicate a departure from the norm with regard to the underlying geology, which could be explained by the anomolies being mineralised bodies. Unfortunately anomalies can also be associated with water, magnetite and pyrite; useful things to find sometimes, occasionally very informative, but perhaps not always of primary interest in early-stage exploration.

GEOPHYSICS

Geophysics was adopted in mineral exploration as early as the 1930s, but it was not particularly successful, as many users did not back their programmes with proper geological evaluation. Thus they surveyed and then drilled ground with limited geological promise, and then were disappointed and even angered by their failure to find economic mineralisation. In due course exploration companies began to pay more attention to the geology of their claims and at the same time technical developments improved the efficacy of the geophysics. Better geology and better technology meant that after the Second World War geophysics and geochemistry became increasingly important tools in the explorer's strategy.

Initially, geophysics was applied at ground level, with lines being cleared in the bush and then seismic or electrical techniques would be used to map the responses from depth along these lines. In North America this activity was described as doodlebugging and had a direct historical link back to the use of divining rods in mineral exploration, as described by Agricola in the 16th century. By the mid-1940s airborne geophysics had increasingly become the norm as it allowed large areas to be surveyed relatively quickly, and with much more sophisticated equipment more useful information was gathered to assist drilling decisions. This improvement, of course, is a continuing process and in recent years, for example, the technique of induced polarisation has become widespread as a further method of enhancing airborne geophysical data.

MAPPING

Another element in an increasingly scientific approach to exploration was the investment in geological mapping made by governments, both local and central, around the world as the 20th century wound on. The work done was usually of a high standard and provided a raft of information for exploration companies seeking prospective ground to investigate. Mapping also provided a solid base on which to make decisions about the most promising areas for geophysical and geochemical work. The use of geochemistry to improve the chances of identifying likely areas of mineralisation dates back to the mid-20th century.

Rudimentary geochemical techniques have been used for centuries where old timers (and ancient timers) have used surface staining, for example, as a guide to possible mineralisation at depth – the development of the Zambian copperbelt was much helped by the observation of staining from oxidised copper. As the years passed, more sophisticated analysis was adopted to search for surface indicators that scientific work had identified as prospective for this or that mineral. For example, in the search for diamond kimberlites today, geologists and prospectors look for garnets (G 10 ones) as a good indicator for diamonds – though glacial movement may have shifted surface indicators a long way from the mineralised source, as happened in the Canadian diamond hunt.

ORE SORTING

Improving efficiency has also been a big motivator over the last 50 years in the application of new technology in mining. One example is in the area of ore sorting. In particular, in the South African gold and platinum mining sectors, ore is extracted from the face either by blasting or occasionally by cutting; in both cases it is impossible to avoid bringing at least some barren or uneconomic ore to the surface for treatment. So early on some mines practised hand sorting to try to avoid waste rock from entering the treatment process. The practice of sorting has a long history but until fairly recently hand sorting by keen-eyed staff was the main method used. The ore was loaded onto a beltway and then sorted by hand with discarded material being diverted to the waste pile and consequently higher-grade material going to the mill. The advantage of sorting was that it increased the grade of ore being milled, boosting operating margins, and it also allowed for a reduced throughput with a smaller mill and thus lower operating and financing costs.

Hand sorting was inefficient in that the human eye was not precise enough, particularly when tired, and too much waste ore was processed. Still the idea was a good one in principle and was just in need of technical improvement. Automated sorting devices go back to at least the 1930s but have become widely used in recent decades. Early devices used a radiometric process and this was a useful advance in the uranium and gold mining industry where radioactive material was present. In due course more advanced equipment was developed using sensors, and then lasers, to measure metal content in the ore and select higher-grade material for the mill. Environmental pressures have also increased in recent years, forcing mines to find ways of reducing waste going into the extremely toxic tailing ponds; enhanced sorting, which allows more rock to be left at the mine, is one way of limiting the size of the ponds.

SHAFT SINKING

Early mining often consisted of digging a hole and mining the ore by hand. In due course miners noticed that the ore often carried on down, but they were restrained by their ability to create deep holes and as we have seen earlier, in the past shafts usually only went down 100 feet or so. The normal method of opening up the shaft was to dig down taking the spoil out by hand in baskets, using ladders to reach the surface. In due course basic hoists were erected to allow dirt to be brought to the surface more efficiently by rope and pulley. This method was also used to bring out ore. In the 18th and 19th centuries steam and then electricity led to a step change in mechanised operation of shaft hoisting. During the 19th century the first incline shafts were being dug, which enabled rail tracks to be laid into the mine for ore to be hauled to the surface; we go into the advance of mechanisation more fully below.

One of the most important advances in shaft sinking was the use of concrete in lining deep level shafts. Although concrete has been known and used for many centuries, the arrival of stressed steel struts, which could be buried in the concrete as it dried, gave it strength and so allowed for shafts to be sunk deeper, without the risk of them collapsing. Before that a variety of methods for lining shafts were developed, including wooden and then brick tubbing, and by the early 19th century iron tubbing was being used in German and Northumbrian coalmines. The main problems requiring sealing of the shaft were sand and water inflows. In the shaft sinking process the use of explosives had been introduced in the late 18th century, which speeded up the process of shaft deepening, especially where the miners were digging through hard rock. Great care had to be taken in terms of the size of charge so as not to damage the shaft already dug and endanger the lives of the miners doing the sinking.

Today shafts can be sunk deep – the South Deeps shafts on South Africa's Witwatersrand go down almost 10,000 feet in one single lift and even with modern techniques these took over six years to sink and another year and a half to equip. The shafts themselves are huge, allowing several functions to be carried out in them – transport of miners to the working faces, ventilation, carriage of supplies and raising of ore to the surface. The physical digging out of shafts has been replaced by the use of power drillers which can drill several holes at once to a depth of around 20 feet enabling explosives to be fitted and set. However, the blown material has to be cleared to the surface to allow the sinking to proceed to the target depth, which is a laborious process.

Another late 19th century advance in shaft sinking was the process of freezing ground, particularly where it held a lot of water, then breaking up and extracting the dirt like solid rock. Although there have been many advances in machinery in shaft sinking, it is one of those tasks in mining which cannot be circumvented – if you want to access deep level ore you have to dig down to it, and as you go you have to dispose of the shaft spoil by hauling it to the surface. There are no short cuts.

MECHANISATION

In the past mining has been an industry where labour has usually been cheap, sometimes due to the lack of alternative employment, sometimes due to historically low wage work in the area – often farming – and sometimes due to the rudimentary, and therefore low-value, skills required of miners. Here we are talking about labour required to work substantial developed mines and not the entrepreneurial miners of the past who, for instance, followed the great gold rushes of the 19th century and expected to stake their own claims and find gold and consequently great wealth. Such prospectors were drawn from all levels of society and worked solely for themselves, either alone or in partnerships.

MECHANICAL LIFTS AND HOISTS

Over the last few decades this has changed markedly, partly for social and partly for technical reasons, and rising wage costs have led mines to develop mechanised mining methods. One of the earliest examples of this was the replacement of ladders as a method of accessing mines with lifts or cages raised by electrically-powered hoists supported by surface head frames. This was particularly essential as mines went deeper; that trait in itself only made possible by the use of refrigeration for deep-level cooling and the development of deep level techniques to drive passageways and install secure roof supports. The onset of inflation after the Second World War and the rising influence of trade unions meant that, in the Western world at least, wage levels rose markedly. In the mining industry, where formerly large workforces had often been deployed, mechanisation was increasingly introduced as a more efficient and safer alternative to raw strength. So mines invested in underground rail systems to haul ore to stations where it could be brought to surface quickly by cage.

DECLINES

Another development technique introduced was in mine design where, depending on depth and the structure of the orebody, declines (roadways) are built which allow large pick-up trucks to drive down into the orebody and take ore directly from the mine face back up to the surface. Although declines have much larger dimensions than shafts, they are easier to excavate and there are considerable advantages on the cost side in labour saving, safety and access time.

OPEN PIT MINES

A further design advance was in the building of deep open pit mines. In the distant past, small open pit operations had been the mining norm and it was only advances in areas such as shaft sinking and refrigeration that enabled deep level mining to be contemplated. Later, the development of large vehicles made it possible to lift and move huge quantities of dirt/ore and therefore meant that mining companies could look at lower-grade surface deposits and, of course, at very remote but high-grade prospects, such as the Western Australian iron ore and Wyoming coal deposits. The ability to design and create stable benches within the open pit as payable ore was removed, enabled the huge haulage trucks and diggers to descend many hundreds of feet following the mineralisation.

Another option was the creation of huge open pits from contiguous and relatively small and shallow old underground mines which had closed, as they could no longer be worked economically. An example of this is the super pit in Kalgoorlie, Western Australia, which has been developed into one huge operation from the old gold mines of the legendary Golden Mile.

MINING MACHINERY

One of the key elements in the advances made in mining over the last hundred years or so has been the establishment and growth of a variety of technologically-driven mining equipment manufacturers around the world. It is their engineering, technical and design skills that have enabled mining companies to meet the fast rising demand for raw materials by exploiting, amongst others, deposits previously too difficult to bring to production. The table shows international mining equipment manufacturers and their specialisms.

INTERNATIONAL MINING EQUIPMENT MANUFACTURERS

Manufacturer	Date of original foundation	Country of incorporation	Main mining equipment activities
ABB	1883	Switzerland	Mechanical drive systems
Atlas Copco	1873	Sweden	Haulage and drilling equipment
BELAZ	1948	Belarus	Haulage trucks
Boart Longyear	1903	USA	Drilling equipment
Brunner&Lay	1882	USA	Rock drilling tools
Bucyrus	1880	USA	Walking draglines
Caterpillar	1925	USA	Large scale haulage trucks
Cummins	1919	USA	Engines
Dyno Nobel	1865	Norway	Explosives
Euclid-Hitachi	1907	USA	Excavators, haulage trucks
F L Smidth	1882	Denmark	Materials handling
Famur	1922	Poland	Underground coal supports
Joy Mining	1919	USA	Loaders, continuous miners
Komatsu	1917	Japan	Excavators, haulage trucks
Kopex	1962	Poland	Underground supports, shaft sinking
LeTourneau	1929	USA	Large wheel loaders
Liebherr	1949	Germany	Excavators, haulage trucks
Nordberg-Metso	1886	USA	Crushers
OMZ	1933	Russia	Crushers, electric shovels, drill rigs
Outotec	1910	Finland	Metals and minerals processing
P&H Mining	1884	USA	Excavators, draglines and drill rigs
Padley&Venables	1911	UK	Drill tools
Rockmore	1948	USA	Drill bits
Sandvik	1862	Sweden	Drill rigs, crushers, loaders
Siemens	1847	Germany	Mechanical drive systems
Takraf	1725	Germany	Continuous mining systems, crushers
Telsmith	1906	USA	Crushers
Terex	1925	USA	Materials handling. Excavators
ThyssenKrupp	1811	Germany	Crushers, mills, materials handling

Source: *Mining Magazine* 2009

Many of the biggest companies were formed in the USA in the 19th century as the economy began decades of rapid growth, supported by the country's then massive raw material resources. The falling away of the US's position in the world of mining over more recent decades has not materially damaged the prospects of companies such as Caterpillar, Boart or Nordberg, as they became international businesses many years ago.

A lot of what the giant mining machinery companies do now is what they have done for most of their corporate existence; they just do it more efficiently, on a greater scale and more reliably. The watchword is often *smart,* with emphasis on improvement in equipment design and operating. After all, in the world of mine development the challenges do not regularly change but the solutions often do. Nordberg's ore crushers, crushing ore into a variety of sizes for the mill, have been central to its offerings almost from the beginning, in 1886. But over the years, starting with its cone crusher in 1926 and followed by its gyratory crusher in 1946 it has upgraded its products to make them more reliable and less likely to need expensive downtime for repairs, but most importantly it has made them faster to allow for greater throughput for much the same cost.

The huge haulage trucks of Caterpillar and excavators of Euclid have transformed open pit mining – concepts like Kalgoorlie's Superpit would be impossible without the huge machines available today. The same thing could be said of the walking draglines of American firm Bucyrus, which have revolutionised open-cast coal mining in places like Wyoming and Queensland. The scale of the manufacturing process and investment, as well as the demands from customers for a range of mining equipment, has also made itself felt at the corporate level in recent years. Bucyrus, having acquired Terex in 2009, has now been acquired by Caterpillar.

Many of the great mining machinery companies started life as manufacturers of equipment for construction, the emerging communications sectors, power and transport – the new industries of the 19th century – and many continue to generate substantial, even dominant, revenues from non-mining activities. The founding of Atlas Copco in Sweden in 1873 was built on the expansion of the railways and although the company these days derives around a quarter of its revenues from sales to the global mining industry, particularly from the manufacture of rock drills, its general engineering activities are financially more significant. The same would be true of ABB and Siemens, specialists in the provision of electrical-drive systems for mining, but much bigger in non-mining areas.

Interestingly, Outotec started life in 1910 as the Finnish mining company Outokumpu, with its large copper mines in Karelia in the east of the country. Over the century it gradually built up its technical processing side as a support for its copper mining activities and then in the 1990s, following the closure of the Karelian mines – which had produced over a million tonnes of copper during their lives – it began to branch out through acquisitions into associated mining support activities, including processing, engineering, grinding, and plant design and construction.

Lower grades and rising demand for metals will lead to a continuing need for mining equipment companies to increase the size of their mining machinery and improve both their grade control and treatment practices. The inexorable march of environmental legislation will also lead to an accelerating demand for treatment and disposal techniques to become ever more efficient when dealing with the issue of pollution during a mine's life and reclamation when it closes. To this end Outotec, for example, has in recent years expanded its minerals processing and service operations mentioned above. Over the next few decades raw materials industries as a whole will be paying far more attention to this part of their activities than may have been the case 50 years ago.

4. DIAMONDS

We have seen how over the centuries metals have played a key part in technological development. This was development of an increasingly sophisticated kind as the advance of the industrial revolution gathered pace in the 19th and 20th centuries. Precious minerals have also been part of this process, in particular diamonds. The common image of diamonds is as jewellery used to adorn beautiful and rich women, which made diamonds desirable, but hardly essential, in the process of human advance. However, along with these gem stones that were fashioned into jewellery, there were a huge number of much less valuable, industrial diamonds, and it was they that had a key role to play in the advance of technology. Later we will see how the diamond industry's giant monopolist, De Beers, courted controversy as it fought to maintain its control over these industrially vital diamonds.

The discovery of diamonds in South Africa in the 19th century heralded the arrival of that country as a major global mineral province and also introduced the world to a group of buccaneers who became known as the Randlords. We earlier saw how these often larger-than-life characters not only financed and developed the diamond fields around Kimberley, but also played an enormous part in the development of the Witwatersrand gold fields. But the amalgamation of the Kimberley mines at the end of the 19th century was in many ways just the start of the diamond story and in the 20th century the story progressed in a way that the Randlords could not have foreseen.

SOUTH WEST AFRICA

In 1908 diamonds were discovered in German controlled South West Africa, known today as Namibia. The discovery was not particularly well timed as the US had experienced an economic slump the previous year and demand for diamonds in the US had collapsed. At the same time the diamond market, although supposedly controlled by the De Beers syndicate established by Cecil Rhodes, was facing an uncooperative major producer in the form of the Premier Mine, the source of the famous Cullinan diamond. This was a South African mine, but was located in the Transvaal, considerably to the north of Kimberley. Premier, which eventually became part of the De Beers group, was then owned by Sir Thomas Cullinan and had its own selling office in London, and despite the poor economic conditions at that time did not curtail its production.

DIAMONDS IN THE DESERT

The discovery of diamonds in German South West Africa (SWA) was made by Zacharius Lewala, a Coloured labourer from the Cape working on the railway that ran east from the port of Luderitz through the desert to Keetmanshoop. He was working under a German, August Stauch, who had acquired prospecting licences from the German Colonial Company for South West Africa (the DKG) and who had asked his workers to look out for any interesting minerals that they might pick up in the course of their work on the railway line.

Stauch, whose wife and two children had been left behind in Germany when he joined the railway and sailed out to SWA, had made a study of the desert around Luderitz to fill in lonely hours. This held him in very good stead when Lewala brought him the discovery stone. De Beers was extremely interested in the news of diamonds at Luderitz and commissioned Hans Merensky, eventual discoverer of platinum in SA but then a young geologist, to go and investigate.

When Stauch announced that diamonds had been found in the desert at Kolmanskop there had been considerable scepticism. The general view was that diamonds were found buried in kimberlite pipes not lying around the desert, but when Stauch, who after Lewala's find had accelerated his land pegging around Kolmanskop, brought a sample of diamonds he had found into Luderitz and started giving them away the townsfolk began to take Stauch's claims more seriously.

This started a pegging rush which came to the notice of the German colonial authorities who themselves sent a team to Luderitz to investigate and to peg land. Merensky's arrival added expert analysis to the frenzied activities of the new diamond miners; his investigations led him to postulate that the diamonds at Kolmanskop came from kimberlites under the sea. He was half right, for we now know that the diamonds came from an inland source and over the millennia had been washed downriver out to sea where the tide and waves had then washed some of them back on shore, to be buried under the ever shifting sand.

AUGUST STAUCH (1878-1947)

August Stauch was born in Ettenhausen, Thuringia, in central Germany in 1878. His family were poor peasant farmers but staunch Protestants and Stauch had to pursue his education between long periods working on his father's smallholding. He had one brother, two half brothers and two half sisters, his father, Andreas, being married three times. In 1892 Stauch joined the German Pioneer Corps where he became involved in military engineering projects. After military service he became a land surveyor with the construction firm Lenz & Co. It was at this time that Stauch met Ida Schwerin and in 1904 married her in Franzburg and started a family; a son and a daughter were born in 1905 and 1906 respectively. Unfortunately, Stauch suffered badly from asthma so he took a posting with Lenz, a railway construction job in German South West Africa where the climate was kinder, although he had to leave his family behind.

Stauch worked as a railway gang supervisor on the Luderitz (on the SWA coast) to Keetmanshoop railway line, which went across the desert. His was a very solitary life, but he was extremely interested in the remote area in which he lived and this was to pay huge dividends in time. In 1908, believing that the area around Luderitz had mineral potential, Stauch purchased two prospecting licences from the DKG. One of his workers found a diamond and Stauch then pegged a number of other leases which threw up more diamond discoveries. To begin with there was considerable scepticism about the finds, but quickly a pegging rush developed and at this stage the German government – in the form of the DKG – stepped in to bring the new diamond fields under its control. Stauch was able to keep his original leases which he had obtained and worked, but he pushed out further into the desert to avoid the DKG's dead hand and that same year made the fabulously rich Pomona discovery.

In 1909 he returned to Germany, was reunited with his family and in due course another son was born. Stauch and his financial backers had become immensely wealthy from the Luderitz diamond discoveries and Stauch decided that he would divide his time between his business interests in SWA and his family in Germany, and in 1912 another son was born. Stauch's diamond discoveries had turned remote Luderitz into a huge contributor to SWA's prosperity – and to the government back in Berlin – and he continued to support, through his leadership and hard work, the growth of the town. Back in Germany Stauch had moved his family to a lavish house in the Berlin suburb of Zehlendorf, but in Luderitz for some years he lived simply in the local hotel. Stauch had also become active in local politics and as a further demonstration of his commitment to SWA he acquired a farm, Haribes, in the Gibeon district inland east of Luderitz in 1911 and he invested heavily in turning the farm into a substantial agricultural enterprise.

With the coming of the war in 1914, Stauch's luck changed for the worse. The Luderitz diamond mines were immediately shut, although with the defeat of the Germans in SWA by South Africa in 1915 a partial re-opening was permitted. During the war Stauch was in Germany but in 1918 he was forced to face the reality of a non-German SWA and the determination of the South Africans, in the form of Anglo American, to acquire the Luderitz diamond operations. In the end Stauch and his associates received £3.5 million in cash and shares in the new Anglo American controlled company, CDM, which acquired the Luderitz diamond operations.

In 1923 Stauch became chairman of AGV, the German railway company that had had a substantial stake in Luderitz diamonds, but his heart was in SWA and he resigned the following year to pursue his business interests there. Over the years these became extensive as he encouraged Germans to emigrate to SWA and settle on land which he had acquired. As an adjunct of this agricultural investment, Stauch set up a cattle and dairy distribution businesses. He also expanded his mining interests into tin, with limited success. In 1923 he brought all his SWA business activities under the corporate umbrella of South West African Trust. Back in Germany, Stauch established the Vox Company, which became a manufacturer of tape recorders and gramophone records. In 1927 Stauch persuaded his family to make SWA their home, settling on his farm at Dordabis, although his children continued to be educated in Germany.

With the 1930s depression Stauch's businesses went into rapid decline and in 1935 he was effectively bankrupt and all his assets had to be sold to pay his creditors. His family had kept Dordabis and Stauch lived there where he started to study science with the ambition to prove Einstein's Theory of Relativity wrong. He failed, but the quality of his work was noted by many in the scientific community. In 1938 Stauch returned to Germany to study science at Breslau University. The war saw Stauch's health begin to fail; he was trapped in Germany, separated from his family and in 1945, wanting to return to SWA, he found he had been de-naturalised. He had stomach cancer and in 1947 entered hospital in Eisenach where he died in May, with only his youngest daughter, Kathe, at his side.

Although Stauch died a poor and lonely man, broken by first the economic slump of the 1930s and then the Second World War, he was one of the giants of the African mining scene. His personal fortune had been measured at over £100 million (in current money terms) in 1914. A sometimes volatile man with enormous energy, he was loved by his family and liked and admired by friends and business associates in both Germany and SWA. It is ironic that Stauch died as poor as when he was born having, as a young man, accumulated wealth which put him on a par with South Africa's Randlords.

THE GERMAN AUTHORITIES TAKE AN INTEREST

To begin with, the SWA diamond discovery did not meet with much interest back in Germany and investors and banks were reluctant to become financially involved. This suited Stauch and other parties in the colony, but as further news of additional discoveries and also rumours that De Beers was anxious to find a way into the developing play, if only to try and exercise control over the new supply heading for world markets, the government in Berlin decided that it should intervene.

Bernhard Dernburg, a top German banker, was appointed Colonial Secretary and sent to SWA to report back on the territory's diamond potential. Although his visit was generally welcomed by the mine owners, his proposals, which he quickly implemented, were far less welcome. A very large area of the desert called the Sperrgebiet was reserved for the DKG – this encompassed the Kolmanskop diamond finds – although those like Stauch who already had legally recognised leases were allowed to keep them and continue mining. Fledgling diamond entrepreneurs in Luderitz were forced to look further afield to the north for new leases.

As part of the new dispensation, the handling of uncut diamonds from the private companies was regulated by Berlin, which also fixed production quotas. In due course a central buying and marketing organisation, the Regie, was established in Germany, which became the only route for diamond producers, both inside and outside the Sperrgebiet, to sell their diamonds. The prices offered by the Regie were often not competitive with those available from the Diamond Syndicate in London, which was controlled by De Beers.

As a further blow to the registered diamond producers in the Sperrgebiet, the state established its own mining company, the Diamond Lease Company, to work its leases; Dernburg also supported the setting up of a German-based company, the South West African Mining Syndicate, to explore for and develop other diamond properties in SWA. However within this tight structure August Stauch came off quite well, for not only did he have his original leases, but he also had a contract from Diamond Lease to mine their claims.

THE POMONA DISCOVERY

Stauch's attention to detail and hard work in promptly registering his claims paid off in spades, as one of his leases contained the fabulous Pomona diamond find. The Pomona discovery was made by Stauch and Professor Robert Scheibe from the Royal Mining Academy in Berlin, a diamond expert, who in 1908 was spending a sabbatical year in SWA. Scheibe had sought out Stauch on his arrival from Germany and the two men decided to explore south of Luderitz in an area named Pomona after the island which lay offshore. Stauch, Scheibe and their party – after a very difficult trek into the desert – set up camp in a valley which Stauch named after his wife, Ida.

When one considers the effort that goes into diamond exploration today, the immediate success of Stauch and Scheibe was like a fairy tale. On their first full day exploring an adjoining valley, to their amazement, they came across a huge area where they found diamonds lying on the floor of the valley. The discovery was one of the largest ever made and over a million carats were *mined* in the first two years. Stauch, ever a stickler for procedure, swiftly pegged several leases and rushed back to Luderitz to register his claims. The story he told led to an immediate rush to Pomona and back in Luderitz the local stock exchange went mad.

Unfortunately, the discovery was so enormous and attracted so much attention that it reminded the British company, De Pass and Co, that it already held leases over the Pomona area. De Pass's claims actually related to silver and lead and when the area was prospected in the 1860s no economic discoveries were made. The De Pass Company simply forgot about the claims but when Stauch and Scheibe made their diamond find Daniel de Pass, chairman of the company, quickly became aware of the excitement and started legal action to enforce the company's rights over the diamond ground.

After considerable legal expense, both in Europe and SWA, an agreement was reached with De Pass whereby an option was granted to Stauch and other interested parties to mine Pomona for diamonds in exchange for a gross sales royalty of 9%. The dispute, which had led to a complete stop on any mining in the area until its resolution, dragged on a little longer as De Pass had Cape Town partners in the original Pomona project. They objected to the deal, as they believed they had rights which De Pass was ignoring. After further expensive litigation a new structure was agreed that gave the Cape Town group a small stake, but Stauch came out the major shareholder with a near 50% interest in Pomona Diamonds and his old partner, Scheibe, had a 21% interest.

What is particularly interesting about the development of the SWA diamond industry was the speed with which it happened. Of course the fact that the diamonds were initially found lying on or very near to the surface was a major help. The discoveries were quickly developed; with no deep shafts required it was merely a matter of designing a basic plant to wash and separate the gravel and diamonds. To begin with these plants were small but as production multiplied they became bigger and more complex.

DE BEERS REACHES A MARKETING AGREEMENT...

The rapid expansion attracted the interest of De Beers, which initially had tended to belittle the importance of the discoveries; a reaction that it was to repeat decades later when Canadian diamonds were first discovered at the end of the 20th century. De Beers was concerned that rising production and difficult economic circumstances in consumer markets would lead to a slump in diamond prices and therefore in profits. It approached the SWA producers and the Regie and after extended negotiation reached agreement for a diamond pool to be set up covering South African and SWA production and for the output to be marketed through the London Diamond Syndicate. The agreement was reached in 1914 but the outbreak of the war later in the year changed everything.

...BUT ANGLO AMERICAN TAKE OVER THE MINES

With the outbreak of the First World War diamond production in SWA ceased and the black and Cape coloured workers were sent home. After South African forces entered German SWA in September 1914, hostilities between them and the Germans continued until Germany surrendered in July 1915 and SWA came under the control of South Africa; a situation that continued until Namibian independence in 1990. Mining on a small scale re-started in late 1915 in SWA, with the output being sold to the National Bank of South Africa; two years later in 1917 SWA diamonds again began to be sold to the Diamond Syndicate in London through De Beers.

After the end of the war, with SWA now in the hands of the South African government, a new deal for the future of the diamond industry was worked out. Unsurprisingly, the idea was to vest all the SWA diamond mines in a new company where a South African entity would be the major shareholder and the original SWA interests, including August Stauch, would receive cash and shares in the new company. Surprisingly the South African entity was not De Beers, who had thought that the mines would naturally fall into its lap, but Ernest Oppenheimer and his Anglo American Corporation.

For many years Oppenheimer, whose early career had been in the diamond industry, had harboured a wish to become a director of De Beers and he had been re-buffed – after all, he was not a miner but a diamond trader. So using Anglo American, which he had partly set up to protect the South African gold assets of German investors whose status had become precarious with the outbreak of war, Oppenheimer gained control of the SWA diamond industry. In 1926 he sold Consolidated Diamond Mines of South West Africa (CDM), the new company set up to mine SWA's diamond fields, to De Beers in exchange for a large stake in De Beers and a seat on the board. Once inside De Beers, he and his family members built up the Oppenheimer stake in the company, enabling him to become chairman in 1929 – he thereby achieved his ambition of controlling De Beers and within his portfolio of interests added diamonds to gold.

SIR ERNEST OPPENHEIMER (1880-1957)

It is impossible when considering mining in the 20th century not to place the Oppenheimer family at the centre of the development of the South African industry, one that is pre-eminent in the production of precious metals. Sir Ernest Oppenheimer played a crucial role in establishing the Anglo American group and, as Chairman of De Beers, in organising the modern diamond-trading cartel, the Central Selling Organisation, now much reformed.

Sir Ernest was born in 1880 in Freidberg, Germany, where his father Edward was a cigar merchant. The Oppenheimers were a large German Jewish family with excellent connections, particularly in the diamond business in England. When he was 16 he went to England and started work as a clerk in the London office of diamond merchant A. Dunkelsbuhler, who was his cousin, and became a naturalised Briton. In 1902 he was sent to Kimberley to run the South African buying office of Dunkelsbuhler where two of his brothers were already ensconced. From there his career took off, but it was gold that provided his initial launch pad. He acquired gold properties in the Transvaal for a group of German investors and in so doing took a participation in the properties, plus an option to increase his stake in the future. During the First World War, to avoid his partners' interests being expropriated, he injected them into a new internationally-financed company, Anglo American, in 1917. He also became Mayor of Kimberley during the war, formed the Kimberley Regiment and was knighted in 1921 by George V for war services in South Africa.

Although born a Jew, Sir Ernest converted to Christianity in the 1920s. He had married Mary Pollock, a London girl, in 1906 and they had two sons. One of them, Harry, was to be Sir Ernest's successor to the family's commercial dynasty. Mary, then Lady Oppenheimer, died in 1932, and in 1934 Sir Ernest married Caroline Harvey.

Fired by his gold successes, Sir Ernest set out to rationalise and then control the South African diamond industry and in particular, De Beers – the creation of Cecil Rhodes. By 1929 Sir Ernest had achieved his aim and became Chairman of De Beers. During the Depression years of the 1930s the world diamond industry came close to collapse and Sir Ernest devised a single-channel marketing concept, or cartel, to bring the selling of the majority of the world's diamonds under the control of De Beers. This cartel survived until the early 2000s, when new discoveries outside the control of De Beers – and the decision of the group to stop being the buyer of last resort – led to the establishment of the Diamond Trading Company as the group's marketing arm. After the end of the Second World War Sir Ernest consolidated the position of both Anglo American and De Beers; in the latter years of his life his son Harry took over the reins of both companies. Sir Ernest died in 1957, having established two of the world's largest mining companies in the short space of 40 years.

Although Sir Ernest never achieved the notoriety of Cecil Rhodes or Barney Barnato, he was a ruthless operator; if he had not been it is unlikely that he would have been able to bring order to the unruly South African diamond industry. Some of his business practices were, however, controversial, particularly his protection of his German investors in the First World War and the establishment of the diamond cartel. His unhelpful, in the view of the American government, attitude towards the US demand for secure supplies of industrial diamonds during the Second World War was another black mark against him, although the ban on De Beers doing business in the US, as a result of the diamond cartel, has now been lifted.

Away from business Sir Ernest was particularly generous in his endowments to Stellenbosch University in SA and to Oxford University. He also plowed large amounts of money into decent housing for mine workers and Africans displaced by apartheid, who were forced to live in townships like Soweto. The Ernest Oppenheimer Trust was established after Sir Ernest's death to help the disadvantaged, providing support in both education and housing. With Anglo American now a British company and De Beers subsumed and much changed within it, Sir Ernest's commercial empire has moved on. However, like Cecil Rhodes before him, Sir Ernest remains one of the great figures in the development of the modern mining industry.

NAMAQUALAND OPENS UP

The discoveries around Luderitz stimulated a wider diamond exploration effort in the years following the end of the First World War. They had also caught the attention of South African prospectors, who reasoned that if diamonds existed in SWA, they might well be found in the desert area south of the border in South Africa, in Namaqualand. Amongst those who were involved in this new diamond exploration effort was Hans Merensky who, following his earlier visit to Luderitz, had made the first discoveries of platinum in South Africa. Merensky engaged the services of Dr Ernst Reuning, a German who had been looking at the platinum discoveries around Potgietersrust, to go to Namaqualand and peg diamond leases. This followed the initial diamond discoveries by a number of prospectors, including William Carstens and Solomon Rabinowitz. Merensky and Reuning were drawn to the area around Alexander Bay where diamonds had already been discovered and they purchased substantial claims in the area.

Early in 1927 Merensky and Reuning formed a syndicate, the HM Association – named after Merensky – to finance a major diamond digging operation at Alexander Bay. As with the mines in SWA, the equipment needed to carry out what was a fairly rudimentary surface pit mining operation was initially quite modest. A significant mine, however, would require much larger capital input. Merensky, through the HM Association, obtained the required finance, ending up a 50% holder in the syndicate. As the operations expanded it became clear to Merensky and Reuning that they had found another potentially massive source of diamonds, something that clearly was going to interest De Beers.

Merensky decided it would be prudent in the light of a developing diamond rush around Alexander Bay to approach the South African prime minister, General Herzog, and urge him to take control of the whole Namaqualand diamond region in order to forestall over-mining; a prudent idea, as the world was facing hard economic times. This Herzog did and in due course Merensky was paid over £1 million when the SA government finally bought out the HM Association. The state ownership of the Alexander Bay diamond operations continues today in the form of Alexcor Ltd, a public company wholly owned by the SA government.

DE BEERS TAKES AN INTEREST

In the meantime, the Alexander Bay discoveries had inevitably attracted the attention of Ernest Oppenheimer and De Beers. Oppenheimer had acquired a large stake in the Association for over £500,000 and also set up the Cape Coast Exploration Company (CCEC) to explore for diamonds. CCEC remained outside the government's control and eventually in the 1940s was incorporated into De Beers. The Alexander Bay discoveries also inevitably set CDM thinking about prospects in the coastal desert on the SWA side of the border, south of the Luderitz diggings.

In 1928 De Beers's exploration team found diamonds in huge quantities, stretching for miles up the coast towards Kolmanskop; diamonds that had been washed out of the interior down river courses and then re-deposited over millions of years back on land. It is believed that since the start of mining in 1928 the CDM operation has mined over 100 million carats. The discoveries formed the base for CDM's operations for the rest of the century and were De Beers' biggest source of profits for decades.

As with the diamonds around Luderitz, the Alexander Bay stones lay close to the surface and were straightforward to mine – the diamonds were captured in gravel terraces, which simplified their extraction. In time the operations were scaled up and literally millions of tonnes of sand had to be moved to get at the diamonds which lay in the deeper terraces. The system CDM used then, and still uses today, was to build huge sand dunes to the seaward side of the gravels to keep the sea at bay, using the overburden to reveal the gravels which increasingly lay below sea level. CDM's operations continue today along these terraces, although as onshore production slips, De Beers is increasingly turning to offshore sea diamonds.

OFFSHORE DIAMONDS

The development of the Namibian offshore diamond mining industry has a patchy history going back to the 1960s when Sam Collins, an American offshore oil-drilling contractor, became interested in the potential for offshore diamonds in what was then SWA. There had been a legal ruling that the monopoly De Beers had over the onshore diamond terraces that stretched from the Orange River border with South Africa to Luderitz, did not apply to diamonds that lay below the tide-line. This ruling meant that offshore diamonds could be mined by independent concerns. Sam Collins and his Marine Diamond Corporation (MDC) pioneered offshore mining using a converted British salvage tug as an exploratory vessel. The mining method tested used suction pipes to suck up diamonds from suitable locations where diamonds might be trapped in gullies. The initial efforts were unsuccessful but after many months of exploration, in 1961, diamonds were finally found offshore.

This discovery led to Collins building a more appropriate barge-type vessel that could suck up and process the offshore gravels. At this stage De Beers was merely watching the Collins project but General Mining and Anglovaal, two smaller SA mining groups, had already taken stakes in Collins's MDC. By 1963 the sea diamond operation was consistently producing small but excellent quality diamonds and De Beers finally became involved by providing Marine Diamonds with a loan for new equipment to expand its operations. Then after a golden patch in the mid-1960s, when as much as 30,000 carats per month were sucked from the sea floor, production began to wane.

In 1965, with Collins's MDC in difficult straits financially, De Beers struck. Collins was unable to resist De Beers's offer for the company and having pioneered the concept of sea diamond mining, he sold out. De Beers continued to mine with decreasing success and in 1971 shut down the SWA offshore operations. However, rising demand and prices for diamonds and improvements in technology saw the now Namibian offshore diamond mining industry start to revive in the 1980s, with De Beers and new independent operators beginning to prospect offshore once again. In due course, more effective mining technology and better surveying techniques led to rapidly rising production and profits; now De Beers gets half its Namibian diamond output from offshore Namibia in a joint operation with the Namibian government.

DIAMONDS IN THE SECOND WORLD WAR AND AFTER

Perhaps the period that most starkly illustrated problems with the De Beers approach to diamond marketing was the Second World War. By the time of the 1930s the Great Depression had led to a collapse in demand for gem diamonds and De Beers was forced to curtail supply drastically, by reducing mine output and closing many mines. In doing that it also reduced the supply of industrial diamonds. However, as the 1930s wore on and consumer confidence across the developed world remained fragile and very muted, many western governments began to crank up their armament industries as the German Third Reich expanded relentlessly across Europe.

Following the end of the First World War, Germany had lost control of its African colonial holdings, two of which, South West Africa and Tanganyika, were already or were to become major diamond producers. With the rise of Hitler and the onset of the war, this loss became something of a headache as the fast-growing German war machine required substantial quantities of industrial diamonds to shape and produce a whole range of precision components to enable mass production techniques to be applied to making weapons. They were also used to produce components for the electronic side of modern warfare. There was the same demand for growth on the Allied side as Britain, its Empire and Commonwealth, and the come-lately US armed forces quickly re-built their defence structure.

The concern for both sides was that De Beers controlled both mine production and the stockpile of diamonds which had built up as demand fell following the worldwide economic slump of the 1930s. The Americans in particular, noting the early successes of the Germans, were nervous that the stockpile of diamonds which were stored in London could fall into the hands of the Reich and so they tried to persuade De Beers to sell them a large slice of the stockpile. Ever worried about losing control of its market monopoly De Beers declined, but instead offered to supply the US with increased regular supplies. The British suggested that De Beers could move some diamonds to Canada but the US was unenthusiastic and by then, with America in the war and Russia increasingly occupying German forces on the Eastern Front, the air of crisis concerning the fate of England, and therefore the stockpile, faded.

The fallout from this bad tempered scrap continued for many decades as US investigators alleged the uncovering of operating practices associated with De Beers's vice-like grip on both the mining and distribution of diamonds. In the light of its problems over securing adequate supplies of diamonds during the war, US investigators also believed they had traced a smuggling route for diamonds from the Belgian Congo, administered during the war by the Belgian government in exile in London, through north Africa and ultimately to armaments customers in Germany, who were paying through the nose for their stones. The route was from a Belgian mine which the Americans claimed that De Beers effectively controlled. However, it is unlikely that Sir Ernest Oppenheimer, with his Jewish family background, encouraged the supply of diamonds to the Reich. The huge profits to be made all the way along the smuggling chain and the clandestine nature of some of the syndicate's activities made it inevitable that Germany would one way or another obtain vital diamond supplies.

American displeasure with De Beers led to a long period when the company's directors and executives were unable to visit the US as government anti-trust action, particularly aimed at alleged price fixing by De Beers of industrial diamonds, rendered them vulnerable to arrest. This situation remained unresolved for over 50 years with De Beers refusing to acknowledge either US legal jurisdiction over its activities or the justice of the price fixing charge. During that period De Beers, using a New York-based advertising agency, NW Ayer, completely restructured its marketing effort in the US and achieved a huge post-war increase in US jewellery sales. Its own direction of this sales drive was done from just over the border in Canada, where its executives regularly met with NW Ayer and US customers. Finally, in 2005, De Beers settled most of the price fixing charges and its executives could once more safely enter the US.

It is often said that war drives technological advances, as weapon systems become ever more sophisticated, requiring equally sophisticated components for their effective operation. After the war the technology remains but new uses for it are found in order for the patent owners to exploit it commercially. As we have noted, diamonds played a critical role in the Second World War as weapon systems became increasingly complex and sophisticated. After the conflict technological advance speeded up, helped by further localised wars like Korea and Vietnam, the space race, the Cold War and a potent combination of rising populations and strong economic growth.

Technology itself played a material part in elevating diamonds to a central position in the production process of the advanced industrial world. Although industrial diamonds were theoretically plentiful, the great industrial economies required substantial and reliable supplies, and whilst supply fluctuations were acceptable, if not welcome, on the gem side, reliability of supply was vital as far as industrial stones were concerned. In the 1950s the technology to manufacture synthetic industrial quality diamonds was perfected and as the years went by, first General Electric of the US and then De Beers began producing a rising quantity of synthetic diamonds for industrial use. Today synthetic diamonds represent around 80% to 90% of the market in industrial stones. The drive behind this work was based on the discovery in the late 18th century that diamonds were pure carbon, but it took another 150 years for the technology to produce synthetic diamonds from carbon to be commercially viable.

Industrial applications for diamonds cover a range of areas. They have particular importance in drilling; from huge drill bits used in natural resource drilling and tunnelling, to high-precision tools used for metals and other hard materials. They also are used for very fine wire drawing and also extensively in machining metals, ceramics and glass, the watch phrase being harder materials and finer tolerances. This trend is likely to continue into the distant future.

THE AGE OF DE BEERS

The principles for mining and marketing diamonds established by Cecil Rhodes were built upon and with the onset of the 1930s depression they were formalised by Sir Ernest Oppenheimer as he set up a single-channel organisation, the Central Selling Organisation (CSO), to oversee the distribution of rough diamonds, the market for which was in danger, as we have seen, of being flooded with SWA diamonds. However, Sir Ernest, despite his lifetime service to the diamond industry, was not, as we noted above, initially a De Beers man.

He had graduated to having a direct interest in the mining industry through the formation of the Anglo American group whose key holdings were in the South African gold mining industry. He, however, had always harboured ambitions to become involved in diamond mining and, like Cecil Rhodes before, he believed that controlling the supply was the secret to maintaining price levels at the distribution end. Retail prices of jewellery were another thing altogether and in the period between the two world wars economic conditions were very volatile, as they had been in the early years of the 20th century, and this created great uncertainty for both retailers and consumers.

MAINTAINING STABLE DIAMOND PRICES

Whilst De Beers had some success in regulating the supply of diamonds to the market following the restructuring of operations in Kimberley, the discoveries in South West Africa had been large enough to cause headaches for the Diamond Syndicate in London as it came to terms with the fiercely independent Regie in Germany, which had the responsibility for buying and marketing SWA diamonds. This continued after the First World War with the discoveries in Namaqualand and just over the SWA border, described above. But De Beers had to work hard throughout the 20th century to maintain stable markets both in the hard times of the 1930s and the good (inflationary) times towards the end of the century.

In aiming for stable markets, and therefore steadily rising diamond prices, De Beers was materially assisted by the fact that many of the discoveries subsequent to those in SWA were made in British colonies, such as British Guiana (now Guyana) in South America, Sierra Leone, the Gold Coast (Ghana) and Tanganyika (Tanzania) in Africa. In every case attempts were made by the owners to sell the diamond output behind De Beers's back in order to obtain better prices than a selling deal with the CSO would achieve. In the end, by exhorting the colonial authorities and the British government to apply pressure on these break-away miners – Britain used to make a lot of money out of London being the headquarters of De Beers's selling operation – De Beers was able to maintain its position as the sole supplier of diamonds to the diamond-cutting trade in Belgium, the US and, after the Second World War, Israel. De Beers itself also bore down on its own customers amongst the diamond traders and cutters if it thought that they were dealing direct with these new sources of diamonds.

THE WILLIAMSON HEADACHE

One new source of diamonds which fell victim to De Beers's tactics was the Williamson mine in Tanganyika. Discovered by Dr John Williamson, an ex-De Beers geologist, in 1943, the Williamson pipe at Mwadui was the largest pipe discovered at the time. Sampling the pipe took two years, but by 1945 Williamson knew he had a major potential producer on his hands, as did De Beers who offered to buy the project out. Williamson declined the offer and in 1946 production began.

De Beers was concerned that Williamson was such a big producer that it, on its own, might be able to destabilise diamond prices during an economic downturn. De Beers therefore applied a two-prong attack on Williamson, firstly by indicating to major cutters in Antwerp that if they bought diamonds direct from Williamson they risked being cut off from De Beers sales. Secondly, pressure was applied by De Beers on the British government, Tanganyika being a British colony then. The line was, as it had always been, that De Beers needed to control this major source of diamonds or the party might come to an abrupt end if Williamson flooded the diamond market. So with Williamson only able to sell his exceptional diamonds to independent cutters and the British government pressuring the colonial government in Tanganyika to nationalise the mine, the economics of the Williamson pipe were severely threatened. There was little that Williamson could do but succumb to De Beers and in 1947 he sold out to Sir Ernest Oppenheimer. Williamson remained a De Beers mine until it was sold to Petra in 2008.

BOTSWANA

As the British Empire broke asunder, the UK battled to structure the approaching independence of its colonial possessions as fairly and efficiently as possible. Not every situation responded well to the decisions taken by London, but in southern Africa where apartheid South Africa was a particular problem, the British were determined to honour their commitment to the three protectorates – Becuanaland, Lesotho and Swaziland – and not allow them to fall into the hands of South Africa. This turned out to be particularly important for the people of Becuanaland, which as an independent nation was renamed Botswana in 1966.

FINALLY DE BEERS FINDS A DIAMOND MINE

It is said, perhaps not entirely unfairly, that in its long history De Beers has often relied upon independent prospectors or other exploration companies to find diamonds and bring the projects to De Beers for them to be developed into mines. In other cases, mines are developed independently and then taken over by De Beers. Very few mines come about as a result of De Beers's own exploration efforts but the two great diamond mines of Botswana – Orapa and Jwaneng, brought to production in 1971 and 1982 respectively – and the smaller Letlhakane and Damtshaa mines, were all products of the group's own exploration effort, led initially by three De Beers geologists – Gavin Lamont, Manfred Marx and Gim Gibson.

The exploration techniques used in finding Orapa, which is situated in northeastern Botswana, are interesting in that they required some very innovative thinking; something that many would not attribute to the somewhat bureaucratic Anglo American/De Beers group. Originally a small number of diamonds had been found washed up by the Moutlouse River and this led to a group of De Beers geologists, directed by Dr Lamont, spending several years in the area of the source of the Moutlouse looking for the originating kimberlite without success. Eventually, Lamont speculated that the historic source of the diamonds found earlier might lie beyond mountains that separated the river from the Kalahari Desert, the upthrust of the mountain range having buried the river's original source.

This was pure speculation on Lamont's part and the theory was difficult to test because of the deep sand cover of the desert. In the end, by examining the huge white ant mounds in the desert, built with mud retrieved from below the sand cover, the geologists discovered that some carried trace minerals, garnets and ilmenite that are often associated with kimberlites. Drilling targets were identified and cores obtained and then analysed. They had hit kimberlite, and kimberlite which had a high diamond content – this is *the perfect storm* for diamond miners, because most kimberlites do not contain diamonds at all. The kimberlite that had been found was named Orapa after the nearby settlement.

DE BEERS' BIGGEST REVENUE SOURCE

Over time the Botswana operations of De Beers, a 50:50 joint venture with the Botswana government, named Debswana, became the group's largest source of diamonds by far; today it accounts for two-thirds of group production. It was therefore inevitable that as the importance of Botswana to De Beers grew, that the government, which gets almost half its revenues from diamond mining, would become increasingly interested in capturing more of the value added for the country. Apart from diamonds, Botswana's main industries are tourism, agriculture (beef) and base metal mining. The population is around 1.8 million with around 200,000 reasonably well educated, and thus trainable, people living in the capital Gabarone where De Beers has established a major sorting and cutting operation.

Interestingly, in moving such an important part of its value-added business to Botswana, to that country's financial benefit, De Beers has impaired the export earnings of its old friend, ally and original mentor, the UK. The business of importing rough diamonds and then sorting them in London and selling them at London sites (sales) was a very nice earner for the UK's balance of payments. The transfer of much of this profitable business to the old British protectorate is yet another example that the sun has truly set on the British Empire.

DE BEERS AND ASHTON MINING

The discovery of the huge diamondiferous Argyle pipe in northwest Australia in the 1970s, unusually in a lamproite rather than a kimberlite setting, was a major headache for De Beers due to the large number of near gem and industrial diamonds contained in the deposit. De Beers found the Australians, in the form of mine developer Ashton Mining, a tough proposition when it came to negotiating sales agreements. In the end, De Beers bought a lot of poor near-gem stones off

Ashton and paid premium prices for them. Ashton kept a sizeable proportion of the small number of gemstones from Argyle, some of which were highly sought after coloured fancies – the pinks were particularly popular – for its own marketing effort. These fancies were marketed by Argyle entirely separately from De Beers' sales arm, the Diamond Trading Company, although with De Beers' tacit if reluctant approval, and so represented the first crack in the monopoly.

LAC DE GRAS

In the 1990s De Beers faced another and greater challenge to its supplier monopoly status; the discovery of diamonds in Canada. Exploration for diamonds started in Canada in the 1960s but De Beers was late into the game; it only accelerated its programme after BHP/Dia Met's 1991 Lac de Gras discovery, now known as Ekati. The discovery was made by the prospectors, Chuck Fipke and Stu Blusson, who had put their prospects into Dia Met and sold a participation in the project to BHP. Their exploration had been a painstaking ten-year effort based on their belief that the great Canadian Slave craton could host kimberlites. The duo's persistence in following the direction of the glacial till in reverse – which carried classic indicator minerals, such as G10 garnets from the kimberlite source – paid off with the discovery of diamond-bearing kimberlites in the Lac de Gras area of the Northwest Territories.

The whole process was a painstaking one and took many years before the prospectors could be sure they had anything of interest. During this period the De Beers PR position denied the likelihood of there being diamondiferous kimberlites in such an unexplored area as Canada. That didn't, of course, stop the group from acquiring claims both near Lac de Gras and in Saskatchewan. In due course, through joint ventures, corporate activity and even exploration, De Beers also became a diamond mine developer in Canada.

CHUCK FIPKE

Chuck Fipke was born in 1946 in Edmonton in central Alberta, Canada, and is the unchallenged father of the Canadian diamond industry. He grew up in the Okanagan in BC, the fertile farming valley adjacent to the US state of Washington. As a boy he was thought backward due to an overactive mind and a chaotic lifestyle, which earned him the nickname Captain Chaos. He was also given the sobriquet Stumpy, a comment on his lack of height. In 1966 Fipke went to the University of British Columbia to study geology. In his final year in 1970 he ran out of money but obtained CAD$300 personally from Walter Gage, the University President, to complete his studies. In 2006 Fipke remembered this act of generosity and he donated CAD$6 million for a new geological research centre at UBC.

After university he joined Kennecott and was sent to Papua New Guinea to explore for gold and copper. He also worked in South America, Australia and Africa, before joining Superior Oil in 1978 to look for diamonds. Whilst in South Africa Fipke came across the work of John Gurney of the University of Cape Town, who was researching the link between indicator minerals such as chromite and garnets, and the incidence of kimberlite pipes, the source of diamonds. Fipke was intrigued by Gurney's theory and put into practice its concepts when Superior sent him to Colorado to look for diamonds in the northern part of the state. He found a number of kimberlite pipes but none were diamondiferous. Fipke postulated that there was a huge craton, an ancient piece of continental plate, where diamonds formed before being pushed to the surface as kimberlite pipes, beneath the northern US and stretching thousands of miles into Canada's North West Territories. He thus joined up with another geologist, Stu Blusson, to form Dia Met Minerals and headed for Canada's frozen wastes.

Fipke's theory was that earlier unsuccessful work in the region done by De Beers had failed to consider the possibility that indicator minerals, in situ millennia before, had been picked up by ancient glaciers and moved hundreds of miles from the source kimberlite. Although it took many years to prove, Fipke's theory was correct as he and Blusson painstakingly surveyed back along the old glacial trail until they alighted on Lac de Gras. In 1991 Fipke was able to announce that Dia Met had found its first diamondiferous pipe. He and Blusson sold a control stake in Dia Met to BHP for around CAD$700 million each, and each kept a 10% direct stake in what was to become the Ekati mine.

Fipke was a larger than life character, a real frontiersman who played as hard as he worked. He married his wife, Marlene, after leaving university and wherever he went around the world his family came with him. It is said, as evidence of his chaotic lifestyle, that he failed to turn up for his wedding the first time and it had to be re-arranged. Certainly Fipke lived well after making his Dia Met fortune and he developed a passion for horse racing. However, exploration was always in his blood and he established a presence in places such as Brazil, Greenland and Angola. He has also invested a large amount of time in researching the subject of indicator minerals to refine the process of searching for diamondiferous kimberlites.

All this frenetic activity eventually caught up with his marriage and in 2000 Marlene divorced Fipke, gaining a Canadian record settlement of CAD$125 million, largely in the form of Dia Met shares. Six months later BHP bid for Dia Met, giving the divorcee a substantial profit. Fipke himself retained a 10% direct stake in Ekati, providing him with the wherewithal to pursue his personal and commercial interests. Currently he is Chairman of Metalex Ventures, a Canadian junior with a number of interesting but disparate diamond projects. Critics say the unfocused nature of the company mirrors Captain Chaos's long-standing problem of organisational unpredictability. However, it would be unwise to bet against Fipke pulling another diamond rabbit out of the hat.

An unusual aspect of the Canadian diamond boom, which really started to fizz in 1992, was the fact that very poor global economic conditions at the time had led to a serious collapse in diamond sales worldwide. Despite that, a string of discoveries followed on the back of the BHP/Dia Met find and this led to a surge in exploration capital raised in Canada and beyond which went into programmes aimed not only at the Slave craton but at other targets in Canada. Ekati eventually reached production in 1998.

After that the large sums of exploration capital raised in the 1990s spawned a number of other diamond mines. These included Diavik, a Rio Tinto project, just ten miles from Ekati, which opened in 2003; De Beers' Snap Lake mine which began production in 2007; and its Victor mine which started producing in early 2008. The first two mines are located in the Northwest Territories and Victor is Ontario's first producer. The next expected producer in 2012 is Gahcho Kue, near Lac de Gras, which is a joint venture between De Beers and Mountain Province.

De Beers is also involved in Saskatchewan with the Fort a la Corne prospect, which first saw the light of day in the 1990s when it was a project of Kensington Resources. There is also the Foxtrot project in Quebec being developed by Stornoway Diamonds, which has also discovered a number of other diamondiferous kimberlites in Nunavut to the north and runs an extensive exploration programme across all of Canada.

In a matter of little more than a dozen years Canada has become the third largest diamond producer by value in the world behind Botswana and Russia. Whilst its big two mines, Ekati and Diavik – the latter discovered by female geologist Eira Thomas – are expected to decline during the second decade of the 21st century, there are other mines on the Canadian drawing board, and with long-term trends showing that diamond supply is expected to lag demand in the future, this new production should secure Canada's position in the global diamond hierarchy.

EIRA THOMAS

The role of women in mining has been expanding rapidly over the last 20 years or so and perhaps one of the most distinguished women in the industry is Eira Thomas, who is currently, amongst other positions, a director of African diamond developer Lucara, after a long spell on the board of Stornaway Diamonds, the fast growing Canadian diamond development company where she ended up as Executive Chairman.

Eira Thomas was born in 1968 in Yellowknife, Yukon, and brought up in Calgary, Alberta and Vancouver, British Columbia. She went to high school in North Vancouver and hoped to go to the University of BC, but eventually wound up across the country at the University of Toronto at Mississauga. Her father Grenville Thomas was born in Swansea, Wales. He graduated from University College, Cardiff, and worked for the National Coal Board in South Wales as a mining engineer until emigrating to Canada in the 1960s. He spent a number of years as an explorationist and eventually set up and ran Aber Diamonds, which Eira joined as a geologist after graduating from the University of Toronto in 1990, her interest in geology having been stimulated by field trips as a child with her father whilst on holiday in the north.

She worked as a field geologist for Aber from 1992 until 1997, when she became vice president in charge of Aber's exploration programme.

During her time as a field geologist she led the team which discovered the Diavik diamond deposits in 1994/5. It was her decision to drill under one of the lakes in the Lac de Gras area, a move opposed by some of the team, that led to the Diavik discovery when a two-carat diamond was found in the drill core. After two years as VP of Exploration, Eira joined the Aber board and remained as a director until taking over as CEO at Stornaway in 2006; the year she also joined the board of oil sands giant Suncor Energy. Her departure from Aber in 2006 was not an amicable parting of the ways and her father Grenville left with her. The story is that Aber got fed up with the way that Diavik was too often linked with Eira's name, and Eira and Grenville Thomas in turn decided that Aber's pursuit of jewellery retailer, Harry Winston, was not where they thought the company should be heading. During her time at Aber she was also president of Navigator Exploration, another company backed by her father.

One of her earliest initiatives at Stornaway, following a friendly merger with Contact Diamonds, was the acquisition of Ashton Diamonds Canada, where Stornaway negotiated a lock-up deal with Rio Tinto, Ashton's majority owner. The bid was strongly contested but Eira battled for months, helpfully with Rio's holding in her back pocket. In 2007 Ashton capitulated to Stornaway, handing Eira a well-earned corporate victory. As well as her role at Lucara Diamond Corp, which has acquired a control stake from De Beers in the major AK6 project in Botswana, Eira Thomas remains on the board of Suncor. More recently it has been rumoured that Eira is part of the Apollo Global approach to BHP which is seeking buyers for its Ekati diamond mine.

All work and no play, as the old nursery adage goes, makes Jack a dull boy and Eira Thomas found time in 2004 to join up with Vancouver investment dealer, Eric Savics, to buy the old Pinot Reach Winery in BC's Okanagan Valley in 2004. A major investment programme at the vineyard saw it replanted and extended, its name also being changed to Tantalus, and importantly its reputation much enhanced. Eira has received a number of accolades from the mining and broader business communities in Canada and her record at Aber and Stornaway suggest that in the years ahead success in these endeavours will lead her to the highest levels in mining both in Canada and worldwide.

DE BEERS REORGANISES

The reorganisation of De Beers was arguably the most radical step in the modernisation of the diamond industry as the 20th century came to a close. There were a number of influences at work at this time, Canada and Russia being two, but De Beers itself was finding its role as the dominant buyer increasingly burdensome. It was also finding its role as a publicly listed company increasingly difficult to justify as well. The problem was that apart from the politically necessary decision to give its largest producer, Botswana, a stake in its worldwide business through the issue of a substantial equity stake in the 1970s to the Botswana government, De Beers had not issued equity for project developments or for financing working capital for many decades.

Of course Anglo American, which was financially active and thus valued its public listing, was De Beers's main stakeholder and needed a listed De Beers for group valuation purposes. However, De Beers was always subject to a considerable amount of scrutiny by the stock broking community and the company did not really enjoy these moments in the spotlight, especially when they led on to events like the humiliating market disaster of 1992 when 'unexpectedly' profits crashed and the dividend was slashed.

It was also the case that because of the nature of the diamond market, and the fact that De Beers was selling a luxurious dream rather than a necessity, the group had sometimes to act to protect that dream from over supply, falling prices and other market problems flowing from its luxury status. Often actions taken, like long-term purchase programmes of low quality Russian and later Australian diamonds, were expensive and undermined shareholder value. Whilst Anglo was able to take the long view, outside shareholders were often less forgiving.

THE BAIN REVIEW

So it was that in 1999 De Beers mandated American management consultant group, Bain & Co, to review its business and structure from top to bottom and make recommendations as to its future shape. Bain's recommendations, which although controversial De Beers accepted, led the company to abandon its role as *buyer of last resort* and instead institute a *supplier of choice* system. What this meant was that De Beers no longer would try and control the diamond market by taking new mine supply off the market into its stockpile of rough stones, instead it would take the initiative in driving demand by encouraging and helping its customers to sell more product. There was, however, a stick to go with this carrot as customers who failed to drive demand could find their *supplier of choice* relationship curtailed. Bain also set in motion a thorough reform of how the group was structured and operated. In due course, following pressure from African countries, the group moved functions such as sorting from London to South Africa and Botswana, although for the moment London still hosts important sales sights.

DE BEERS GOES PRIVATE

In 2001 there was further upheaval as Anglo American made a bid for the shares in De Beers that it did not own. When this takeover had been completed Anglo re-structured De Beers's capital, with itself keeping a 45% stake and the Oppenheimer family having 40%. The remaining 15% was retained by the government of Botswana. After over 100 years De Beers had been returned to private ownership, its public listing finally a thing of the past. This also made it easier for the corporate re-structuring recommended by Bain & Co to proceed without undue public scrutiny, although De Beers's major shareholder, Anglo American, remained firmly in the public eye as an important constituent of the FTSE 100.

As the world financial system came under enormous pressure, triggering the major economic recession of 2008/09, diamond demand collapsed and De Beers needed rescue funds which Anglo provided. The cost to Anglo was huge and for the first time in living memory it passed its dividend, despite record 2008 earnings. This caused great controversy, particularly amongst Anglo old timers like ex Deputy Chairman Graham Boustred whose rant at Anglo's American CEO, Cynthia Carroll, caused outrage amongst South Africa's feminist groups.

THE END OF THE OPPENHEIMERS

But the switch from London to Africa of many of De Beers's activities was arguably not the most revolutionary thing to happen to Cecil Rhodes's beloved company in the new century. In 2011 as the diamond industry began to boom again, following the mid 2000s financial crisis, Nicky Oppenheimer, son of Harry Oppenheimer, the architect of the post-war diamond industry, announced that the Oppenheimer family were selling their 40% stake in De Beers to Anglo American for £3.2 billion. The motivation for this sale related to the great difficulties the family had in supporting their stake in De Beers during the financial crisis, as well as the realisation that Nicky Oppenheimer was getting close to retirement age and there was no natural successor. Some of the pain that this decision must have caused the Oppenheimers was mitigated by the fact that buyer Anglo was the child of Nicky Oppenheimer's grandfather, Sir Ernest Oppenheimer, and had been the largest shareholder in De Beers for many decades. Nonetheless, the severing of the Oppenheimer link with De Beers was a dramatic and historic event and truly did mark the end of an era.

THE DIAMOND REVOLUTION

The new millennium saw other huge changes instituted in terms of the organisation and operation of the worldwide diamond industry. This was in part due to the rise of mining giants, BHP and Rio Tinto, as major diamond producers following their discoveries in Canada. However, in 2012 both companies, citing changing corporate strategy and the relative modesty of diamonds in terms of contribution to earnings, intimated that they might withdraw from the sector if a suitable buyer for their operations could be found. Also, the break up of the Soviet Union saw the rise of Russian business oligarchs in natural resources, which opened up mining and diamonds to private investment. One of the most aggressive new diamond magnates was Soviet born Lev Leviev, a major player in Israel on the cutting and polishing side and in both Angola and Russia on the production side. Although still the biggest name in the world of diamonds De Beers will clearly not dominate the sector in the future as it has done over the previous century.

5. COPPER IN CENTRAL AFRICA

The developing Industrial Revolution meant that during the 19th century leading industrial powers were anxious to obtain secure raw material supplies for their fast growing manufacturing base. We have seen that the United States was extremely fortunate in having abundant local sources of metals and minerals, and this put it at an advantage over one of its main rivals, Great Britain, which apart from coal was fast running out of economically exploitable raw materials. Britain, however, had its Empire, which supplied it with all sorts of agricultural commodities and had the potential to do the same with metals and minerals. Of particular importance here was Africa, where Britain's colonies in the west and central southern parts of the continent were directly controlled from London.

As the 20th century wore on, secure raw material supplies became critical because of Britain's early involvement in the two World Wars. Twice the US hung back from entering these conflicts and gained immeasurable advantage in its attempt to overhaul Britain as the West's most powerful political and industrial power. This determination to push Britain into second place was probably as important in framing the US's negative attitude to Britain's colonies as was the moral position that the US evinced, sometimes unconvincingly, when 19th century US foreign adventures such as The Philippines were taken into account. As long as Britain's colonies could provide it with cheap and secure raw materials, the US ambition of overtaking Great Britain would be more difficult to achieve.

RHODES IN CENTRAL AFRICA

Before the potential of Africa could be unlocked, the formal colonisation process had to be completed. Cecil Rhodes was important in establishing British colonies in central southern Africa, two of which, Southern and Northern Rhodesia (now Zimbabwe and Zambia), bore his name. Whilst the Portuguese, who had been in Africa for centuries, had historically established trading posts, the British preferred territorial administration and Rhodes was pushing the concept of British hegemony from the Cape to Cairo, certainly in terms of a transport link, hard at the British government.

As the 19th century drew to a close deals were being done all over Africa by the Europeans, not least in Southern Rhodesia where Lobengula, King of the Matebele, in 1888 granted Cecil Rhodes, through his representative Charles Rudd, exclusive rights to all metals and minerals within his kingdom. From this the United Concessions Company was formed. Lobengula claimed that he had been hoodwinked into signing the deal when others approached him later with attractive alternative propositions.

By 1893 the inevitable happened; the colonists found themselves at war with Lobengula, who was swiftly overcome. In due course, with the southern part of what was now called Rhodesia fully annexed, Rhodes entered into negotiations with Lewenika, Paramount Chief of the Barotse, who controlled large areas of the western part of Northern Rhodesia, including what was to become the Copperbelt. In 1890 Lewenika signed an agreement with the British South Africa Company giving the company mining and other commercial rights over his territories; in return he was granted the protection of the British Crown and an annual payment from the company. Lewenika repudiated the agreement but to no avail – the British were firmly in central Africa and the search for minerals could begin in earnest.

BELGIUM VERSUS BRITAIN, WITH A LITTLE AFRICAN HELP

At this stage it is important to understand one thing: although sophisticated European and American mining men did make greenfield mineral discoveries in Africa – the gold mines of Johannesburg would be one example – in the Copperbelt bridging the border between Northern Rhodesia and the Belgian Congo ultimately many of the mines established were based on previous activity by indigenous African miners.

This activity also clearly went well beyond simply mining enriched super-grade ore; lower-grade copper which required multiple treatments to unlock the metal from its ore was mined and the multi-stage treatment of the ore was recorded by 19th century explorers. There was naturally quite a lot of accompanying mumbo jumbo from tribal doctors, which went along with the treatment of the copper ore. This consisted of a variety of rituals such as sprinkling sacred water and chanting which occurred at each stage of the treatment process, but despite this the processes were firmly based on science rather than witchcraft. Although this demonstrates a perhaps unexpected level of technology on the part of indigenous Africans, the record seems to suggest that none of their operations went very deep, maybe as a result of being uncomfortable with mining at any material depth (superstition perhaps). This, of course, simplified the Europeans' search for mineable deposits, as there were many abandoned mining sites to be examined.

Two quite distinct developments went on in central southern Africa, one involving the Belgians in the Congo and another the British effort in Northern Rhodesia. Whilst the preliminary ground for exploring for mineral deposits was covered in the dying years of the 19th century, the establishment of the modern large-scale Copperbelt copper mining industry is unquestionably a 20th century event. In the early days there were disputes with indigenous tribes like the Angoni and with Rhodes's British South Africa Company whose interest was fired by other corporates anxious for a piece of the action and by prospecting adventurers. In due course, the British South Africa Company's writ became established in Northern Rhodesia and this position of control continued until the part nationalisation of the copper mines by the Zambian government in the 1960s.

In the Belgian Congo, as in Northern Rhodesia, the Europeans had little trouble in identifying prospective areas to look for mineralisation, thanks to the extensive historical workings of the indigenous inhabitants of the Congo's Katanga region. Some of these workings were huge; the Star of the Congo mine was an ancient pit almost a mile long and a sixth of a mile wide although quite shallow – it was maybe no more than 30 feet deep, in keeping with indigenous mining tradition. The Katanga province, the centre of copper mining in the Congo today, is thought to have produced 100,000 tonnes of copper before being *discovered* by the Europeans in the 19th century. Indeed, two mines being revived today, Kolwezi and Dikuluwe, were operating when the Europeans arrived in the Congo in the late 1800s.

One of the major differences between the development of the mining industry of the Belgian Congo and of Northern Rhodesia was the motivation issue. The Belgian Congo became the personal fiat of King Leopold II of the Belgians and from 1879 when he began colonising the Congo until 1907 when he sold it to Belgium, the assets of the colony were exploited for the benefit of the King and his friends. The colony itself received nothing from the exploitation of its mineral and agricultural resources and this did not really change after Belgium took control.

Northern Rhodesia's Copperbelt was developed for commercial reasons, providing Britain with much-needed raw materials and the participating mining companies with an often more than adequate return. Although Britain was a hardheaded colonial power, it did re-invest some of the economic spoils from its commercial exploitation of its colonies back into those colonies, and Northern Rhodesia benefited from this over the medium term.

THE IMPORTANCE OF COPPER

The two-pronged European search for copper was driven by rising usage in the rapidly industrialising world. The arrival of electricity was one of the key drivers of this, underlying our basic theme that it is through metals and minerals that civilisation advances. It is hard to envisage a more important development in the evolution of the modern world than the discovery of electricity. Today's world would be entirely different without electric power and life would be very much more precarious than it is now. It is hard to see how worldwide communications, healthcare and transport, for example, would have advanced beyond the basic level that they had reached by the start of the 19th century without electricity.

Even with the process for making electricity having been discovered, little use could have been made of it without the means of transmitting the power to users. We first came across copper in the Bronze Age, but treating ore to produce copper metal for weapons is a quite different task from capitalising on copper's qualities as a conductor of electricity, to refine it and then produce high-quality copper wire for electric transmission cable and wiring.

Visionaries are by definition rather thin on the ground in any age and with the class system apparently set in stone worldwide during the Victorian era, few might have forecast the popular boom in Western housing that followed the First World War and which has continued up to this day. Allied with industrialisation, this not unnaturally led to a huge increase in the usage of copper. If electricity was to fulfil its life enhancing potential, huge new sources of copper would be needed. Indeed, as the 20th century dawned, over 60% of copper mined was going towards the manufacture of copper wire.

As the US began to flex its industrial muscles around the turn of the century, Great Britain needed to respond and for this reason its African colonies were important as potential sources of cheap raw materials to fuel its industrial reply. The acquisition of prospecting rights in Northern Rhodesia was therefore an important step forward.

THE DEVELOPMENT OF NORTHERN RHODESIA'S COPPERBELT

The old adage that if you want to find a new mine the best place to look is where there has already been one is particularly apt in the case of the Copperbelt. We have already mentioned that some of the old indigenous copper mines in the Congo are being revived today and the same thing is happening in Zambia where one of the most profitable *new* mines, Kansanshi, was first developed in 1908 and was eventually flooded in 1957. The old mine was a typical Copperbelt high-grade underground operation, the new Kansanshi is a large, low grade surface mine.

As the 20th century dawned Edmund Davis, later knighted, a director of the Bechuanaland Exploration Company (BEC) was beginning to establish a stable of companies to explore for and develop copper mines in central southern Africa. Among the group companies were Northern Copper, Rhodesia Copper, Rhodesia Copper & General and Rhodesia Broken Hill Development – the latter primarily a lead and zinc mine. BEC was also agent for Kafue Copper and Bwana M'Kubwa Copper, two familiar names in the modern Zambian copper industry. The latter, which was an ancient native mine, was re-opened in the 1990s after a long period of dormancy. The biggest landholder in the region was the British South Africa Company, from whom the BEC obtained exploration rights which eventually extended to 700 square miles in area and were later acquired by Northern Copper and then re-distributed to other companies. This shuffling of the asset pack was a feature of that period and it is not surprising that the British South Africa Company was behind much of this activity, as it was eventually to become part of the Anglo American group (itself a past master of asset shuffling).

Another early century group involved in acquiring exploration licences in Northern Rhodesia was Tanganyika Concessions, which at one time controlled Kansanshi and, as well as copper, also had interests in coal and bismuth. However Tanks (as Tanganyika came to be known during its long history as a listed UK registered mining company), was also involved in the Congo with the Belgian metals group, Union Miniere. Eventually, having established Rhodesia Katanga Junction to hold its Northern Rhodesia interests, Tanks pulled out to concentrate on the Congo.

The First World War also had a profound effect on mining activity in the region. The Germans in Tanganyika later on in the war attempted to disrupt activity by threatening the Broken Hill lead mine and the railway that transported copper to Beira on the east coast from the mines on both sides of the Congo/Northern Rhodesian border. There was also a major attempt made to lift the level of mined output from the region, as Great Britain was facing extremely high metal prices as a result of the war effort and the need to buy American copper, with the US not entering the war until 1917. Around that time the Nkana claims were pegged and then traded on a few times over the next few years. Widths were reasonable and grade was quite high – 27 feet of 3.3% copper and 45 feet of 2.8% copper in one sample – but although mined today on the Zambian Copperbelt in the form of cobalt rich waste dumps, in 1919 Nkana's ore, which was oxide, could not be mined and treated economically.

Two small mines located south of the Copperbelt, Sable Antelope and Silver King, produced throughout the war but closed down in 1923 as metal prices, following the wartime boost, collapsed. Kansanshi and Bwana M'Kubwa were the other two copper producers of note, but none of these mines were anything other than modest producers. For instance, Bwana M'Kubwa produced around 2,800 tonnes of copper in the final two years of the First World War and the other mines produced on a similar scale, with none of them making much, if any, profits. Other mineral deposits found, coal and gold in particular, were either of poor quality or modest in extent.

BANCROFT'S EXPLORATION PROGRAMME

In 1924 the British South Africa Company ceded administrative control of the two Rhodesias to the British Colonial Office, but retained its mineral rights. As the 1920s progressed, two major mining figures, Sir Ernest Oppenheimer of Anglo American and Alfred Chester Beatty of Selection Trust, became increasingly involved with the Northern Rhodesian mining industry. It was said with justification that since the latter part of the 19th century when prospectors were active in the field, very little of anything of interest had been found. All the prospectors did was re-examine the ancient workings of the natives which had been abandoned for the very good reason that they had been worked out, or at least worked out if you only considered surface mineralisation.

The time had come for a more extensive, modern and efficient exploration programme to be put in place. At this stage the Canadian geologist, J. Austen Bancroft, entered with a mandate from the British South Africa Company to head the largest geological exploration programme in the world at the time, covering thousands of square miles of northern Rhodesia. Bancroft was the first economic geologist and it is a tribute to this change of emphasis in exploration that led to the development of the modern Zambian Copperbelt, now reviving after years of drift and neglect.

At this stage, two separate groups were beginning to emerge as regional leaders as Bancroft's programme began to crank up: Anglo American and Selection Trust.

Ahead of that, the Nkana Concession – which came to host the Nkana, Mufulira, Chambishi, Chibuluma, Roan Antelope, Mindola and Baluba mines, some of which still operate today – was reserved in 1924 for Copper Ventures, in which Selection Trust, one of Chester Beatty's companies, acquired an interest. Copper Ventures then sold on its rights to Edmund Davis's Bwana M'Kubwa Copper.

An even bigger concession, ultimately known as Rhodesian Congo Border, had been reserved for Copper Ventures in 1922 and in 1928 yet another, this time amalgamated, concession was incorporated into the Loangwa Company where both Edmund Davis and Sir Ernest Oppenheimer were directors. In 1926, what was to become one of the great names of the Copperbelt, Nchanga Copper Mines, was registered. The corporate structures necessary to take advantage of Bancroft's programme, which began in 1928, were slowly falling into place.

The 1930s were a difficult time for world economies and for metal prices and demand. Bancroft's exploration programme continued, although circumstances did mean that the number of field teams and the level of work did fluctuate from year to year. A lot of work was done on the Kansanshi mine which had closed down in 1914, but even though several years of development work turned up reserves of 10 million tonnes grading 4.3% copper, it was not enough to get the mine re-opened and Kansanshi fell back into dormancy until a short lived revival in the 1950s.

Despite the economic and financial problems of the pre-war decade the mining industry in Northern Rhodesia expanded steadily, although, as we saw with Kansanshi, this progress was not always smooth. The Broken Hill mine, opened in 1906, in particular proved important not only as a provider of key metals for Britain's war effort but also as an operation which pioneered important technological developments to overcome, in particular, the problems of water at depth in the country's mines. In the 1920s mining ceased at the 225 feet level; by the early 1950s with new technology and investment, mining was possible down below 1,000 feet with the main pump chamber located at the 1,580 feet level. However Broken Hill, or Kabwe as it became after Zambian independence, was an ageing mine as the metal price booms of the 1960s and 1970s unrolled. It became increasingly marginal and eventually closed in 1989.

J. AUSTEN BANCROFT (1882-1957)

The name of Joe Austen Bancroft, a Canadian born in North Sydney, Cape Breton, is synonymous with the exploration and development of what is now known as the Zambian copperbelt. The exploration programme that he oversaw in then Northern Rhodesia in the 1930s was probably the most extensive scientifically-based programme seen anywhere up to that time and from it was born one of the largest copper mining provinces in the world. Bancroft pioneered the science of *economic geology* in the first part of the 20th century; at the time such a term would have been considered an oxymoron but now it is the driving force behind most commercial geology.

He was born in 1882 one of eight children and his father was a Methodist minister. The early part of Bancroft's adult life, after graduating first in his class from Acadia University, Nova Scotia in 1903 and being awarded a Yale fellowship, was spent studying and then teaching geology. He joined the faculty of McGill University in Montreal in 1905 and took post-graduate courses at Leipzig University, Georgius Agricola's alma mater, and Bonn University between 1908 and 1910. It was here that he met his French wife, Jeanne Poirier, who he married in 1911; they had no children. Bancroft also gained his PhD from McGill in 1910. Three years later he became Professor of Geology at McGill and then Head of Department, but he continued to keep his practical geological skills up to date during the university breaks, mapping for the Geological Survey of Canada.

As the years went by Bancroft began to develop a strong interest in the economic aspects of geology, then a much neglected part of the science, and took leave from his university duties to act as Consulting Geologist for Granby Consolidated, a copper miner in British Columbia. In 1929 he finally left McGill to organise a massive exploration programme in Northern Rhodesia. Although considered a fine teacher and an exceptionally gifted academic, Bancroft found his real niche in heading this extremely ambitious venture.

Following the British South Africa Company's success in acquiring huge tracts of land in central southern Africa for mineral exploration in the latter part of the 19th century, a somewhat disorganised effort was undertaken by individual prospectors to uncover copper, gold and other deposits. Few significant mines were developed, but the prospectors did frequently stumble across very old native workings suggesting that the region was highly prospective for minerals. This aspect of these early efforts intrigued Bancroft and in his seminal book *Mining in Northern Rhodesia* he makes extensive and approving comments on the evidence gathered together by many explorers and prospectors as to the relatively sophisticated mining and, in particular, treatment methods used by the natives over a very long time span to produce finished copper.

Bancroft's team consisted of almost 170 geologists and 30 prospectors drawn from around the world, and he organised a systematic observation and mapping exercise covering no less than 160,000 square miles of Northern Rhodesia. Whilst there were mutterings that Bancroft had involved himself too deeply in what was something of a technical exercise, the reality was rather different with the Nkana, Nchanga and Bancroft mines being discovered as a direct result of Bancroft's team, and this laid the foundation for the modern Copperbelt. This success led to Bancroft becoming Consulting Geologist for Anglo American in Johannesburg, where he not only continued to be responsible for the Copperbelt but also oversaw De Beers' diamond interests and the development of the Orange Free State gold field after the Second World War. He forged a close relationship with Anglo's founder, Sir Ernest Oppenheimer, and in 1957, just before he died, was honoured with the Bancroft Mine being named after him.

Bancroft was a man who was able to get along with people at all levels, from mine worker to mine magnate, and who was respected and liked by all who worked with him,. These are major advantages when carrying out an exploration programme as large as his Northern Rhodesian project. In 1956, a year before his death, Bancroft received the Gold Medal of the Institution of Mining and Metallurgy in London in recognition of his central role in advancing the science of economic geology. Before Bancroft, it was prospectors who found mineral deposits, but after Bancroft the task of finding new mines fell increasingly to geologists.

THE NCHANGA DISCOVERY

The development of the Nchanga mining area, still an important part of the Copperbelt today, began with an extensive exploration and drilling programme by the Rhodesian Congo Border Concession Company, which started in 1923. The complexity of the area and the relatively unsophisticated nature of drilling programmes in those days meant that the process was somewhat drawn out, but by late 1926 the results were sufficiently encouraging to lead to the flotation of Nchanga Copper Mines in London. The first annual report of the company indicated what we would today call resources of between 12 and 15 million tonnes of 3% copper; a useful but not startling figure. Often using surface copper staining to identify prospective drilling targets the programme made satisfactory if unspectacular progress.

However, in 1928 a drill hole pulled 4.56% copper over 93 feet in a previously examined area of copper mineralisation on the Nchanga lease. Bancroft knew he was on to something and the drilling programme was intensified. Early in 1929 Bancroft was able to start reporting some wide, rich intersections with 65 feet grading 7.96% copper from 795 feet and directly below that another 30 feet grading an astonishing 20.94% copper. The great Nchanga mine had been found and the modern Zambian copper mining industry had been born.

Unfortunately nothing has ever been straightforward on the Copperbelt and in September 1931 with a small mine having been established at Nchanga that year, a massive inflow of water into the workings led to flooding to within 100 feet of the surface. Fortunately only one miner was killed, but the water problems were judged to be so bad that the mine was shut; the de-watering cost being judged too great at a time of worldwide depression and low copper prices. Nchanga's ore reserves, however, at over 100 millions tonnes grading between 3.5% and 7% copper, were a substantial resource and in 1936, with global economies recovering, it was decided to reopen the mine on a much larger scale than before.

In 1937 the Nchanga Consolidated Copper Mines Company (NCCM) was formed and Rhokana Corporation, the holder of the Nchanga leases, ceded them to the new company in exchange for shares. During the Second World War production was restrained by a shortage of supplies, but its output was nonetheless important in the Allied war effort. After the war for a number of years there were problems with obtaining coal from Wankie in Southern Rhodesia, but production rose steadily and by 1951 had reached 58,000 long tons, worth almost £10 million.

THE BRITISH RETREAT BEGINS

In 1951, following the payment of its maiden dividend, the management and control of NCCM was moved from London to Lusaka, Northern Rhodesia's capital, where the local tax laws allowed capital expenditure on new mines to be written off against profits, unlike in the UK. This move was also significant as it was one of the earlier signs that Britain, after the exhaustion caused by fighting two World Wars in 40 years, was beginning to fade as an economic force and a centre for international finance. This deterioration continued for years, leading to further departures of originally British financed mining companies from London. Of course in recent years things have swung back dramatically and now many of the world's largest mining companies are once again incorporated in the UK.

As a consequence of this initial move, Nchanga, whose mine management contract had been with the British South Africa Company for years, came increasingly under the influence of Johannesburg, the home base of many of Nchanga's directors, such as Sir Ernest Oppenheimer. This closeness to South Africa may in hindsight seem strange, but black African independence was still a full decade away and though the Nationalists were in power in South Africa, apartheid was still very much a work in progress and of little interest to the world then. Also King George VI was Head of State and South Africa was in the sterling area currency bloc.

One thing that should be mentioned here is that the rising tide of exploration and mine development from the 1930s onwards carried a considerable environmental cost, which Bancroft noticed. In particular, huge swathes of forest were felled to provide fuel for the mining and treatment facilities, as well as for domestic use amongst the rising number of people who worked on the mines and in the towns growing up to support the mines.

Rhokana Corporation, which succeeded Rhodesia Congo Border Concessions, and whose exploration programme was also under Bancroft's direction, had by the 1930s developed the Nkana prospect and Roan Antelope; the former being of a similar size to Nchanga, although the grade was lower at around 4% copper. This was a time of low copper demand and low metal prices and, like Nchanga, Nkana struggled. Rhokana reduced production, but later in the decade as war loomed production began to pick up and when war was declared in 1939 all of Northern Rhodesia's production was put at the disposal of the Ministry of Supply in London. During the war Rhokana produced 570,000 long tons of copper and also paid total dividends of 34s 6d (172.5p) – a satisfactory outcome for both Britain's war effort and shareholders.

In 1953 Bancroft Mines was formed and it acquired mining rights from Rhokana, which eventually led to the development of the Konkola and Kirila Bomwe orebodies. The resulting mine was named after J. Austen Bancroft and had reserves of around 100 million tonnes, grading between 2.5% and 4.8% copper. Bancroft's major exploration programme had paid off handsomely and the naming of the mine after him (today it is simply called the Konkola mine) recognised his efforts and the successful development of a number of large mines on the Copperbelt.

The early 1950s was a time of considerable excitement in base metals, due to demand stimulated by the Korean War. However, after the war an extended period of growth set in for the advanced economies and during the 1960s, despite Cold War tensions, such as Cuba, economic progress was more than satisfactory. As the 1960s advanced, the US became embroiled in Vietnam at the same time as public spending began to soar, due to President Lyndon Johnson's Great Society initiative. Inflation and metal prices began to rise. At the same time, the wind of change blowing through Africa brought independence to many former British colonies and, as commodity prices strengthened, newly independent governments began to take an interest in the commercial assets that had remained in the hands of foreign investors after independence. This was, after all, the age of Salvador Allende and Fidel Castro, who were heroes throughout the dark continent.

ZAMBIA NATIONALISES THE COPPER MINES

In 1964 the UK, well into the process of dismantling its Empire, granted Northern Rhodesia its independence as Zambia, following the failed experiment of amalgamating Northern Rhodesia with Southern Rhodesia (Zimbabwe) and Nyasaland (Malawi) into the Central African Federation (CAF). It is a moot point whether the federation ever had a chance of success, bearing in mind that white settlers in Southern Rhodesia constituted a far larger group than did their white counterparts in Northern Rhodesia and Nyasaland. The black majorities did not like the set up either as the CAF was not independent but remained essentially a British colony. This made the governmental structure complicated and as the idea in Britain turned to granting independence to its colonies, the black leaders in Northern Rhodesia and Nyasaland became increasingly uncomfortable with the federation.

For their part, white politicians from Southern Rhodesia, although the most powerful group in the federation, became increasingly concerned about the future direction of the CAF. These tensions led to the dissolving of the federation in 1963 and the granting of independence to Northern Rhodesia. The first independent Zambian government was headed by Kenneth Kaunda, a fierce proponent of the idea that Zambia could go it alone.

The British had left Zambia with just a handful of indigenous people educated to university level and for the first few years of independence, business and commerce remained firmly in the hands of the settler community and the largely British firms who had established subsidiaries in Northern Rhodesia. What the British had left behind, though, was a bulging treasury of foreign currency, worth hundreds of millions of pounds. These currency reserves allowed the new government to build up its educated, professional workforce. Although, ultimately, large quantities were wasted on prestige projects like eastern European designed commercial buildings and a national airline, the early years of independence were not without promise. Then, in 1968 the Kaunda government, encouraged by Greek Cypriot businessman Andrew Sardanis – who was a close friend and adviser to Kaunda and not naturally well disposed towards the British and Zambia's settler community – decided to take a 51% stake in the country's larger commercial businesses and, of course, the copper mines.

At that stage the copper mines were managed by two groups, Anglo American of South Africa and Roan Selection Trust which was controlled by American Metal Climax (Amax). The Anglo mines were amalgamated into Nchanga Consolidated Copper Mines and Anglo created Zambia Copper Investments to hold its 49% stake and Climax's 49% was vested in Roan Consolidated Mines (RCM). The government paid for its stake in the copper mines through the issue of bonds by Zambia Industrial and Mining Corp (ZIMCO) – the government's vehicle for holding its stake in the mines. Following the part nationalisation, these bonds, which were repayable through a mechanism taking into account underlying profits credited annually to holders, traded in the market at well below their redemption value.

In 1971 Sir Peter Tapsell, a Conservative MP and old friend of President Kaunda, persuaded the Zambians to start buying back the ZIMCO bonds in the stock market in London, using James Capel, the firm that he was a partner of. The Zambians continued to do this and steadily redeemed the compensation bonds at attractive prices. In 1973, the government was persuaded by a merchant banking consortium led by Barings, and, it is alleged, encouraged by Lonrho's Tiny Rowland, a big holder of ZIMCO bonds, to nationalise the copper mining industry properly, by buying out the ZIMCO bonds at their full face value and terminating the management contracts of Anglo and Amax. This value was considerably above the prices that James Capel had been buying the bonds in the market, but the Zambian government was persuaded that this was the right thing to do, as it would finally allow them to run the mines themselves and begin the process of having indigenous Zambians taking senior management and operational positions.

This state of affairs continued until 1978, when the government increased its stake in the mines to 60% by capitalising most of the loans that the industry owed it. In 1980 management was vested in a new company, Zambia Consolidated Copper Mines (ZCCM), which brought all the mines under a single corporate umbrella. Foreign interest in the consolidated industry was divided between Anglo, who indirectly controlled 27.3% of ZCCM's 'B'shares which had no votes, and private investors with 12.7% of the 'B' shares. The Zambians were now completely in charge and a gradual decline on the Copperbelt set in.

The Zambian government, having capitalised the industry's old borrowings, cynically began to use the cleansed balance sheet to borrow large sums of money in the international banking market, using ZCCM as collateral to fund government spending. The industry began to starve through lack of capital needed to maintain production levels; in the late 1970s copper production had peaked at around 700,000 tonnes of metal, but 20 years later it had fallen back below 200,000 tonnes.

There were two attempts made by the government to privatise the copper industry following the fall in 1991 of Kenneth Kaunda, who had been the Zambian president since independence. The first attempt took up much of the 1990s, as the government vacillated over the form it wanted privatisation to take – the sale of ZCCM as a corporate entity, or the piecemeal sale of ZCCM's assets. In the end, the route taken was to sell off the individual mines to foreign interests. ZCCM kept a 10% to 20% stake in each operation, but the foreign owners were responsible for financing and operating the mines. Although not perfect, the new regime has led to a huge surge in copper production, with output back above 700,000 tonnes and two new mines, Kansanshi and Lumwana, opened. Originally optimists had believed that Zambia's copper output would exceed 1 million tonnes; despite the improvement, their hopes were rather too high. Huge sums of money are also going into developing the North orebody and the Deep orebody at Konkola, a mine first opened 84 years ago.

ZAIRE'S OUTPUT COLLAPSES

Across the border in the Congo, after being given independence by Belgium in 1960, the copper industry performed quite well following its nationalisation in 1967, with the historic owner Union Miniere acting as operator for a number of years. By the second half of the 1980s it was producing close to 500,000 tonnes of copper a year on behalf of Gecamines, the state copper company that had taken ownership of Union Miniere's Congo operations. By the mid-1990s Zaire, as it had become known, saw production collapse to little more than 20,000 tonnes as first its notorious President, Mobutu Seko Sese, stripped the national treasury of its money and following that a wide-scale civil war broke out. By the early part of the 21st century, both the political and economic situation had stabilised and with Gecamines being basically bankrupt, foreign mining companies began to return and a number of mine resuscitations and developments began to crank up. This recovery process, however, lags well behind Zambia's, and has already attracted some unwelcome government infighting and disputes over the standing of some of the foreign companies exploitation agreements and work contracts, which has led to confiscations and subsequent asset sales to new parties. Plus ça change!

6. OTHER NATIONALISATIONS IN THE 1960S AND BEFORE

Above we have seen how central Africa in the shape of Zambia and the Congo (Zaire) nationalised their copper mines and we have also seen how in both cases it was a commercial failure, although the process did see a major expansion in the number of indigenous people employed at the higher technical and managerial levels, particularly in Zambia. But African copper was not the only mining industry affected by nationalisation in the 1960s, nor was Africa the only region involved in the practice of returning strategic industries to public indigenous ownership during that decade. However, radicalism in Latin American mining was not simply a 1960s phenomenon and in Bolivia nationalisation came early.

BOLIVIA

Although now one of the least prosperous South American countries, Bolivia held an important position during the transition from industrial to technological revolution in the early decades of the 20th century, due to its role as one of the largest producers of tin in the world. Used for centuries in bronze and pewter ware, tin's properties as a corrosion-resistant coating for metals such as steel and as a solder in the rapidly-expanding industry of electronic circuitry emphasised its key role in the advance of technology.

The 20th century tin mining industry in Bolivia was dominated for decades by the Patino, Aramayo and Hochschild families, with significant support from British capital in the early days. Bolivia became a reliable source of metals, particularly for Great Britain, in the second half of the 19th century when many new mining areas were opened up, often as a result of the construction of British-financed railways. One of the most famous historic mining areas was around the city of Potosi in the south of the country, where silver had been mined at least since the 17th century, and indeed probably before that. In the early years of the 20th century silver exports ran at around 6 million ozs annually. Other silver mining centres included Chocaya, Colquechaca and Huanuni, but as the 20th century progressed tin took over from silver as the prime metal being produced.

One of the richest tin mines was the La Salvadora mine in Uncia, Potosi, acquired by the legendary Simon Patino in 1894; he then bought out the British-owned Uncia Mining in 1910 and the Chilean-owned Llallagua Mining in 1924. This made him dominant in this southern stronghold of tin mining, with over 50% of Bolivia's output under his control. Thereafter, in 1916 Patino took control of Williams Harvey in England, the world's largest tin smelter, giving him vertical integration in the tin industry, and by the time of the Llallugua purchase he was living in Europe, concentrating on expanding his non-Bolivian mining and industrial interests. As Patino developed his tin interests Mauricio Hochschild and Carlos Aramayo were building up their tin mining activities.

As to the mines themselves, they were largely underground operations and working conditions were poor, with deaths from mine accidents and related conditions like silicosis unfortunately high. Historically the workforce in Bolivia's mines had consisted of slave labour, first indigenous Indians and then Africans, and although the mines had received considerable investment they remained dangerous, and the workforces and their unions became increasingly radical as the 20th century progressed. The most tragic manifestation of this was the Catavi massacre just before Christmas 1942 when Bolivian troops shot dead what was alleged at the time to be several hundred miners and their families as they marched on the Catavi tin mine in support of a strike at the mine. Catavi then was owned by Patino, but clearly the mine, despite its size and richness, had a death wish independent of its ownership, because in 1967 the Bolivian military again massacred a large number of protesting miners at the mine, which by then had been nationalised.

Bolivia's other mining activities included copper, where the Anglo-French Corocoro United Copper was the largest producer in the early 20th century, with First World War demand pushing its output up to 20,000 tonnes of copper by 1918. Some lead was also produced during First World War, as was antimony, bismuth, zinc and tungsten, with Great Britain the major market for all these metals. As the decades wore on Bolivia's mineral output was increasingly dominated by tin, with silver in relative decline, and by the middle of the century Bolivia was responsible for producing from 5% to 10% of the world's tin output with the Siglo XX mine of the Patino Group the largest contributor. In the years following the end of the Second World War Bolivia's tin mines struggled to maintain production and a combination of age and falling grade put the industry under pressure.

In 1952 the radical government of President Estensorro, having overthrown the military government, nationalised Bolivia's mining industry and the once prosperous tin mines became state owned. If there had been considerable public controversy over the way that the mines had been run in the past, they had at least made huge profits for the original owners. The age of the mines in the 1950s meant that with metal prices in decline in the years following the Korean War, state ownership was accompanied by escalating losses.

Even where the mines were relatively modern, losses ensued. Patino's Siglo XX mine was one example, where the clash between the radical mine unions and the mine management and technicians now working for the state corporation, COMIBAL, led to increasingly inefficient operations and mounting losses. The government allocated few resources to exploration and was nervous of committing capital to opening up new mines, the Mina Matilde zinc project near Lake Titicaca being a case in point. Hochschild had ownership of the project until the 1952 nationalisation, then in 1966 an operating lease was sold to US Steel and Engelhard, and a mine was developed only for the government to expropriate it again in 1971. This chaotic style of industry management has dogged Bolivian mining right up to the present day, where private sector and state exist somewhat uneasily together, and where labour relations remain poor.

MAURICIO HOCHSCHILD (1881-1965)

Moritz (Mauricio) Hochschild was born in Biblis near Frankfurt in Germany in 1881. The extended Hochschild family was Jewish and large; Mauricio, however, was an agnostic. He went to Freiberg University where, with his two brothers, he studied mining engineering, graduating in 1905. Following graduation he went to work for Metallgesellschaft, the German metal trader. In 1906 he was posted to Spain and in 1908 left Metall to run a gold mine in Australia, thus beginning a long spell abroad. The gold mine closed and Mauricio headed for Latin America.

He first worked in Chile trading copper and visited Peru and Bolivia, the latter country being where he eventually built the base of his fortune. With the outbreak of the First World War he returned to Germany to serve but did not see action as he worked in Austria sourcing metals for the war effort. He married in 1918 but his wife, Kathe, died in 1924. They had one son, Gerardo, who was born in 1920. In 1921, after being awarded his doctorate in mining engineering at Freiberg, Mauricio returned to Chile. Leaving his brother to continue the copper trading business he then moved to Bolivia where he started to deal in tin. All mining in Bolivia was small scale and independent miners sold their output to dealers who aggregated small amounts of ore into marketable parcels which could be traded for treatment – the process was called *rescate*. Hochschild's Bolivian tin business was based on this process. So successful was his venture that he incorporated Mauricio Hochschild & Company and this became the vehicle for his Latin American trading and mining activities.

Building on these activities Mauricio began to assemble a Bolivian tin mining group, which brought together a number of small mining operations, and in due course became a major force in Bolivia alongside the mining interests of the Patinos and the Aramayos. Whilst he enjoyed cordial relations with Carlos Aramayo he was often at odds with Simon Patino, who was a bigger force in Bolivia than he was and often a shareholder in some of his mines, where he harried Mauricio.

Mauricio's success came at a price as he became unpopular with many in political power in the country. In 1939 as fascist sentiments swirled around Bolivia, fed by the rise of Hitler in Germany, President Busch ordered Mauricio's death by firing squad. This order was reversed but in 1944 President Villaroel imprisoned him and only after political pressure was brought to bear on the dictator was he released. His unpopularity with the Bolivian authorities was in part due to his activities in the late 1930s in bringing German Jews to Bolivia to work for him and so saving their lives; many of his own family took this route. Amongst them was cousin Philippe whose wife Germaine had an affair with Mauricio and married him after she and Philippe were divorced in the 1930s.

Mauricio left Bolivia in 1945 at the end of the war as his position as a Jew and mine owner fatally undermined his status in the eyes of his social peers. Over the next few years Mauricio worked at building up his other Latin American mining interests in Chile, Peru, Mexico and Argentina, a wise move as in 1952 an increasingly radicalised Bolivian government led by new President Victor Estenssoro nationalised the tin industry which by then was in irreversible decline. Fortunately some compensation was paid and this enabled Mauricio to re-invest his share in his other Latin American mines, which included Mantos Blancos in Chile, now part of the Anglo American group. After Bolivia he lived in Paris until his death in 1965, during which time he built up his non-Bolivian interests and administered the Hochschild Foundation, which he had endowed in 1951.

It is ironic that though the Hochschild story is far less well documented than those of the two other great Bolivian tin mining families, the Patinos and the Aramayos, Mauricio Hochschild's activities in Latin America laid the base for the establishment of the Hochschild Mining Company PLC which is now listed in London and part of the FTSE 250. Both the Patinos and the Aramayos have now disappeared from the mining scene, whilst Hochschild is a growing force in precious metals. Interestingly, one of Mauricio's uncles was a founder of the

American Metal Company which became the US giant Amax, now lost within the bowels of Freeport McMoran.

Mauricio himself was a gracious man who lived a modest life and showed great loyalty to both family and employees. It is much to his credit that he survived so many difficult years in the 20th century, laying the foundations of a mining business which still thrives today.

CHILE

One of the most audacious expropriations was by the far left Chilean government of Salvador Allende. Allende was an experienced radical Chilean politician who had served in congress many times. He had also unsuccessfully run for president on three previous occasions before becoming a minority president in 1970 in a close three-way race. In his three years in office before committing suicide following an army coup in 1973, Allende was constantly in dispute with Chile's Congress, where his party was in the minority.

Still able to stir the hearts of left-wing radicals over 30 years since his death, Allende was a curious figure with a penchant for rubbing up potential allies with his radical economic ideas. He was also very happy to use the privileges of his office to live considerably more lavishly than the ordinary Chilean was able, and he used government residences in the hills above Santiago to entertain an international group of admirers, dispensing his favourite tipple Chivas Regal Scotch whiskey as they debated the coming international triumph of the left deep into the night. UK admirers included Labour Minister Judith Hart and inevitably Tony Benn, who probably drank tea rather than Scotch.

Allende's nationalisation of the Chilean copper mines in fact was not contentious within Chile; the legislation needed was supported by all the political parties in congress. Also it merely took a step further a process that had been developing since the 1950s when the government took a majority stake in the mines, essentially Chuquicamato and El Salvador operated by Anaconda, and El Teniente operated by Kennecott. So in 1971 Allende announced a full take out of the 49% foreign-owned portion of the mines with no compensation. The terms were based upon Allende's view that the US mining companies had extracted profits way beyond what was reasonable in the past and because of this nothing was owed for the 49% residual stakes. Whatever the truth of his assertions, and we discuss these further on, this was not a wise stance for Allende to take. Richard Nixon, the arch anti-communist, was in the White House and the US was increasingly concerned about Allende's close relationship with Cuba's Fidel Castro and also the financial support Chile was getting from the Soviet Union.

Expropriation without compensation was the last straw as far as the US was concerned and the CIA increased its support for anti-Allende groups within the country, particularly in the army. There was deepening unrest and division in Chile – Allende's party had increased its vote in the elections for congress but was still in the minority and congress, concerned by deteriorating economic and financial conditions within Chile, sought Allende's overthrow. In August 1973 General Augusto Pinochet was appointed commander-in-chief of the army and in September backed by congress the army staged a coup and Allende committed suicide.

During Pinochet's rule and thereafter the copper mines, incorporated as Codelco, remained under state ownership and the army was entitled to around 10% of the profits. Although subsequent Chilean governments have considered the privatisation of Codelco, as many key industries were privatised after Allende's overthrow, state ownership of the old mines remains popular within Chile. However, Codelco lost its monopoly status with the coming of Pinochet and eventually a large number of foreign-owned mines were developed. Well known to UK investors are BHP Billiton/Rio Tinto's Escondida mine and Antofagasta's Los Pelambres mine, both developed after Allende. In today's Chile Codelco produces around one-third of the country's copper with two-thirds coming from the private sector, as we will discuss later.

PERU

Peru has a long history of foreign involvement in the mining industry, starting in the early 20th century and continuing through to the late 1960s, when a coup led to the installation of a military government which then set about nationalising a large part of Peru's mining industry. The 1950 Mining Code, which was very friendly to the industry, had led to a major upsurge in mining developments by foreign interests, particularly in the base metal field.

At that time the industry was split into two informal blocs, the major mining projects and the smaller ones. The big mines represented about two-thirds of national metals output by value and were largely owned and run by foreign interests. The smaller mines covering the other third of output value were primarily locally owned. This ownership structure, although favouring foreign miners, had a more active local input than in many countries, particularly in Africa where there was little or no local participation in mining. This may have been due to many African countries being colonies until the 1960s, decades after the colonial period had ended in South America.

One of the mines to escape nationalisation was the Toquepala copper mine. Its owner was US-controlled Cerro de Pasco, which was established in 1901 by, amongst others, J. Pierpoint Morgan, to develop the original Cerro de Pasco area base metal deposits. The reason Toquepala was not nationalised was that Cerro de Pasco had merged its interests in Toquepala and the huge Cuajone copper deposit into Southern Peru Copper, where fellow shareholders were ASARCO (51.5%), Phelps Dodge (16%) and Newmont (10%). The Cuajone deposit was being developed on the back of 50 banks on five continents and even the radical Peruvian government, with its small and poor country, acknowledged that nationalisation would be unwise bearing in mind the forces arrayed against it. However, the nearby La Oroya smelting and refining complex, which Cerro de Pasco owned, and its original base metal operations, were both nationalised in 1974, and vested in the government-controlled mining company Centromin. Although nationalisation inevitably led to inefficiencies and lost opportunities there were some successes. The Cerro Verde copper mine was opened, and treatment complexes for copper at Ilo and zinc in the Rimac Valley were built.

At the time of the coup the notorious Maoist group, the Shining Path, was founded but kept its head down until 1980 when civilian presidential elections were held, leading to the start of change of attitude towards private and foreign economic activity in the country. For the next ten years or so Shining Path was a thorn in the side of the government and its terrorist activities acted as a brake on foreign investment interest, including mining investment, as it had control of large parts of the countryside. Its leader was captured in 1992 and its activities and control subsequently collapsed. At this stage the government's increasing interest in liberalising the economy and encouraging foreign investment acquired momentum.

By the end of the 1980s the Peruvian mining industry found itself in serious straits, losing up to $100 million a year and with production stagnating and labour productivity suffering for lack of investment. Metal price weakness did not help either and it was clear that the growing burden on the state that mining had become needed relieving. It was at this stage that Peru introduced the Mining Investment Promotion Law in 1991 and the process of de-nationalisation began in earnest.

By the start of the new millennium Peru had privatised over 90% of the state's mining interests. Following the passing of the General Mining Law in 1992, exploration and development of mines by the private sector became much easier, in line with worldwide trends following the fall of the Soviet Union. This has led to over a hundred foreign mining companies of all shapes and sizes becoming active in Peru and has led to the investment of billions of dollars into the industry. Of course not everyone has been happy with the return of foreign capital. At the local level there have been complaints that new mines have not brought major benefits to the still impoverished Peruvian countryside and that government revenues derived from mining have been only modestly spent in the areas where they originated.

CUBA

The nationalisation of primarily US-owned mines in Cuba by Fidel Castro in the early 1960s was prompted by much wider issues than some other nationalisations at that time. Castro had been a thorn in the side of Cuba's government, headed by Fulgencio Batista, for many years and when, after a number of failed attempts, he overthrew Batista in 1959 there was a short period where the US might have been able to steer him away from his natural inclination to cuddle up to the Soviet Union. This opportunity passed and Castro, spurned by the US, started to align Cuba both politically and economically with Moscow. The US was furious and started to apply sanctions against Cuba, which inevitably led to the expropriation of all US businesses on the island, including the 80% US-owned mining industry.

One of the first US companies to suffer expropriation of its Cuban investments was Freeport Sulphur, with its new laterite nickel and cobalt mine at Moa Bay. Freeport had spent around $120 million developing the mining operation in Cuba and on a treatment complex in Louisiana in the late 1950s, and Castro confiscated the mine in 1960. Freeport got no compensation from the Cuban government and had to write off the whole Cuban investment. Fortunately such a body blow did not endanger Freeport's ability to survive corporately as most of the loans for the Moa Bay mine came in the form of customer loans (mainly from US steel companies anxious for nickel feed) and these were secured against the project itself not against Freeport's balance sheet; though common now, this sort of finance was unusual in the 1960s. Indeed just a few years later Freeport pushed into volatile Indonesia to establish the Ertsberg copper mine using similar project financing techniques.

Freeport also ran the Nicaro Nickel plant for the US government, which had financed the plant in 1942 to supply US wartime needs. At the time of its nationalisation, Nicaro was operated by Nickel Processing Corp, joint-owned by National Lead of the US and private Cuban interests. Castro was careful about taking it over since in essence it was a US government business, indeed he was prepared for Freeport to run it, and also Moa Bay, but only if special tax breaks were ended. The Americans refused to agree on the grounds that such a move could endanger profitability and Castro swooped.

Nicaro required a special process to treat complex nickel ore, which the Cubans were unable to master, so ironically they later sold a 50% stake to Canada's Sherrit

Gordon in exchange for its technical expertise, much to the anger of Washington. Another of the nationalised mines was the Matahambre copper mine, originally financed by Manuel Luciano Diaz and opened in 1916. For many years it was one of the US's largest sources of foreign copper. Eventually it was taken over by American Metal Company and with the revolution fell into Castro's hands.

There were other somewhat smaller mines producing copper, iron ore and a variety of industrial minerals in both American and Cuban hands and they all were confiscated and placed in one of the Cuban state mining companies formed by Castro for the purpose.

The actual outcome of Cuban mining nationalisation over the years has not been as disastrous as it proved for Zambia and the Congo/Zaire. This was not really because of indigenous technical skills, although Cubans were well trained, but rather because of Soviet capital and management assistance. The Soviet finance did not survive the fall of the Soviet Union but before that time Cuba had begun to build a relationship with Canadian mining interests whose advanced technology was able to treat the difficult Cuban nickel ores and this relationship was expanded further.

In the 1990s when Castro slightly loosened economic controls, a few foreign mining companies began to explore Cuba's potentially significant mineral areas. Exploration for minerals, being an essentially speculative activity and thus alien to the bureaucratic mind, had been rather downgraded during the earlier Castro decades. However, we are unlikely to see a new surge of foreign mining activity until Cuba's dispensation post-Castro becomes clearer.

7. THE AUSTRALIAN NICKEL BOOM

The Australian mining boom of the late 1960s was given the generic title of *the nickel boom*, although it can be argued that nickel was, in economic terms, a relatively minor part of a period of exploration and new discoveries that saw the genesis of the giant iron ore industry in the northern part of Western Australia and the discovery of uranium in the Northern Territories.

In terms of nickel there were three major events – the discovery of nickel by Western Mining at Kambalda in Western Australia in 1966, the sensational but ultimately disappointing Poseidon discovery at Windarra to the north of Kambalda in 1969, and in 1971 the Selection Trust group's Agnew nickel discovery, which was further north still.

A FINANCIAL EVENT

Although Australia had spawned a number of mining booms in its past, the 1960s boom at times was as much a financial event as a mining event. As far as stock market activity was concerned, the surge of interest in Australian mining shares followed an extended worldwide boom in industrial, technology and financial shares, and was symptomatic of an era when confidence was high and investors, buoyed by profits elsewhere, were in the mood for speculation. The boom itself was precipitated by strengthening metal prices and in the case of nickel by an extended strike by Canadian miners in 1968, which pushed the metal's price to record levels as shortages bit. An extended economic upturn stimulated by US deficit financing to accommodate accelerating spending on the Vietnam War and a consumer boom underpinned rising metal demand.

Inflation too was in the air as the American economy geared up to provide both guns and butter, putting an increasing strain on price levels, thus inadvertently laying the ground for the revival of gold in the 1970s, and extending investment interest in hard assets (commodities) for a further decade. In addition to the war in Vietnam there was another key event that drove metal demand; the phenomenal expansion of the Japanese economy.

In 1952 Japan languished as a defeated and, in economic terms, lesser-developed country; 20 years later after two decades of 8% annual growth it was amongst the most powerful economies in the world. Such growth, mirroring that of China today, spawned a large and rising demand for raw materials, thereby creating the climate for an expansion in Australia's then ageing mining industry. Given the size of the Australian iron ore industry of today it is extraordinary to think that before the discovery of the Pilbara mines Australia actually operated an export ban on iron ore, so limited it was thought were indigenous supplies.

POST-WAR EXPLORATION

Whilst the Aussie boom can be said to have started with the Kambalda nickel discovery in 1966, prospectors in Australia had been busy since the 1950s and with some considerable success. Indeed the successes of the earlier years almost certainly whet the appetite of nascent mining entrepreneurs, providing the impetus for the late 1960s frenzy that was most famously exemplified by Poseidon. In the early 1950s tungsten ore (scheelite) was discovered at King Island off Tasmania. The Mary Kathleen uranium deposit near Mount Isa was found in 1954. Huge bauxite deposits were discovered in 1957 in Western Australia, allowing Australia to cease importing the mineral for the Bell Bay aluminium smelter.

This was followed by the Weipa bauxite discovery in the early 1960s, about the same time as the giant Pilbara iron ore region was being opened up. By the mid-1960s the Queensland coal basin was under development. Other large bulk mineral deposits were discovered around the same time, such as manganese in the Gulf of Carpentaria and the Duchess phosphate deposit in western Queensland. It is little wonder then that these local successes attracted the attention of increasing numbers of international mining groups, large and small alike, and from metal processors such as US Steel, Sumitomo Metals and Union Miniere.

As we have seen, nickel was not always a desirable metal to find and was often cursed by old prospectors. However by the 1960s its use in stainless steel making and other technologically advanced applications had transformed its image – it was highly sought after and Australia did not have the metal. Western Mining, incorporated in 1933 to search for gold in Western Australia, had become a sizeable gold producer around Kalgoorlie, but by the 1950s the industry was in rapid decline and Western Mining began to diversify its metal interests. It had some considerable luck in that regard, discovering the Darling Range bauxite deposits near Perth and also the Koolanooka iron ore mine near Geraldton, Western Australia.

The company's discovery of nickel near Kambalda, south of Kalgoorlie, was, however, something of a fluke. In the early 1960s Western Mining was involved in a regional mapping exercise, mainly but not exclusively for gold, and by 1964 was working around Kambalda. This led an old prospector, John Morgan, who had himself been working in the area 20 years beforehand, to show Western Mining some old samples from a dark gossan outcrop which, whilst containing no gold, had shown some anomalous nickel values on earlier examination at the Kalgoorlie School of Mines. After conducting a surface exploration programme in 1965 Western Mining commenced a drilling programme at Lake Lefroy and in January 1966 struck nickel, to be precise 10 feet of 8.3% nickel at 500 feet. The Australian nickel boom had been ignited.

THE KAMBALDA DISCOVERIES

Western Mining's initial success was followed by a number of other discoveries and indeed, though there have been major corporate developments concerning nickel mining in Western Australia over the last five years or so, new nickel deposits continue to be found today. The Kambalda nickel deposits are sulphide in nature, high grade, often quite shallow, and usually formed as lenses and thus mineable by decline methods. Their relative shallowness and high grade enabled Western Mining to develop mining operations at Kambalda rapidly.

From the initial Lake Lefroy discovery in early 1966 it took a mere 13 months for the company to produce its first ore from the Lunnon Shoot. Thereafter other shoots were discovered including the Fisher, Juan, Otter, Durkin and Long Shoots; some of these operations are still producing today although WMC, as it became known, exited mining in Kambalda in the early years of the new millennium. Production levels climbed above 50,000 tonnes in the 1980s and 1990s although they only represented around 10% of world output at best and were a quarter of the production levels of Canadian giant Inco. Nevertheless the Western Mining discoveries were important from both a geological and historical point of view.

The discoveries were even more important for their impact on the Australian stockmarket, for Western Mining entered the sixties with its shares trading listlessly at around A$1.20 (15 shillings in old money converted at the current sterling/A$ rate). By the time the Kambalda discoveries had begun to produce, and Western Mining had been able to re-start dividend payments at the end of the decade, the share price had already peaked at A$95. This was a stupendous performance, but one that was comfortably topped, and in a much shorter period, by arguably the most phenomenal mining share of the 20th century – Poseidon. Although Canadians might argue for their own champion, the late 1990s shooting star Bre-X.

THE RISE OF POSEIDON

The great Poseidon was a classic example of a company being in the right place at the right time, and with the right product, or in its case the right prospect. The story itself can be told relatively quickly, although in due course it probably deserves a book all to itself so stunning was its impact on the securities markets of the UK and Australia. Poseidon NL, which was named after a famous Australian racehorse rather than the mythical King of the Sea, had started as a wolfram producer in 1952.

By the time Western Mining began to uncover the Kambalda nickel deposits in 1966 Poseidon was destitute and its shares traded at 2c(A). The Kambalda discoveries alerted a number of Australian entrepreneurs to the possibilities of the mining sector as a source of opportunity. Poseidon, quite inadvertently, attracted the attention of two separate players, Boris Ganke and Norman Shierlaw. Ganke had built up a 25% holding in Poseidon in 1967 but was diluted down when Shierlaw vended the Bindi Bindi nickel prospect in Western Australia into the company. Bindi Bindi proved interesting but was not big enough or rich enough to be economic and it was dropped. However those in Poseidon at that time saw their shares rise from 10c(A) in 1967 to a 1968 peak of A$3.45.

Poseidon, during this run, had hired Ken Shirley, an experienced prospector and friend of Shierlaw's, to look for other prospects for the company, and he came across some interesting ground at Mount Windarra northwest of Laverton in Western

Australia. An attempt was made to joint venture the ground with Consolidated Gold Fields of Australia but they thought it held little promise and declined. Poseidon then decided to seek the view of its consulting geologists, Burrill and Associates. Burrill, in contrast to CGFA, liked what they saw and with Shierlaw's permission bought Poseidon shares at around 60c(A) in April 1969. The shares doubled but then sank back to 80c(A) following the release of Burrill's positive report. Poseidon then commissioned an exploration programme on Windarra in August, and in September began percussion drilling. The rest, as they say, is history, and dramatic history at that as the share price rose from 80c(A) in September to A$280 five months later.

At the beginning of October the first Windarra drilling results were announced and they included a 40-foot intersection assaying 3.56% nickel starting at 145 feet – a relatively shallow level – with lower grade intersections above, all the way to the surface. Later on considerable doubt was thrown on the accuracy of these figures but in anticipation insiders had been buying the shares aggressively. With the announcement Poseidon moved up to A$12 but its run had hardly started. A week later they had reached A$20 and as October wore on the market began to be influenced by London interest and by the end of that month the shares had reached A$37. In mid-month there had been some disappointing results from two other holes and the shares had temporarily stalled, but the market had the bit between its teeth and was not going to let a good story spoil. At the same time the excitement generated by Poseidon infected the rest of the Australian mining sector, and mining shares, particularly junior explorers, rose across the board.

In the middle of November Poseidon made another announcement. Though the orebody was growing in size there was no repeat of the earlier 3.56% grade. The average nickel grades came in between 2% and 2.5% and there was copper and even some platinum and palladium. Poseidon's shares took heart and by the end of November stood at A$57. Poseidon was also being supported by a rocketing nickel price, stimulated by a labour strike at Inco's giant nickel mine in Sudbury, Canada.

December saw Poseidon's share price continue to progress, and by mid-month it had reached A$84 in anticipation of the AGM when further drilling news was expected. By the time of the AGM on 19 December, attended despite its closeness to Christmas by investors and analysts from all over the world, the share price had reached A$110. Curiously the main issue at the AGM was a long and at times angry debate about an earlier placement of Poseidon shares at a price way below A$110. Shareholders had to wait two hours for news of progress at Windarra, but seemed well pleased by confirmation of the 2% plus grade and the enthusiasm of the board concerning prospects for the continuing drilling programme.

CRAZY CHRISTMAS 1969

The week following the AGM Poseidon's shares went mad. On Monday they closed at A$175 and on Christmas Eve they rose to A$182. In London the market on Christmas Eve was open for only half a day but witnessed some of the heaviest share trading ever seen in London, with most of it centring on the Australian pitch. In hindsight 24 December 1969 marked the peak of the Australian mining boom, although the tail was a long one, with the Agnew nickel discovery of Selection Trust not occurring until early 1971 and other discoveries, many of them in metals other than nickel, happening in between.

Poseidon also was not finished as the New Year loomed and on New Year's Eve it broke through A$200, closing the old year at A$210, a year it had started at below A$1. Its course was almost run, although by early February it had reached a new peak of A$280. Some of that further buying related to research work by London brokers Panmure Gordon who had prepared a major report on Poseidon that valued the company's shares at between A$300 and A$382.

THE BOOM BEGINS TO EBB

As summer loomed in the northern hemisphere Poseidon went into a sickening dive, falling back to A$64 by mid-year. Then, inspired by buying from Mineral Securities, the nascent but ultimately doomed Australian mining finance group, the shares doubled to A$144. Minsec's interest, however, was not matched by others and by December Poseidon had slumped to A$39. One of the most extraordinary events in modern mining history was over. By 1976 the company had de-listed from the Sydney Stock Exchange, and although it returned to the market as a gold producer within the Normandy group in the 1980s it was literally a shadow of the phenomenon it had been.

Whilst Poseidon is remembered for its stock market fireworks, a mine was finally developed at Windarra. However a number of excellent opportunities to finance the mine on reasonable terms were passed over. Also Poseidon was undermined by a number of problems, largely of its own making. In particular it was reluctant to enter into a joint venture with a nickel user like Amax of the US who would have appreciated the secure supply, and even worse, as it dithered it missed the zenith of the nickel price and nickel demand.

Production at Windarra was finally achieved in 1974, the mine having become a joint venture with Western Mining in the middle of the 1970s economic slump when attention had switched almost entirely to gold. Windarra also turned out to be a rather modest mine producing less than 10,000 tonnes of nickel a year before Poseidon NL folded in 1976, which compared with expectations, as a result of

Panmure Gordon's famous research note, of around double that figure per year. Poseidon's life as a producer was short and after it went under its half of the Windarra mine was bought by Western Mining, where it was always a modest part of the group's nickel production profile.

Apart from Western Mining's Kambalda discoveries and Windarra, a number of other nickel finds were made during the boom. Metals Exploration and Freeport Sulphur developed two mines, the small Nepean mine in Western Australia and the much larger Greenvale nickel laterite mine in Queensland, which was always a marginal producer. Great Boulder/North Kalgurli, Carr Boyd Minerals and CRA Group/Anaconda were others with small Western Australian nickel mines. There was also the American consortium Union Oil/Homestake/Hanna that had a deposit just south of Windarra which was eventually developed in conjunction with Mount Windarra. None of these operations remotely matched Kambalda which, in turn, as we have already seen, was itself a modest producer in terms of world nickel output levels.

Right at the end of the nickel boom in 1971 London-based Selection Trust and diamond producing subsidiary Consolidated African Selection Trust (CAST) announced the Agnew nickel discovery. This was a high-grade discovery located north of Windarra. If it had been found in 1969 when the nickel boom was at its height it would undoubtedly have been like pouring petrol on a fire in its effect on the mining market.

As it was there was keen interest initially in the discovery and the shares of both Selection Trust and CAST performed strongly for most of 1971, as speculation about the size of a possible mine swirled around the market. After that a weakening nickel market and the serious, drawn-out business of developing a mine at Agnew led to a deflating of interest. Ultimately a mine was built there but rather like Windarra it was on a smaller scale than originally hoped and a dull nickel price ensured that the mine was rather marginal. In the end Agnew became part of the ubiquitous Western Mining.

A MODEST MINING EVENT

Poseidon was a stunning event in financial terms, able still, almost 40 years later, to amaze even denizens of the 1999/2000 TMT boom, but from a strictly mining point of view it was a relatively small event. Despite this its importance cannot be overstated for it underlines the attractions of the mining industry to risk capital, the lifeblood of the industry over the centuries. It is the prospect of being involved in another Poseidon that drives so many investors to provide mining juniors with the funds to explore and hopefully to find. Without this access to capital it could be argued that the mining industry would waste away, though no doubt there will be some who would be prepared, against all the evidence, to put their faith in a publicly-funded mining industry relying on bureaucrats to allocate both risk and development capital. Mineral-rich Venezuela, for example, has embarked on this course.

Another consequence of the Australian mining boom was the establishment by stockbrokers of integrated mining teams, particularly in London, to follow what was considered to be a sector requiring specialist coverage. In London this was a relatively new phenomenon, indeed in the largely English speaking stock markets in the 1960s, dominated inevitably by the US, the rise of sector specialist research analysts was a new development, let alone specialist teams covering sales, corporate advice, research and trading as an integrated unit. This cross-fertilisation of investment disciplines, which started with mining, prepared the ground for the development of sophisticated solutions to corporate project financing that we enjoy (sic), and often suffer from, today.

However, as we have seen earlier, these speculative activities do act in an important way as oxygen for the industry, and in the case of the nickel boom and its aftermath they also encouraged the Australian government and the capital markets to tighten the regulatory regime, something that continues to this day around the world. It is also interesting to observe that one of the largest brokers in Australia in the 1960s, and at the centre of the nickel boom, Patrick Partners, behaved in such a risky manner, essentially operating outside its competence, that it brought the company down. The investment banking/broking world of today suffers, some would say, from the same dangerous addiction to risk, with its own fatalities!

8. THE MINING PROMOTERS

Although this book is primarily a history of mining and miners we should not ignore the activities of promoters and financiers, a breed which have been prominent throughout mining history, and in particular thrive during periods of high excitement like the Aussie boom. Their efforts are, as often as not, unsuccessful, for their investors that is; many seem to be in business largely to separate investors from their money. The following description from London mining analyst Rob Davies's *The Mining Promoter's Handbook* sums things up succinctly:

> "Never forget that the object of the company is to raise money for the benefit of the directors. It is not the aim of the company to find anything. That makes life far too complicated and the management might have to manage something."

In this chapter we will look at a few of these characters starting with a couple of Englishmen who found ripe pickings in Australia.

HORATIO BOTTOMLEY

Long before the Aussie nickel boom one of the most famous of all promoters of largely worthless Australian mining paper was the legendary Horatio Bottomley, an Englishman and also a Liberal MP at Westminster. Born in 1860 and orphaned at age four, Bottomley worked his way up from office boy to partner in a legal shorthand firm. He was a highly persuasive man and was very skilled in avoiding legal sanction, particularly during his days as a promoter of Australian mining stocks between 1893, ahead of the great Kalgoorlie gold rush, and 1906. He conducted his scams from London, promoting over 50 Australian mining stocks over the period. Sometimes he used umbrella-financing vehicles like West Australian Joint Stock Trust and other times it was direct *explorers* like North Eastern Associated Gold.

He also made large amounts of money out of genuine projects like Great Boulder Mines to add respectability to his business. Underneath he manipulated the accounts of many of his promoted companies to siphon off cash into his pocket. Another of his tricks was to raise money for a company and then use some of that to pay a first dividend – with shareholders relaxed and happy he could then begin to skim the company. In the end Bottomley exited the Australian mining field and turned to British stocks where he was unsuccessfully tried for fraud. As a result of this he had to resign his parliamentary seat. During the First World War he was a prominent recruiter for the armed forces, but eventually was successfully prosecuted for fraud over a Victory Bond trust he set up in 1918 and was sent to prison. He died penniless in 1933.

CLAUDE DE BERNALES

Other colourful mining promoters who operated in the Australian industry included Claude Albo de Bernales, an Englishman like Bottomley, but who based himself in Australia in the 1930s. De Bernales's activities were mostly above board and many of his promotions like Wiluna Gold were substantial companies. However he did frequently manipulate situations to his advantage and this eventually led to his downfall when the London Stock Exchange suspended a number of his companies, forcing them into eventual liquidation. Although investigated over many years by the British authorities de Bernales was never charged. He died in 1963 in London, a recluse.

But mining promoters were not always dubious characters! William Robinson, one of the major Australian mining financiers after the First World War, was synonymous with the rise of Melbourne's Collins House Group that included companies of the stature of North Broken Hill and Western Mining. More recently Lang Hancock, who we will meet later, was the leading promoter of the prospects for iron ore in Western Australia.

HAROLD LASSETER (1880-1931)

Over the centuries gold prospecting has become arguably the most glamorous of all mining exploration activities. Literature and the cinema has underlined this with stories of greed and daring in the pursuit of this rare, beautiful and valuable metal. Great stories have grown up around gold, typical of the genre is that surrounding Australian prospector, Harold Lasseter, whose name is synonymous with one of the most intriguing tales of gold exploration and discovery – the search for Lasseter's Reef.

Lasseter was born in Victoria in 1880 and ran away to sea as a youth after getting into trouble with the law. Sailing off the coast of Queensland in the late 1890s he heard of ruby deposits near Alice Springs and left his ship to become a prospector. He found no rubies but decided to push west hoping to find work in one of the mining camps. The story was that during this trip he found extensive shows of free gold in the desert on the border of Western and South

Australia. He almost died of exposure but was rescued by a camel driver who took him to a camp run by a government surveyor called Joseph Harding.

In due course Lasseter claimed that he and Harding went back to the area where Lasseter had originally found the gold and further rich gold shows were recorded with values up to 3 ozs per ton. The two attempted to interest investors in Western Australia to back a full-scale expedition to the gold deposit but there were no takers. Disappointed, Lasseter then went back to sea and pitched up in the US where he got married to Florence Scott in 1903 and took out US citizenship; later he and Florence had a daughter and a son. He returned to Australia in 1908 and settled in Tabulam, New South Wales. He worked on the roads and bridges being constructed in the NSW and also transported horses to the Sydney sales.

He claimed during this time that he attempted to return to his gold discovery but the group of men he hired had no stomach for the trip and it was aborted. In 1914 the war broke out and Lasseter took the family south to Melbourne where he enlisted. Away from home during the war he became an inveterate womaniser and is said to have had several children by various women. After the war he became a contractor and worked on the Sydney Harbour Bridge.

His desire to locate Lasseter's Reef again had not been dimmed by the passing years and in 1930 Lasseter approached the Australian Workers Union to help him equip another expedition to the Central Australian Desert. The story he told of the rich gold reef intrigued John Bailey of the union and he engaged two experienced mineral men, George Sutherland and Fred Blakely, to assess Lasseter's story. In 1930 a new company was formed, Central Australian Gold Exploration (CAGE), to finance an expedition, which included both Sutherland and Blakeley, to locate Lasseter's Reef. The story of the expedition so excited the imagination of the public that backing was easy to obtain this time.

Lasseter was wary of his co-explorers and planned to peel off from the main expedition and join up with a dingo scalper, Paul Johns, to press on to the Reef. Problems arose over equipment, supplies and water and the expedition made slow progress. At one time Fred Blakeley, with whom Lasseter had a poor relationship, had to return to Alice Springs for supplies, but Lasseter and Johns remained behind to continue the search on camel. During this period Lasseter went off on his own for a couple of days during which time he claimed that he had relocated the

Reef. The main CAGE expedition got bogged down with problems with trucks and planes, and in the end it was disbanded.

By now Lasseter was operating on his own as Johns had also returned to Alice Springs for supplies and new camels. In 1931 the old members of the expedition worried about what had happened to Lasseter and launched a search for him. Lasseter's body was eventually found in the protection of an aboriginal community who had found him close to death in the desert and then nursed him until he died some weeks after. The aboriginals insisted that his body lie in peace where they had buried him to assuage the spirits; in 1957 his body was exhumed and finally buried in Alice Springs.

Did Lasseter's Reef ever exist?

Some were sceptical, thinking that Lasseter's story was uncannily similar to one told of a major gold discovery supposedly made by an expedition in 1895/96 led by a Western Australian bank official called Earle. Indeed over the years more than 15 expeditions were undertaken to find what was known as 'Earle's Reef' which was thought to be located in an area close to the Western and South Australian border. There is another theory that in fact Lasseter was not the discoverer of the Reef but Joseph Harding was, and he told Lasseter where the reef was to be found. Lasseter misunderstood Harding and thought the reef lay to the west of Alice Springs whilst it actually lay to the east.

It remains a moot point whether there ever was a Lasseter's Reef. Perhaps it was just an elaborate promotion; a number of prospectors similarly mounted expeditions in the years after Lasseter's death to try and locate the Reef and failed. Despite the sceptics the legend of Lasseter's Reef survives and has come to embody the romance of gold exploration, and the hope that out there in the vast Australian outback there still lie gold deposits that can make a man fabulously rich.

OTHER PROMOTERS INCLUDING DISRAELI

Promoters were not the exclusive preserve of Australian mining, Cecil Rhodes in the 19th century was a major promoter of South African gold and diamonds, as was his rival Barney Barnato. In Canada Murray Pezim is a fairly recent example of that country's rich promotional tradition with his efforts over the Hemlo gold camp in Ontario and BC's Eskay Creek and Snip gold projects in the 1980s.

One of Canada's most colourful and aggressive mining promoters is American-born Robert Friedland whose scalps include the Voisey's Bay nickel mine in Labrador, the Fort Knox gold mine in Alaska and the Oyu Tolgoi copper and gold prospect in Mongolia. Less gloriously Friedland has also been involved with promoting mining investment in Burma and his Galactic Resources Company developed the Summitville gold mine which was closed in 1990 when its waste dam collapsed. One might also add the name of Benjamin Disraeli, who we discussed earlier, to any list of mining promoters; his mining pamphlet on Latin American prospects was pure puff.

A FEMALE MINING PROMOTER

Most mining promoters were men but women have also occasionally engaged in the practice. One of the earliest was Julia Moffett who in the late 19th and early 20th centuries very profitably promoted mineral prospects in the US for herself and Hopper & Bigelow, the New York stockbrokers she worked for. These included Victoria Chief Copper, of which she was Treasurer, and Sierra Consolidated. Mrs Moffet, who was actually the Bigelow in the broker's name, claimed that she personally investigated prospects for the firm and its corporate clients.

After extensive field visits she and senior partner Robert Hopper then promoted the stock to the firm's investment clients. The claims were largely worthless and eventually the authorities had Hopper & Bigelow closed down. With legal suits pending Mrs Moffet disappeared and was not heard of again until her death in 1913. She had not gone far though; she had lived on East 125th Street, not a million miles away from the Broadway offices where she had helped hatch Hopper & Bigelow's mining frauds.

9. GOLD COMES TO THE FORE AGAIN

Whilst the 1960s had seen a remarkable recovery in base metal demand driven by technological innovation, huge growth in Japan and the Vietnam War, precious metals and particularly gold had attracted little attention. As the 1970s unrolled gold began to push itself to the forefront with investors beginning to worry about inflation and rising political, financial and economic instability. Gold had remained unchanged in price since the American depression of the 1930s, and another commodity, oil, had also been stagnant in terms of price since the 1950s. Both were to undergo a massive lift in position and reputation.

GOLD AND INFLATION

The huge increase in gold supply in the wake of the Spanish conquests in South America in the 15th and 16th centuries was a major reason for the surge in inflation in Europe at the time, particularly in Spain. Since gold was money in those days it meant that a massive increase in gold and thus money in the Spanish economy pushed up prices as output in the pre-industrial age was completely incapable of keeping pace, so prices inevitably rose sharply to balance the surge in money supply. In the 1930s during the economic slump when there was a need to stimulate demand by arresting the fall in the price of goods and services, the US government took complete control of the internally-held gold of American private citizens, enabling it to push up the price of gold, without any private market interference, in the hope of stimulating prices in general and agricultural prices in particular.

The rise in the gold price in the 1970s was a result of different influences again. Inflation had begun to take hold in the advanced world in the late 1960s and accelerated throughout the following decade and into the 1980s. Gold output, dominated by South Africa with the Soviet Union also a large producer, was stagnating and with production costs rising rapidly a supply squeeze loomed. When the US closed the *gold window* in Washington in 1971 – where holders of overseas US dollars could exchange these dollars for gold at the fixed price – attention was drawn to the fact that, in particular, the French had been switching into gold from the dollar, and the US no longer could support this leaking away of its gold bullion reserves. Demand for gold started to increase as people began to see it as a rare asset that acted as a hedge against inflation.

In these three instances we see gold as firstly a cause of inflation, then secondly as a tool to kick-start price growth to avoid deflation and economic slump, and thirdly as an asset to be used as a hedge against inflation. It can be legitimately argued that gold has a very close relationship with the financial phenomenon of inflation but what is not so clear is what that relationship is. Today we are most likely to see gold as a real asset and therefore a hedge against inflation and increasingly as an alternative to the dollar and other currencies suffering from financial turmoil, which means, as it did in the 1930s, that gold can also provide protection in times of deflation.

SOUTH AFRICA

Returning to the 1970s, South Africa was by far and away the largest gold producer, followed at some distance by the Soviet Union. Considerably behind these two were Canada, the US and China, and other smaller producers included Ghana, Australia and Peru. As the new age of gold dawned the market was in the embarrassing position that the apartheid pariah, South Africa, was not only the largest producer but had the only substantial listed gold mining companies. Also South Africa had a number of new mines, or still-expanding older ones, which further underpinned Johannesburg as the quality place for international gold share investors. Forty years later the situation has changed dramatically. The table shows figures for gold production by country in 1970 and 2011.

GOLD PRODUCING COUNTRIES

	1970 (tonnes)	2011 (tonnes)
South Africa	1,000	187
Russia (Soviet Union)	218	198
Uzbekistan (Soviet Union)	78	88
Canada	75	110
US	54	235
China	47	350
Ghana	22	99
Australia	20	268
Peru	3	150
Indonesia	n/a	118
Others	n/a	869

Source: Goldsheet Mining Directory; Gold 1981: Consolidated Gold Fields (estimates)

The South African gold mining industry had started in the 19th century in the area which is now known as the Central Rand but as the decades passed mining pushed out east, west and then southwest. This created the present crescent shape of the total Witwatersrand goldfield. At the same time the easy rich surface and near-surface gold deposits of the original discoveries had been worked out. The new fields such as the Orange Free State, Evander and Klerksdorp/Vaal Reefs fields developed after the end of the Second World War, and the West Witwatersrand field developed in the 1930s, were deep level, large-scale undertakings. They were also, certainly by today's standards, rich, with grades up to 32gms (1oz) per tonne in the 1960s and early 1970s at high-grade mines such as West Driefontein. The table shows South African gold production and grades at points from 1900 to 2009.

SOUTH AFRICAN GOLD PRODUCTION AND GRADES (1900-2010)

	Tonnes	Average grade gms/t	Percentage of Western world output
1900	11	n/a	3
1910	234	11.6	34
1920	254	11.3	51
1930	334	11.2	51
1940	437	7.2	35
1950	363	6.4	44
1960	665	10.1	62
1970	1,000	13.3	79
1980	675	7.3	70
1990	605	5.1	35
2000	428	4.5	20
2010	192	3.0	8

Source: SA Chamber of Mines

Since many of South Africa's mines operated at depths down to and even below 10,000 feet and had very large workforces with limited mechanisation, and the gold price had been fixed at $35 per oz since 1934, grades had to rise during this period in order for the mines to make an economic return on their very substantial investment.

However as the gold price finally began to rise following the closing of the US gold window in 1971, mines were able to lower their grade and thus access ore that at a higher gold price could be upgraded to the mineable reserve category. This led to a steady decline in South African gold output, but not in mine profits and taxes. This also coincided with a strengthening of worldwide opposition to the country's apartheid political system. This meant that traditional exports of raw materials, farm products and basic industrial goods were restricted; for example Canada banned SA wine imports and Denmark stopped buying coal, although Japan went on importing SA coal.

Since South Africa needed to import high-value technical goods and often this had to be done in an expensive roundabout way using third parties to avoid trade bans imposed by some countries, the revival in the gold price couldn't have come at a better time. Of course there was frequent talk of trying to ban South African gold but with the Swiss, who do not mix morals with business, acting as the clearing house for SA metal there was no chance of such a idea getting off the ground.

20-YEAR GOLD PRICE HISTORY IN US DOLLARS PER OUNCE

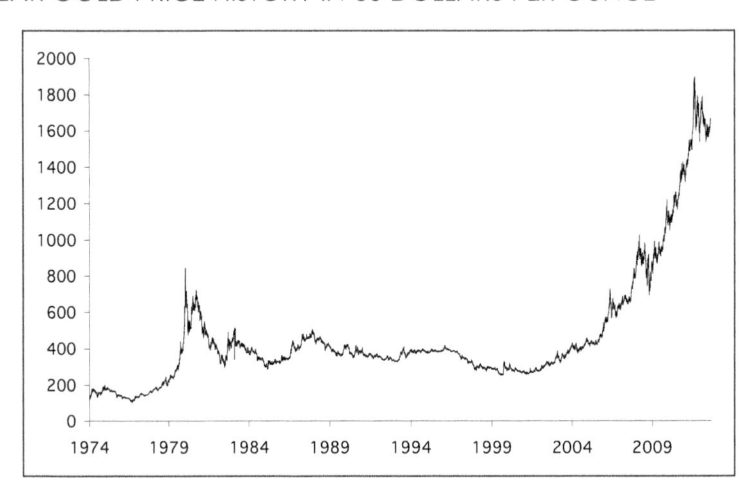

The SA gold share market remained active during this increasingly tricky time for the country with financial markets paying only lip service to external disapproval of apartheid. Ironically, when apartheid was dismantled in the early 1990s trading in SA gold shares fell away steadily and eventually dramatically as industry rationalisation, a weak gold price and squeezed operating margins undermined investor interest in the sector. To illustrate, the bull markets in SA gold shares since 1971 have occurred over the following periods – 1972/74, 1977/80, 1983/86, 1989/90 and 1993. Each of these runs took place against the background of a National Government, during the apartheid era, although in 1993 talks with the African National Congress (ANC) over the shape the new dispensation would take after the elections in the following year were well underway.

The 1990s also saw South African production begin to slip, as we noted earlier, and production from other countries such as Australia, Canada and the US began to close the gap with South Africa. Indeed the last 20 years has seen very few new gold mines developed in South Africa – South Deep and Target are the two biggest. The trend has been one of retreat, as mines have grown old, and, following substantial amalgamations and asset swaps, old shafts in uneconomic or exhausted areas have been closed. This has led to a major reduction in the workforce on the gold mines. Another irony is that, with the end of apartheid, job opportunities in the gold mining industry have been much reduced, although long years of a weak gold price rather than political change in South Africa have caused this.

The South African industry has also changed in that the giant personalities of the founding era of the late 19th century, the Randlords, have been largely replaced by far less exuberant characters. Whilst mining executives of the post-Second World War decades such as Harry Oppenheimer of the Anglo American group, Robin Plumbridge of Gold Fields, Barry Hersov of Anglovaal and Ted Pavitt of Union Corporation were able mining men, none of them would have described themselves as colourful in the style of Cecil Rhodes or Barney Barnato.

Today's mining executives are arguably even more anodyne, although clearly very competent to run what have become giant enterprises. Perhaps the size of the modern mining groups precludes too flashy and individualistic a style of management. But in South Africa one man did, in the period bridging the new millennium, give us a flash, perhaps the last flash, of what a Randlord might have been like – that man was Brett Kebble.

BRETT KEBBLE (1964-2005)

One of the most sensational events in recent South African mining history was the shooting in 2005 in Johannesburg of Brett Kebble. Perhaps one of the last of the old style *freebooting* mining entrepreneurs, Brett Kebble was not a miner by training but a lawyer. He was born in Springs on the East Rand and went to school at St Andrews in Bloemfontein and then to the University of Cape Town from which he graduated in 1986. He stayed in Cape Town to practice law for a few years before joining forces with his father, Roger Kebble, who was a practising mining man of the old school. In 1990 he married his wife Ingrid and they had four children. In the mid-1990s the two Kebbles came together with a distinguished group of mining practitioners and investors, including London's leading gold guru Julian Baring and Adam Fleming of the eponymous banking family, to turn the cosy South African gold mining industry on its head through the acquisition of Rand Mines.

Over the next ten years Brett Kebble became a major force in SA mid-size mining companies including Randex, Randgold Resources, JCI, CAM, Harmony, Western Areas and Durban Deep (DRD). At one stage there was an attempt to amalgamate all of what were generically known as Kebble companies into a new offshore company listed in Canada. Mark Wellesley Wood, a Brit who later clashed with the Kebbles, was hired to head the new entity. The deal fell through, leading to a number of companies in the group effectively going their own way, including Randgold Resources to list in London and Harmony to expand into Papua New Guinea.

Like so many past SA mining entrepreneurs Brett Kebble found it easy to fall out with previous colleagues and confidants. Perhaps the most damaging of the fall-outs was with Wellesley Wood who became CEO of DRD, the marginal Central Rand gold producer. A number of alleged irregularities were uncovered, including the whereabouts of assets owned by DRD's Australian subsidiary, and Wellesley Wood pointed the finger at Brett Kebble. Accusations and counter accusations flew between the two sides, enriching the lawyers if not the shareholders. In the end Kebble found that his somewhat cavalier approach to the assets of companies he controlled, or had a major stake in, severely damaged his reputation. As a result he began to lose directorships and attracted the attention of both the SA authorities responsible for securities and the JSE itself.

Kebble was also a highly faulted character in terms of his corporate strategy. In the early 1990s he claimed to be a big gold bull and was resolutely opposed to hedging. When Western Areas (WA) needed long-term finance for its South Deep mine he arranged some of the most toxic hedging tactics in the industry, which WA shareholders paid dearly for and which led to the company losing its independence. He also moved cash and assets around Kebble companies without clearance from his board colleagues; occasionally these were transparent loans, often they were unrecorded. Later that decade he became a strong supporter of the ANC as a result of the change of government in 1994, and in the new century he embraced black empowerment and Africa-wide investment – initiatives that unfortunately did little for shareholder value.

In the end Kebble's innovative practices caught up with him and control of his remaining three interests – Randex, JCI and Western Areas – was wrested from him and he was sacked as a director. Not long after this he was shot whilst apparently standing by his car having a conversation with 'person or persons unknown'. By then Brett Kebble had made so many enemies that a score of motives were put forward for his murder/assassination.

A year after the murder Glenn Agliotti, a businessman alleged to have links to organised crime and also contacts in high places in the new South African establishment, was charged with Kebble's murder. Agliotti and his hired hit-man Michael Schultz were sent for trial in 2010, although by then the mystery had deepened with rumours that what had happened had really been an assisted suicide. In 2011 Agliotti was cleared of Kebble's murder. Kebble, like so many earlier Randlords, had a brilliant mind but he was fatally flawed and his end was more like that of a mafia 'don' than of a successful mining magnate, a role he had aspired to but failed to achieve.

PROBLEMS AHEAD FOR SOUTH AFRICA

As to the future of gold mining in South Africa, the outlook is not very good. Production has slumped, new mines are not being developed and costs on average are rising faster than the gold price. A huge surge in the gold price might be helpful in that, all things being equal, it could, if costs can be restrained and financing obtained at an attractive level, lead to development of the deep-level resources in the southwestern area of the Witwatersrand crescent. One of the leading groups in that area is Witwatersrand Gold.

Another geological theory being pursued on the Witwatersrand is that the crescent shape of the gold-bearing reefs is in fact a saucer but one that has dipped to the southeast, leaving the current goldfields in the north and northwest nearer the surface, albeit lying at considerable depth. The main group investigating this possibility is Afriore, a Canadian company now wholly owned by Lonmin. Since the depths at which gold might be found could well be much deeper than even the mineral resources of Wits Gold, the gold price would have to be well beyond $2,000/oz and/or the grades uncovered would have to match the earlier grades of the original fields found over 100 years ago. The chances of this are slim which means that South African gold production is probably going to keep on sliding, allowing China to cement its position as the largest gold-producing nation.

CAPITAL RAISING

There has also been a need to assess the historic method of raising capital for gold mines. The normal pattern in South Africa was for one of the mining finance houses like Anglo American or Gold Fields to explore gold leases, often held by the house for many years, and if they demonstrated potential for economic development the house would float a minority stake on the Johannesburg stock exchange to obtain capital for bringing the mine to production. Since the mining house retained both equity and operational control, and collected annual fees during the mine's lifetime, they got a very good financial deal out of the arrangement. They were also in a position to trade their own shares in the mine from a position of inside knowledge and with no capital gains tax in South Africa until the 1990s the fact that their original *in* price was so low made these historic portfolios of gold shares, when sold, a veritable treasure chest of risk-free earnings.

CONSOLIDATION IN SA

For most of the 20th century the South African gold mining industry consisted of a substantial number of individual gold mines (between 40 and 50 in the 1970s) largely controlled by half a dozen mining houses. Then in the space of no more than four years at the end of the century the number, driven by ferocious corporate activity, was reduced to less than ten. There were two reasons for this dramatic change, the first was that global investors wanted to hold very large, million oz plus gold producers not small producers, as many of the ageing SA gold mines were becoming.

The second reason related to a revolution that had occurred in the mid-1990s, described above, when a group of primarily European investors led by Julian Baring, the leading gold fund manager in London, took control of Rand Mines and ended the old system of individual mines being charged an inclusive administrative fee by the controlling mining house. Following this the industry plunged itself into a wholesale restructuring of the mines, with ownership changing as the mining groups swapped or sold mines amongst each other to establish a more rational ownership structure.

During this process many of the old mines lost their individual identity as shafts were rationalised, many being closed, and cost economies led to mine mergers within the newly structured houses. At the end of all this in the early 2000s the main SA gold groups were AngloGold, Gold Fields, Harmony and DRD. Following this some smaller gold mining groups like African Rainbow have been formed as part of South Africa's policy of empowering black investors by providing them with a stake in the country's mining industry. Whilst perhaps politically necessary this trend is unlikely to arrest the SA gold mining industry's decline.

Elsewhere gold mine exploration and development before the 1980s was carried out in a traditional manner, in the sense that equity rather than bank finance was used, but the whole process was much more individual because often the mines were controlled by the original exploration company not a large mining house. The need also to target high-grade deposits to counter the fact that gold's price did not really change between 1934 and 1971, and indeed had not changed much for the previous century or more preceding the 1934 price hike instituted by the US government, and this meant that ore tonnages mined were often not huge.

The South African mines were different in that they became very deep-level operations and the gold-bearing reefs, although often not very thick, were persistent which allowed for large-scale mining. The idea being that if you had to go down to great depths to mine the ore you needed to extract large tonnages in order to earn an economic rate of return. Small, relatively shallow gold mines, the norm outside South Africa, did not have quite the same issues and therefore did not need large tonnages for mining to be economic. One of the unique features of the SA gold mines was the requirement to mine ore based on the average payable (economic/profitable) grade. If the gold price went up then the grade was reduced, if the gold price fell or working costs rose sharply the grade could be raised to maintain profitability. Gold mines elsewhere usually mined the orebody to extract maximum profit from the beginning, which meant that they often did not have the flexibility to adjust their mining plan if the gold price fell or costs went up.

Over the last 20 years the financing of gold mines has moved towards debt finance, but supported by the use of derivatives to hedge the gold price. The first derivatives to finance mines used a simple forward selling technique, borrowing gold at very low interest rates from bullion banks against future production when the mine had been developed, and then selling the gold into the spot market using the proceeds to construct the mine. One of the first mines to do this was the Pancontinental-owned Paddington in Western Australia in the mid-1980s.

GOLD ELSEWHERE IN AFRICA

Whilst gold production in South Africa declined in the later decades of the 20th century it began to rise elsewhere on the continent, particularly in west and east Africa. The three countries that have led this trend have been Ghana, a well established producer, Mali and Tanzania. In recent years Burkino Faso and the Congo (DRC) have begun to develop gold mining sectors, although the latter has been a major base metals and diamond producer for many decades.

GHANA

After South Africa, Ghana is the largest gold mining country in Africa, producing around 86 tonnes of gold annually. Production in Ghana was around 80 tonnes in 1948 but steadily declined to as low as 8 tonnes in the early 1980s, which, despite almost twenty years of rising prices, took it back to output levels not seen since the depressed 1930s. Perhaps the best known mine is AngloGold Ashanti's Obuasi mine in Ashanti province in the south of the country, which started production in 1897 and over its life has produced more than 750 tonnes of gold. In 1897 Edwin Cade, a London merchant, acquired Obuasi, which had originally been called the Ellis Mine, and floated it on the London Stock Exchange as Ashanti Goldfields. It remained listed in London until it was acquired in the late 1960s by Lonrho, a UK incorporated African trading company which was headed by the charismatic Tiny Rowland who had been encouraged to make the acquisition by a Ghana government concerned by the mine's decline. Ashanti was de-listed from London and owned in tandem by the government (55%) and Lonrho (45%).

In the 1990s after years of economic drift the Ghanaian government decided to introduce liberal, free enterprise principles as promoted by the World Bank at the time. This led to a programme of de-nationalisation and encouragement of foreign capital which culminated in the re-listing of Ashanti Gold and the launching of an ambitious expansion programme at Obuasi. Unfortunately Ashanti fell under the spell of Goldman Sachs and purchased a whole string of derivatives to protect it against a falling gold price in 1999. The gold price actually rose as a group of central banks announced an end to gold sales for five years. As the gold price soared, Ashanti's derivatives went toxic threatening to bring the company down and its share price crashed from $25 to $4.50. In 2003 Ashanti, which had never properly recovered from its derivatives debacle, was acquired by AngloGold, then still an important part of the Anglo American group.

Other major gold groups engaged in the Ghana gold revival include Gold Fields, with its Tarkwa and Damang mines, and Newmont with its Ahafo mine. There are a number of junior companies also active including Noble Resources, Golden Star, Keegan Resources, PMI and Pelangio.

MALI

Mali's re-emergence as a gold mining country towards the end of the 20th century started with the development of the Sadiola gold mine of then Canadian junior IAMGOLD and South Africa's AngloGold. Mali's historic gold story is traditionally linked with the ancient desert city of Timbuktu, which for centuries was a trading centre for all manner of commerce including gold. During the long French colonial period of the 19th and 20th centuries Mali essentially disappeared from view although French and Soviet geologists did a substantial amount of non-commercial prospecting in the country in the 1960s and 1970s, and uncovered a number of areas of gold mineralisation. But its rise to be one of Africa's larger gold producers began in 1991 when it produced around 4.6 tonnes following the start of production at Sadiola, its first material gold production for a long time.

Gold production has risen steadily over the past twenty years to a current level of around 50 tonnes annually and is expected to continue to grow. One of the key factors in this growth has been the successful development of three new mines by FTSE 100 company, Randgold Resources. The emergence of Randgold Resources (RR) is an interesting story which starts with the ambitions of then South African parent Randgold & Exploration (R&E), part of the Rand Mines group of Brett Kebble, to locate to the UK in 1997, as a number of its South African peers like Anglo American, SA Breweries and Old Mutual were doing at the time.

The reason for R&E wanting to move to London was to gain access to foreign investment capital for its RR division which had been formed to explore for gold in Africa. The route which R&E chose was through a reverse takeover, to be engineered by UK stockbrokers Société Générale Strauss Turnbull, of the UK incorporated African trading group African Lakes PLC. Lakes was very much a relic of Britain's colonial past but it retained the valuable asset of a full UK stock exchange listing. Whether it was the approach of R&E's board or whether the South African authorities, still then wedded to stringent foreign exchange controls, simply did not like the idea of R&E's SA shareholders ending up with foreign listed equity, the takeover was blocked by Pretoria.

Following this setback it was decided that RR, which already had a number of seed capital shareholders following its formation in 1995 (including Rio Tinto's Australian subsidiary North), should float itself on the London market. This it did in 1997 during a very difficult period for mining and gold shares. RR had purchased BHP's interests in West Africa, centred on Mali, and amongst the assets acquired was the Syama gold mine which BHP had opened in 1990. RR struggled to overcome a number of operating problems and was forced to shut the mine in 2001, the mine eventually being sold to Resolute Mining. RR then set about developing new gold mines in the country following a successful exploration programme. Morila developed with AngloGold, Loulo – with its open pit and underground sections – and Gounkoto were opened between 2000 and 2011, and as a result of this group gold output and profits soared leading to RR's admittance to the FTSE 100 share index in London in 2008.

Whilst RR has become the largest gold producer in Mali other companies operate mines in the country including IAMGOLD and AngloGold, Resolute, Nevsun and Avnel, and Mali has become one of the top addresses for gold exploration in Africa.

TANZANIA

Another fast rising gold producer is Tanzania, an African producer with a centuries old tradition of artisanal gold mining which supplied the commodity needs of Arab traders passing down east Africa. During the period of German occupation of Tanzania from 1884 to 1918 gold exploration was carried out and eventually mines were developed in the area around Lake Victoria, which remains Tanzania's main gold mining area today. The first was the Sekenke Mine in 1909. In 1918 as a result of First World War reparations Great Britain became responsible for what was then called Tanganyika and gold mining continued to prosper with new mines being opened in the north west of the country, and peak production of around 3 tonnes in 1938 was achieved. With the outbreak of the Second World War gold mining was severely curtailed.

After the war gold mining re-started but when Britain granted Tanzania independence in 1961 industrial activity came under government control and gold mining declined rapidly, until in 1967 no gold production was recorded, a situation that broadly continued for thirty years. Starting in 1998 with the opening of the Golden Pride mine of Resolute Resources gold output has risen steadily, reaching around 45 tonnes in 2011 and making Tanzania the fourth largest producer in Africa. Amongst the leading producers are AngloGold at Geita and African Barrick at Bulyanhulu, North Mara and Tulakawa, with a substantial number of junior explorers, many of which are to be found with programmes in the Lake Victoria region.

RISE OF THE OPEN PIT

There has been a marked change in gold mining techniques over the last two or three decades with the rise of open pit mining leading to the mining and treatment of high volumes of low grade ore. Many of these pits like the Dome in Timmins in Canada can reach many hundreds of feet in depth and require small workforces using powerful capital equipment in the form of diggers and trucks. Indeed the deepest open pit operation in the world is the Russian diamond mine, the Mir, which goes down to 2000 feet, although Mir is exceptional.

This capital-intensive method of mining has its drawbacks as it is damaging to the environment and requires extensive reclamation, particularly if the mine is not in a remote location. Underground gold mines need much higher grades, like Goldcorp's Red Lake operations in Canada, but sometimes an open pit mine can become an underground operation when the pit begins to mature and there is a continuation of economic mineralisation at depth. Randgold's Loulo gold mine in Mali is one example of this. Perhaps though the most well known open pit, which has been developed from underground mines, rather the reverse of Loulo, is the Big Pit in Kalgoorlie in Western Australia, a structure so large that some claim that it can be seen from outer space.

Although some high-grade deposits are still being found they are rare and often quite small; the trend now is to develop very large low-grade (below 3 gms/tonne) open pit mines. These can be new operations or, like Kalgoorlie's Big Pit, mines taking advantage of massive low-grade mineralisation around existing operations. Another example of this is the Pamour open pit mine of the Porcupine Joint Venture near Timmins in Ontario, controlled by Goldcorp. This contemporary concept of large-scale mining of low-grade ore is interesting as its environmental effect can be much more detrimental than underground mining, although it is far less dangerous for the miners themselves.

In recent years as the gold price has risen, but production from established gold mining areas such as South Africa has fallen, the search for gold has increasingly concentrated on these large, low-grade prospects where the economies of scale and the ability of mining companies to access low-cost capital has encouraged the development of capital-intensive operations.

THE AUSTRALIAN REVIVAL

One of the most spirited revivals in world gold output has been that achieved by Australia in the last 20 years or so. When the gold price began to move away from its official level of $35 in the 1970s Australian gold output had fallen to a mere 18.7 tonnes per year, a level it maintained until the 1980s when output began to revive on the back of a strong gold price.

Much earlier Australian governments had become so concerned about supporting local gold production that they had in 1924 exempted local gold producers from paying taxes and royalties, an exemption that was withdrawn in 1991 despite much industry protest. A subsidy paid on the value of gold production had been withdrawn earlier, in 1973. Australian production had peaked in 1903 when annual output reached 118 tonnes; in that decade total production fired by the finds in Western Australia almost doubled to over 1,000 tonnes, 20 years later it had slumped 80%, leading to the granting of tax exemption.

The table on the next page shows Australian production by decade from 1851 to 2000. It can clearly be seen how production rose significantly in the 1980s and 1990s from the low levels it had been at since the 1920s.

AUSTRALIAN GOLD PRODUCTION (1851-2000)

Decade	Total production (tonnes)	Largest state producer	State production (tonnes)
1851-59	693	Victoria	619
1860-69	619	Victoria	504
1870-79	461	Victoria	311
1880-89	358	Victoria	215
1890-99	598	Victoria	212
1900-09	1055	Western Australia	551
1910-19	597	Western Australia	358
1920-29	196	Western Australia	148
1930-39	317	Western Australia	239
1940-49	308	Western Australia	226
1950-59	320	Western Australia	246
1960-69	268	Western Australia	202
1970-79	205	Western Australia	107
1980-89	782	Western Australia	582
1990-2000	3218	Western Australia	2423

Source: Australian Gold Net 2002

The 21st century opened with the Australian gold mining industry in good health as it built upon the successes experienced in the latter decades of the previous century. By the end of the first decade annual output stood at around 230 tonnes, below the plus 300 tonnes achieved in the mid-1990s but back on a rising trend.

NEW ZEALAND

Australia's close neighbour across the Tasman Sea also began to build a respectable gold mining industry with the discovery in 1878 of the Waihi district on the North Island, around 70 miles south east of Auckland. Waihi's first mine at Martha Hill had an early setback and financing from London in the shape of newly formed Waihi Gold Mining led to a change in the treatment method to cyanidation, which enabled the mine's profitability to be secured. Waihi was at the southern end of the Hauraki goldfield where the gold was found in vein structures which were extensive and in the case of Martha Hill widened at depth where further gold lodes were found.

In the first decade of the 20th century Martha Hill became one of the biggest and most profitable gold mines in Australasia, trumping many Australian producers. In 1905 it made a profit of a little over £725,000 from output of 171,000 ozs, the ore grade coming out at 0.63ozs (19.6gms) per tonne, a highly satisfactory result. Just under half the net profits were distributed in dividends, as was the practice then. The orebodies also contained useful amounts of silver – as much as 7 ozs to 1 oz of gold in the deeper sulphide ore.

The Martha Hill mine produced 5.6 million ozs of gold and 38.4 million ozs of silver over its life before being closed in 1952 as a result of falling margins caused by a stagnant gold price and eroding grades. But that was not the end of the Martha Hill operation. Soaring gold prices in the 1970s and 1980s led to the mine being re-opened in 1987 as an open pit operation and in 2004 a decline was driven to access deeper ore which had lain unknown to the original Martha Hill miners. Around the same time the Macraes mine in eastern Otago on the South Island was re-opened by Oceana Gold which also re-started mining the old Reefton orebodies.

In the late 1860s New Zealand produced around 1.1 million ozs annually from alluvial operations in Central Otago where the Gabriel's Gully discovery started New Zealand's first major gold rush. There were also operations on the west coast of the South Island and hard rock mines were developed near Reefton, also on the South Island. After that the rich alluvial operations rapidly petered out, although there were some successful dredging operations at the start of the 20th century and then again 50 years later. The peak of hard rock mining came in the first decade of the century with the development of mining operations to exploit the rich quartz veins found on the Coromandel Peninsula on the North Island where New Zealand's first gold had been discovered by Charles Ring in 1852.

Coromandel and nearby Karangahake were technical innovators in terms of gold mine treatment, with the former using stamp batteries to crush the hard quartz into powder to aid recoveries and the latter being the first user of cyanide to extract the gold from the crushed ore. Despite these early successes New Zealand has never been a first-rank gold producer, and indeed though its annual overall minerals output provides substantial foreign currency earnings for the country, it nowhere near matches that coming from agriculture and tourism. This places the country well down the list of mineral producers.

GOLD RUSHES TODAY

In the modern world mineral exploration tends to be conducted in a more scientific manner than formerly, the required scale tends to preclude the old prospector who often worked alone, and instead requires a more measured and professional approach, even when the company is small. Governments also tend to want a more organised structure. However in lesser developed countries it is still not unknown, although the genesis may be quite different from past rushes, for large gold rushes to occur – Brazil has seen a number over the years and in 2006 experienced a new one at Eldorado do Juma in the Amazon jungle near the town of Apui.

THE BRAZILIANS AGAIN

The Apui gold rush was very 21st century in terms of its genesis: the rush was started by a story which appeared on a Brazilian maths teacher's website and which told of locals mining large quantities of surface gold. No more is the hysteria of a rush necessarily generated by a thirsty prospector in from the outback showing the gold he has found to drinkers in a bar – rumours are now spread by the internet! Since the find was close to the Juma River the prospectors, who numbered several thousand, in some contrast to their 19th century predecessors arrived swiftly by motorboat. Apart from that the Eldorado do Juma site was a throwback to those earlier rushes with thousands of diggers working on tiny plots having completely cleared what a few months before had been dense and pristine jungle. Also, in time-honoured fashion, there were personal fortunes to be made by some – one digger mined $19,000 worth of gold in 17 days – but most made little or nothing.

Also, as with most gold rushes, a makeshift town quickly grew up around the diggings, with bars, restaurants, food shops and most importantly mining supply stores. If past gold rushes are any guide these services very likely made the owners of the businesses more money than most diggers did and even the people actually providing the services are likely to have done extremely well. There were also the inevitable land ownership disputes (the federal government claims ownership), overworked police trying to keep some semblance of order and the odd *godfather* figure collecting tithes from the diggers. And just to complete the link with gold rushes of long ago, disease including malaria became rife in Apui amongst the diggers. In time activity has dropped materially at Apui; gold rushes often have limited lives as easy-to-access ore is worked out.

NO CLASS DIVIDES

One of the things that is particularly interesting about the gold rushes of the 19th century was how classless in a world driven by social class they often were. Aristocrats, businessmen and tradesmen vied with each other to obtain passage to the gold fields of the Klondike or California. The cost of making the journey tended to preclude the very poor unless they lived near the rush – unlikely in the case of Klondike but possible in the case of California or the Victorian and NSW rushes in Australia – or unless they had some skill that a rich prospector or syndicate was willing to pay for.

Despite the example of the Apui rush, today it would be very difficult to re-create the circumstances of past gold rushes, particularly in the developed world. If a sensational discovery were to be made by a prospector, such as the Canadian diamond discoveries of the 1990s, it would still take a properly structured and financed programme to exploit the find. In the developed world there would also be material health and safety issues that would preclude the chaotic conditions of past gold rushes. There is also the problem that most of the obvious surface gold deposits (and indeed those of other metals) were discovered decades ago (although deep in the Congo jungle or in Australia's inhospitable outback there might be something that has been missed or never stumbled across).

Apart from the geological aspect there is also the socio-economic angle to consider today. It is unlikely, given the relatively prosperous circumstances in the developed world, that educated people would be prepared to go through the privations of yore in the hope that they could find both gold and a fortune in a modern rush. It is of course a different matter in the developing world where genuine and widespread poverty still exists – the lure of instant riches in a gold rush such as in Brazil described above can be irresistible when you have nothing.

There is also the fact that the aeroplane and helicopter have made even the most inaccessible places relatively easy to find and certainly very quick to get to. This means that well-organised exploration companies have more than a head start today if a prospector uncovers a new mineral area. The prospector can also more easily leverage his good luck by selling out for cash and a royalty interest to such a group. Today's lay mining speculators prefer to remain in the comfort of their own homes and hope to make their fortune through the stockmarket; most of them fail in that task just as did the gold rush participants of old.

THE NEW WORLD GOLD GIANTS

The modern age for gold began in the early 1970s when President Nixon closed the gold window. At that time the world's largest gold producers were virtually all South African. As the years went by this began slowly to change as a rising gold price encouraged a major increase in exploration and new mine development.

We now find that South African gold miners occupy only two of the slots in the table of the ten largest gold producers worldwide and these two producers both produce over half their gold outside South Africa (see the table). This trend is almost certain to continue with South African output falling steadily and SA gold groups increasing their non-SA output. Indeed Harmony Gold, which falls just outside the top ten world producers, by the end of the decade should have become a primarily Papua New Guinea gold producer.

TOP TEN GOLD PRODUCING COMPANIES (2010)

Company	Country of incorporation	Production (tonnes)
Barrick Gold	Canada	242
Newmont Mining	USA	168
AngloGold Ashanti	South Africa	140
Gold Fields	South Africa	102
Goldcorp	Canada	78
Newcrest	Australia	73
Kinross Gold	Canada	68
Navoi	Uzbekistan	63
Freeport McMoran	USA	53
Polyus Gold	UK (Russia)	43

Source: GFMS World Gold

10. PLATINUM

Today the world's largest producer of platinum is South Africa. However South Africa's pre-eminence has only occurred over the last 50 years despite the huge discoveries of the metal made in the Transvaal north of Johannesburg in the 1920s. These in fact were not developed until ten years later when Rustenburg Platinum was formed to mine the Bushveld Igneous Complex discovered by Hans Merensky in 1924, who we met earlier in his role in the diamond developments in South West Africa in 1908.

HOW PLATINUM'S ROLE AND UNDERSTANDING OF THE METAL HAS CHANGED

Although platinum is now both a precious and a strategic metal, its historic role was as a precious metal, though not, as gold and silver were, a monetary metal. Platinum has been around for centuries though; perhaps its earliest recorded use was in the Casket of Thebes which was placed in the tomb of Shepenupet, daughter of the King of Thebes, in 700 BC. Much later on the Conquistadors in Latin America came across the metal when looking for gold. Sometimes the platinum was found with gold and was difficult to separate and therefore became something of a curse. Indeed some explorers thought that platinum was a sort of unripe gold, discovered before it had had time to become gold. This idea of metals being live like plants was quite widespread in the ancient world and even in the Middle Ages. Thus base metals could, some thought, eventually mature into gold, gold being the most desirable of all metals. Some also thought that metals would *grow back* following their extraction from an orebody.

As the scientific age dawned from the 18th century onwards scientists began to research whether platinum could have some practical use beyond its role as a white metal for jewellery and adornment. The big problem was that platinum had a very high melting point and ancient metallurgists could not achieve high enough temperatures in their furnaces. However in 1782 Antoine Lavoisier, the French chemist, used oxygen to melt platinum properly for the first time, although it was to be another two decades before the process had been refined sufficiently to produce meaningful quantities of the metal. In 1802 William Wollaston and Smithson Tennant, two English scientists, developed a process for refining platinum and also discovered palladium – the first of the other platinum group metals (PGMs) with which the modern world has become familiar.

Two years later they uncovered further PGMs – rhodium, iridium and osmium – from the platinum they were working. By the middle of the century another English scientist, William Grove, had produced the first fuel cell using platinum, but this was very much a laboratory discovery and work continues today to develop the first commercially economic platinum fuel cell. Around the same time Percival Johnson and George Matthey developed a process to refine pure platinum and separate out the other PGMs, and their success led to the establishment in 1851of the partnership still in existence today in the form of Johnson Matthey PLC which is one of the largest refiners of precious metals in the world.

A METAL OF THE 20TH CENTURY AND BEYOND

Despite the discoveries of the 18th and 19th centuries platinum is very much a metal of the 20th century and indeed beyond. Its early use was substantially driven by jewellery demand, and in the 1920s and 1930s it was popular for a time as a substitute for gold in jewellery fabrication. Although it remains a popular jewellery choice today, particularly in the Far East, growth in demand is now highly influenced by industrial demand for things like autocatalysts, specialist glass and chemicals, medical products and the future promise of a huge rise in demand when economic commercial fuel cells are developed. There is also a rising demand for platinum coins and bars for investment, although platinum has a long way to go to achieve the investment status that gold enjoys.

In the first part of the last century the largest producer of platinum was Canada which had taken over from Russia following the discovery of nickel and platinum at Sudbury, Ontario, in 1901. Russia itself had taken over as the largest producer from Colombia in the first half of the 19th century when platinum and palladium were discovered in the Ural Mountains. South Africa, the dominant producer today, came somewhat late to the scene, and it was only after the end of the Second World War that it overtook Canada as the largest producer in the world. This was prompted by a huge rise in demand for platinum jewellery in Japan and the use of platinum as a catalyst in the petroleum industry. In 1974 as inflation led to soaring commodity prices the US introduced autocatalysts to combat noxious exhaust gases, which led to a further surge in demand for PGMs, with platinum prominent.

This growing demand for platinum could only be satisfied by the exploitation of the huge resources of South Africa's Bushveld. The first platinum mine established on the Bushveld was Waterberg, which opened in 1926 after having been discovered by Adolph Erasmus in 1923. The mine operated for a few months before closing down, an astonishingly short life that was caused by complete failure to control grade properly, to do formal metallurgical testing on the ore and to build a plant that could treat the ore. Such sloppiness seems extraordinary today, but then Waterberg belongs to an age when mines were financed by the equity of often-unquestioning private investors rather than more rigorous bank lending that is ubiquitous today.

RUSTENBURG PLATINUM

After this false start, in 1931 Rustenburg Platinum was formed under the control of Johannesburg Consolidated, part of the Oppenheimer mining empire, to exploit the Bushveld platinum deposits discovered by Hans Merensky. Unfortunately Rustenburg got off to a disastrous start in 1932 having to be closed within months as a result of collapsed demand in the face of the Great Depression. Production restarted in 1933 and in 1938 the first of many production increases occurred. During the Second World War Rustenburg's production was around 40,000 ozs per annum.

HANS MERENSKY (1871-1952)

Hans Merensky was born in 1871 in Botshabelo in the Transvaal. His father Alexander, a German, was an ethnographer interested in the scientific study of local African culture; he was also resident missionary in the area. In 1882 the family returned to Germany where Merensky finished his schooling and then went to the State Academy of Mining in Berlin to study mining geology and engineering, and then took a doctorate in geology at the Royal Technical College of Charlottenburg. His course professor in Berlin remarked on Merensky's sixth sense for ferreting out mineral deposits. Following that he worked in the coal mines of Silesia before joining the Department of Mines in East Prussia.

In 1904 Merensky returned to South Africa on sabbatical from the Department to do some geological field studies and it was here that he made the first of a suite of major mineral discoveries in southern Africa. Working in the Transvaal he discovered tin near Pretoria and then became associated with Premier Diamonds. His growing reputation as a highly skilled and successful geologist led to a number of mandates from mining companies in South Africa. Following this success he resigned from his position with the Department of Mines back in Germany and settled permanently in South Africa, setting himself up as a geological consultant, a role that kept him busy.

In 1909 he was commissioned by De Beers to travel to South West Africa, then a German colony, to look at the prospects for diamonds there following the first discovery near Luderitz. Merensky correctly believed that the diamonds discovered came from the sea and predicted further substantial discoveries. For the next few years Merensky experienced a major downturn in his fortunes as he was made bankrupt by the economic depression of 1913 and then interned as an alien during the First World War. During this period he was financially assisted by General Mining founder Sir George Albu and after the war his career resumed its upward path with a run of major mineral discoveries including the Namaqualand diamond fields in 1926 and two years before that the platinum reefs of the Bushveld, the main reef being named after Merensky.

Although following these discoveries Merensky bought a farm at Westphalia in the north of the Transvaal and began the next phase of his life as a farmer following sustainable techniques, he continued to do work as a consulting geologist and was involved in the discovery of the Orange Free State gold mines, the phosphates of Palabora and huge chrome deposits at Pietersberg. Indeed this run of world-class mineral deposits discovered by one man is probably unique in mining history. So is the fact that Merensky was not only a brilliant exploration geologist but also a dedicated environmentalist.

Merensky's later life passion, his farm in the Transvaal, led to him investing in environmentally-sound practices covering agriculture, plantations and forestry, with a particular emphasis on soil erosion and water conservation. He never married so wrote his will in trust to benefit a large number of causes in South Africa over the years through the Hans Merensky Foundation. One of his earlier endowments was the Hans Merensky library in 1937 at the University of Pretoria. Other philanthropic causes included bursaries to students studying farming and mining. Today, 60 years after his death, the foundation has over 4000 employees and is one of the largest timber and fruit growers in South Africa with Merensky's belief in sound sustainable cultivation still the driving philosophy – surely a fitting legacy for an outstanding life.

POST-WAR PLATINUM BOOM

Rustenburg Platinum's production was doubled in 1947 and then doubled again in 1951, with the company acquiring Union Platinum in the interim in 1949. Fired by oil industry demand Rustenburg increased output in 1955 and again in 1957 but that was followed by a sudden drop in demand from the oil industry. Production had to be pared back but fortunately the hiccup was only temporary and both demand and output resumed their upwards march through the 1960s with Rustenburg production reaching 1 million ozs by 1970. Two years later further expansion was undertaken with the development of the Amandelbult section. All the while the prime customer for Rustenburg was Johnson Matthey in the UK and the two companies worked hand in glove on processing technology and other treatment improvements, culminating in the early 1980s with investment to process and sell the by-product base metals in Rustenburg's ore.

OTHER PRODUCERS EMERGE

Whilst Rustenburg was the only platinum producer until the end of the 1960s, highly prospective ground was held by other players and in 1964, in the wake of Rustenburg's continued expansion, Impala Platinum was formed to exploit part of the Bushveld. It started production in 1969. Four years later Western Platinum, part of the Lonrho group, started production. To the irritation of Rustenburg and Impala, Western Platinum published regular financial details about its production and earnings from its mine. In those days both Rustenburg and Impala published only the barest of details of their operations on the grounds that too much information would hand vital advantages to the groups' main market competitor the Soviet Union (something that had never bothered South Africa's gold mines, who likewise had a sizeable rival in the Soviet Union). By the middle of the 1970s South Africa produced around two-thirds of world output of platinum, a figure that has subsequently risen to about 80%.

In recent years other producers have appeared as the tight hold on Bushveld ground dominated by Anglo Platinum, the successor company to Rustenburg, and Impala has been loosened in the post-apartheid era. One of the largest newcomers is Aquarius but it ranks below Gold Fields offshoot, Northam, a company that has suffered operating and metallurgical difficulties over the years, and has not really fulfilled its early promise. Other rather smaller producers include newcomer Royal Bafokeng, Eastern Platinum, ARM Platinum, Platinum Australia and Platmin.

Platinum is a classic example of a metal with a key role to play in the advance of the human race as it confronts the issues of the environment and dwindling conventional energy sources. It is perhaps appropriate that the fortune that Hans

Merensky made from his role as the father of modern platinum was left in trust to South Africa to be used to support and protect the environment. The importance of the platinum-using fuel cell in providing clean fuel for transportation neatly squares the circle.

11. IRON ORE

When the modern iron and steel industry was in its infancy in the 19th century, iron ore was predominantly sourced locally. The idea of shipping this low-value product any great distance was hardly considered, neither was long-distance shipping of coal, the fuel that fired the steel-making process. The giant bulk carriers that today ply the oceans with coal and iron were decades away. The great steelworks of the UK and the US were therefore built near sources of both iron ore and coal. In the case of the UK these also happened to be near ports, which helped the distribution of the steel products of the mills and in due course, with the coming of bulk carriers, allowed the importation of raw materials from overseas when local supplies ran low or became uneconomic to mine.

The US was particularly blessed with its massive Mesabi Range iron ore deposits in Minnesota, found in the mid-19th century, the output from which was shipped east via the Great Lakes to the smelters and mills of Ohio and Indiana whilst coal from the huge fields in Pennsylvania came west to power these facilities. Over a period which lasted from the 1880s until the period after the Korean war in the early 1950s, the US became an international economic and then political power sweeping its old colonial ruler, Great Britain, from its top spot.

This feat was materially assisted by America's huge natural resources in both metals and energy, resources that Britain no longer possessed and had, therefore, to obtain from its Empire. At the end of the Second World War an exhausted Britain under pressure from the US began its retreat from Empire, thereby steadily losing its access to cheap raw materials. This was in contrast to the US where, for a time anyway, cheap commodities and spectacular technological advances allowed it to show a clean pair of heels to Britain, Europe and the Communist Bloc.

However in the East a great economic power was beginning, after the humiliation of defeat in the Second World War, to stir. Japan was on the move again and its appetite for imported raw materials was to stimulate a massive expansion of mining activity worldwide, particularly in the area of iron ore.

The war, though less costly in terms of military lives than First World War, was hugely more damaging on the civilian front and the accompanying destruction of infrastructure led to a massive economic rebuilding programme around the world. Unscathed in a physical sense by the war, the US nonetheless instituted a huge infrastructure investment programme itself to provide a land fit for heroes, as well as to accommodate, as the 1960s rolled into view, the consequences of the post-war baby bulge. The Japanese economy in due course derived great benefit from this development through massive growth in its manufacturing export sector, as did those who supplied Japan with raw materials.

AUSTRALIA

Amongst the big winners were Australia's new iron ore mines in the Pilbara region of Western Australia. However the story of Australian iron ore goes back to the early years of the 20th century when iron ore mines were developed in South Australia in the Middleback Range to feed the new steel plants being established by BHP on the New South Wales coast. In those days Australia was a modest exporter of iron ore, BHP being able to absorb virtually all locally mined ore. In 1938 the Australian government decided that it had a delicate diplomatic problem arising from an iron ore deposit being developed by the Japanese company, Nippon Mining, on Koolan Island in Yampi Sound on the northern coast of Western Australia.

The Japanese, having invaded China, were feared to have other regional ambitions, including possibly Australia, even though its security then was guaranteed by Great Britain. Not wishing to have Japanese workers within Australia – there were some Japanese technicians who worked on the Koolan Island project – but also not wanting to inflame Japan by expelling these Japanese workers, the government simply put an end to Nippon's project by ruling that Australia had so little iron ore that it could not allow exports of the metal anymore. The Nippon development was brought to an end and the Japanese disappeared; no iron ore was exported again for 30 years.

It was not until the 1960s that BHP established a steel mill at Whyalla on the South Australian coast to use Middleback ore. By then Australia had repealed its ban on exporting iron ore, which opened up the huge new finds in the Pilbara. One of the major figures of this period in Western Australia was Lang Hancock, a pastoralist who had first examined the Pilbara ranges in the early 1950s when iron ore exports were still banned. Hancock's farm, in the Hamersley Range, was the location of the massive Mount Tom Price iron ore deposit, which was part of a huge swathe of mineralisation that Hancock had stumbled across ten years previously. Hancock kept quiet about his discovery until the early 1960s when the export ban was lifted.

LANG HANCOCK (1910-1992)

Lang Hancock was arguably the central figure in the development of the massive Pilbara iron ore province in Western Australia in the 1960s. His discovery of huge iron ore deposits in the early 1950s may perhaps not have been the first time that the potential of the area had been appreciated, but his discoveries when finally revealed provided Australia with probably the biggest economic boost in its history.

Hancock was born in Perth, Western Australia, into a well-to-do, land-owning family who farmed at Ashburton Downs near Paraburdoo in the northwest of the state. After being educated at the private Hale School in Perth he went to work on the new family spread at Mulga Downs near Wittenoon to the northeast of Paraburdoo. He was a pastoralist but was always interested in minerals and he found blue asbestos on the farm when he was a boy. Latterly as a young man he returned to the asbestos find and in 1938 established a mine; five years later he entered a joint venture with Colonial Sugar Refiners but in 1948 sold out to CSR when he realised that he no longer had any say in the operation.

In the interim he had married for the first time in 1935 but his wife, Susette, never really embraced rural living and after a number of years returned to her native Perth. Hancock had taken over the station from his father in the mid-1930s and ran it with his boyhood friend, Peter Wright. During the war Hancock served in Australia's internal defence force and in 1947 he married again, his second wife being Hope Margaret. They had a daughter, Gina, Hancock's only admitted child, though there have always been rumours that in his youth he had relationships with aboriginal women on the family station.

After sighting iron ore whilst flying over his station in 1952 Hancock, with prospector Ken McCamey, returned to the site on many occasions, mapping out the huge iron ore deposits over more than 100km. It was only when the government ended the iron ore export ban in 1961 that Hancock, with Peter Wright, pegged the Pilbara iron ore tenements and then began to market the concept of a new world-class iron ore province. In one sense the rest is history. Hancock and Wright entered into a deal with RTZ's Australian arm, CRA, and Kaiser Aluminium of the US, and the giant Mount Tom Price mine was born. Over these years Hancock discovered a large number of additional iron ore deposits, including McCamey's Monster, Parabardoo, Marandoo, Brockman and

West Angelas, all of which were developed by large international companies.

Hancock's royalty stake in Tom Price and other mines made him a very rich man and as a prominent Western Australian he became politically active, although he never sought elected office. He supported the concept of Western Australian independence but was more of a thorn in the side of Canberra than a direct threat to the Australian Commonwealth itself. It was nonetheless the case that central government could not ignore Hancock and his local influence, and also the case that the Western Australian government ignored him at its peril.

As well as a deep interest in local politics Hancock also retained his life-long *eye for the ladies* and when his wife Hope died in 1983 he employed as housekeeper a young Philippines girl, Rose Porteous, whom he married in 1985. The marriage was controversial and eventually volatile, although it survived until Hancock's death in 1992. Hancock's daughter, Gina Rinehart, was never reconciled to her father's third wife and on his death sought an inquest believing that Rose was in part responsible for nagging her father into an early grave. The inquest's decision was death by natural causes.

Hancock's contribution to the development of Western Australia's iron ore industry lives on in the shape of Hancock Prospecting, run by Gina Rinehart, which today is developing some of the iron ore tenements that Hancock had kept under his private control, including the giant Hope Downs mine with Rio Tinto and the Roy Hill 1 prospect. Literally billions of tonnes of iron ore still lie in the Pilbara ensuring a life for the iron ore mining industry stretching into the next century and beyond. The name of Lang Hancock will forever be associated with this extraordinary Western Australian success story.

OPENING UP THE PILBARA

The 1960s saw a huge iron ore development programme in Western Australia, the largest mines being in the Pilbara. Apart from Tom Price, major new mines developed there in the 1960s included Mount Newman (Amax, BHP, Colonial Sugar, Selection Trust), Mount Goldsworthy (Consolidated Gold Fields, Cyprus Mines, Utah Development) and Robe River (Cleveland Cliffs).

One of the key issues that had to be addressed in terms of opening up the Pilbara was transport. The problem that first exercised Hancock and the international miners drawn to the Pilbara was how these fabulous riches were to be transported to market. The potential size of the deposits meant that the majority of the customers for the ore would come from overseas, with Japan being the obvious first port of call. The steel industry in Australia itself, dominated by BHP, was well served by the traditional supplies coming to it from South Australia where expansion was underway in the 1960s.

The suite of discoveries made by Lang Hancock could not all be developed at once, the market just was not there, but Japanese appetite meant that three, Tom Price, Newman and Goldsworthy, were developed. New ports at Port Hedland and Dampier had to be built and standard-gauge railway lines laid to allow the huge ore trains to haul millions of tonnes of product, bound for Asian markets, to the massive bulk carriers at the ports. Port Hedland hosted the Goldsworthy and Newman lines, the longest line from Mount Newman being over 300 miles long, and Dampier was the destination of the original Tom Price railway. Later, ore from Robe River was hauled to Cape Lambert, close to Dampier on yet another line. The trains could be up to two miles long and have over 300 ore trucks.

INDUSTRY GROWTH CONTINUES

Australia is now the third largest producer of iron ore in the world behind China and Brazil, and its exports in volume terms are on a par with Brazil. The Pilbara mines have also materially widened their customer base, exporting not only to Japan and South Korea, but also to India and China. China is particularly interesting being such a major iron ore producer in its own right, but the qualities of the high-grade, haematite Pilbara ore complements China's own much lower-grade producers. The original Pilbara producers have since the 1960s increased their output many times, and companies such as Fortescue Metal have also made further substantial discoveries outside the Hancock belt. In its expansion plans for the next few years Rio Tinto hoped to increase Pilbara output to 220 million tonnes by 2012, whilst BHP also has plans to expand output from its 2008 level of 132 million tonnes. By way of comparison, in 1974 total Pilbara iron ore production was around 80 million tonnes.

Disputes have also arisen between the state government in Perth and the old producers about access to their railway lines for the new generation of iron ore miners. The argument being used by the historic producers is that they have plans for substantial expansion of their Pilbara mines and that they therefore cannot offer capacity to new producers. This has meant that new producers will have to build their own transport links to the WA ports and Fortescue has led the way. Another initiative outside the Pilbara is the Geraldton Iron Ore Alliance which brings together seven small and medium-sized companies with deposits in the central west of Western Australia, including Weld Range, Jack Hills and Wiluna, well to the south of the Pilbara. The long-term plan is to construct transport links to the coast where a new deep-water port at Oakajee to the north of Geraldton would be built.

THE POLITICAL SIGNIFICANCE OF THE PILBARA

The rise of the Pilbara iron ore industry has had a profound effect on the economies of the Far East, and has been a key element in Australia's engagement with its regional neighbours. In the decade or so after the end of the Second World War Britain and Australia remained close political and trading allies, almost neighbours despite the huge distance separating the two countries. However under pressure from the US to embrace the European Common Market on its doorstep Britain began to negotiate for a new political future. As it did so it began the retreat from Empire and also began to send messages to British Commonwealth Dominions such as Australia that it was seeking new friends much nearer home.

At the same time Far Eastern economies such as Japan, Taiwan and Hong Kong were beginning to grow fast and needed raw materials to fuel this growth. The huge mining boom that rattled on through the sixties underlined to its Pacific neighbours that Australia was a serious contender to be the region's raw materials supplier of choice. The opening up of the Pilbara in the late 1960s and the huge output of iron ore from the new mines that began to flow then confirmed Australia as a serious regional economic power with new friends and interests in South East Asia.

Although the massive Pilbara developments were from every angle a game changer economically and politically for Australia and Western Australia, there were other iron ore openings outside the region. Western Mining developed a mine at Koolanooka Hills around 190km east of the coastal port of Geraldton and 300km north of Perth. In the north of the state BHP opened the old Koolan Island mine in Yampi Sound in 1965. Further afield in 1967 the Savage River iron ore mine in Tasmania owned by a consortium of local and international companies started exporting to Japan.

BRAZIL

At its conception as an independent state in 1822 Brazil was essentially an agricultural economy, although it had been Portugese incompetence over gold mining, in what is now the state of Minas Gerais, which had triggered Brazilian demands for independence from Portugal. The state of Minas Gerais also was believed to have considerable quantities of iron ore (as much as 3 billion tonnes) in the Doce River Valley, the potential of which was restrained by the lack of significant customers due to the nascent nature of the iron and steel industry locally and the distance to the then major industrial markets of Europe. The legal position of mineral rights in Brazil was also slightly complicated with Brazilian landowners owning surface mineral rights, but foreign groups being able to mine underground.

Further geological work on the iron ore deposits led to the establishment in 1909 of the Brazilian Hematite Company, inevitably with British capital, which then merged with the Vitoria-Minas Railway Company (CEFVM). CEFVM ran between Cariacica and Alfredo Maia and in 1910 began investigations into the establishment of a line running to the iron ore deposits, a process which took many years. Two years later in 1911 the American entrepreneur, Percival Farquhar, took over the company, which was then re-named Itabira Iron Ore, and an extensive examination of the deposits was undertaken. Many years passed before a mine was developed at Itabira – the first shipments to the coast at Port Vitoria began in 1940.

INDIGENOUS CONTROL

With the start of iron ore mining political pressure within Brazil built up for some local stake in the American-owned mines, and in 1941 Percival Farquhar divided Itabira Iron into Companhia Brasileira de Mineracoa Siderurgia and Companhia Itabira de Mineracoa in partnership with Brazilian businessmen. In the following year the Brazilian state formed Companhia Vale de Rio Doce (CVRD now called Vale) and took over Farquhar's two companies and the Vitoria-Minas railway. The first president of the company was Israel Pinheiro. This was part of an agreement between Brazil, the US and the UK to establish an iron ore mining and export complex in Brazil with US financial aid and in exchange Brazil declared war on Germany. In 1945 a special dock for mineral exports was completed and opened at Vitoria.

By the end of the 1940s CVRD was responsible for 80% of Brazil's iron ore mining and Eximbank of the US, the major lender, made attempts to influence board policy over the operation of the company and the mines. With the iron ore mines up and running CVRD entered into a supply agreement with Companhia Siderugica Nacional (CSN), Brazil's first steel maker, which also operated rich iron ore mines of its own at Casa de Pedra in Minas Geiras. In 1952 the Brazilian government took over complete control of CVRD and in the following year the company made its first shipment of iron ore to Japan. By the end of the 1950s Port Vitoria was further expanded, with the completion of the Paul Wharf enabling CVRD to start regular bulk shipments of iron ore to international customers in Japan and Germany.

The company by then had established a major geological surveying group and over time made a number of significant mineral discoveries in iron ore and also in bauxite and gold. Also active in trying to establish an iron ore exporting operation around that time was Hanna Mining of the US, which had taken over UK-owned gold miner St John del Rey in order to develop St John's large iron ore reserves in Minas Gerais. Whilst an iron ore operation was established in the 1960s by Hanna, the company kept on running into problems with the Brazilian government which was never really happy with a US presence in the strategic iron ore industry. Eventually Hanna sold out to CVRD.

CARAJAS IN THE AMAZON

The discovery of the massive iron ore reserves of Carajas in the Amazon basin was of particular importance – they were to become CVRD's largest source of iron ore. The Carajas discovery was also one of the largest iron ore finds ever made, with estimates of 18 billion tonnes and grades of 66% Fe. However, as so often happens in mineral exploration, the initial find was somewhat fortunate with the story being that US Steel geologists surveying the Carajas region were forced to land their helicopter on a bare hill to refuel from onboard containers. Whilst the refuelling took place the geologists on the helicopter were curious as to why, in such a lush part of the world, the hill was bare. The reason, they discovered, was because the hill was made of high-grade iron ore, not conducive to vegetation growth. The original development was a joint venture between US Steel and CVRD, but in the 1970s US Steel sold out to CVRD, concerned that the energy crisis rendered the operations at Carajas uneconomic.

Around this time CVRD also established a joint venture with Alcan to develop bauxite reserves in the Trombetas River region, which ultimately led to CVRD building up a significant integrated aluminium business. Also over the years it became a material force in gold, paper and pulp in Brazil, and had interests in steel making through its large historic stake in CSN. The latter two interests were sold off when CVRD's privatisation was completed in 2001 and its gold interests also were significantly reduced.

With the decision to develop the Carajas iron ore deposits being made in 1978, construction of a new railway to the Atlantic port of Sao Luis was begun. The Brazilian government also allowed a number of small iron ore smelting operations to be built along the railway line to which a small proportion of the ore on its way to Sao Luis was diverted. This led to a deforestation problem along the line as large quantities of trees were cut down for charcoal for the mini-smelters. Vocal opposition to this, from Europe in particular, led the government to abandon its support for the smelters and they were closed down. The railway line then reverted to simply carrying Carajas iron ore to Sao Luis for export. By then Brazil had become the largest iron ore exporter in the world, a position it holds to this day, thanks to CVRD and the guidance of its iconic president at the time Eliezer Batista.

ELIEZER BATISTA

Eliezer Batista was born Eliezer Batista da Silva in 1924 in Nova Era, a town in the mining state of Minas Gerais in Brazil. His parents, Jose and Maria, were emigrants from Portugal and his father, a saddle maker, built up a substantial business in Minas Gerais. In due course Jose went into agriculture and ranching in the region and was able to comfortably support his two sons and four daughters.

Batista was educated first in Nova Era and then went to secondary school in Ouro Preto and St Joao del Rei, respectively to the south and west of Belo Horizonte. He was a bright but difficult pupil and his academic excellence was treated suspiciously by his monastic teachers who were largely Dutch. His interest was engineering and it was to study this that he took himself off to the cosmopolitan Federal University of Parana in Curitiba in the south of Brazil, graduating in 1948. Batista was a rebel in those days and after university he travelled extensively, helped by a natural bent for languages.

On one visit to his family in Nova Era, Batista contacted the growing iron ore group, CVRD (now known as Vale). He was interested in the modern outlook of CVRD and got a job working on the construction engineering side, very important then as the company began to develop its large iron ore reserves. He worked on the construction of the extension of the Vitoria-Minas iron ore railway line and then supervised the maintenance of the track. In due course he was sent to the US for further training. When he returned he became chief engineer of the railway, acting as a link between the CVRD mine and the port at Vitoria.

Up to this moment Batista's contact with mining had been purely through his engineering role on the railway serving CVRD. However in 1961 he was appointed President of CVRD, a role he undertook for just one year until in 1962 he was appointed Brazil's Minister of Mines and Energy, a post he held until 1964. Political change in the form of a military coup led him to leave the government and he was almost imprisoned in a purge of the old regime by the military rulers. He then became president of iron ore developer MBR (Caemi Group) a position he held until 1968 when he returned to CVRD and was sent to Europe to rebuild the company's markets there, taking his family of seven children and his German wife with him.

In 1979 he returned to once more take up the reins of president and this time also became chairman of a number of CVRD's operating companies; he held the post of president until 1986. During this period he oversaw the development of the giant Carajas iron ore deposits where CVRD was a partner with US Steel. This was a time of great pressure for Batista because it saw CVRD expanding out of its original Itabira operation in Minas Geiras, causing concern in the state, and getting embroiled in the environmental controversy over the development of Greater Carajas – a government-promoted scheme which Batista did not personally support – separate from the Carajas mine. In 1992 he returned to government, taking the position of Minister for Strategic Affairs.

Batista was a great believer in the philosophy of linking raw materials with their end use, particularly in the field of high technology – this was a result of his upbringing as a civil engineer rather than a mining engineer. He saw that huge, high-grade iron ore deposits far from market had absolutely no value unless transport logistics such as rail, port and ships were in place to take the product to its end-user, not something that was commonly accepted in the chaotic, populist nation that Brazil was in the years following the Second World War. Batista was also a great believer in beneficiation and upgrading the product to enhance its price, such as pellets for the steel industry's direct beneficiation process.

Batista, now in his mid-eighties, continues to take an interest in business; he is on the board of a number of Brazilian companies including those of his son Eike, who has built a reputation and fortune in gold and more recently oil, making him Brazil's richest man. He also continues to advise the president of Vale – now the world's largest iron ore producer – and is also active on the board of a number of Brazilian development quangos. In his twilight years Eliezer Batista da Silva continues to be an enormously influential figure within the Brazilian economy as he has been throughout his life.

VALE AND FRIENDS

Although Vale has always been the largest iron ore producer in Brazil, the industry's development was not solely down to Vale, even though the company in the early 1990s still produced over half of the country's output. Mineracoes Brasileiras – part of the US mining group Hanna, Mineracoa de Trinidade, Ferteco Mineracoa and Samarco Mineracoa were four of the larger independent Brazilian iron ore companies. However, after the process of privatising CVRD was completed in the early 2000s the company began to rationalise the Brazilian iron ore industry and one by one it bid for each of its competitors, eventually attaining control of all Brazil's iron ore exports.

Today the industry remains dynamic and amongst a group of independent companies planning new mine developments is Bahia Mineracao, now a subsidiary of Kazakhstan's London listed ENRC, which has a large but low-grade deposit in Bahia state. The privately financed group Ferrous Resources is another future and major producer, with huge resources in Minas Gerais, which should underpin Brazil as a global iron ore giant.

IRON ORE IN THE EAST

A local iron ore mining industry is the ambition of many developing economies as they seek to increase their industrial base. For advanced economies the need for indigenous supplies is less pressing – some like Japan have such huge requirements for iron ore that these can only be satisfied by imports, local sources being tiny. This does not materially inconvenience Japan or other large importers in the East like China and India; the value added that is earned from the export of finished steel-based goods, more than offsets the cost of iron ore imports.

The latter countries, unlike Japan, are big iron ore producers, and China is in fact the largest in the world, although it requires both local and imported sources to fuel its industrial expansion. As we have seen earlier, China was a technically-advanced civilisation in the early centuries of the 1st millennium AD, and before that, and it developed and mastered the blast furnace to convert iron ore into cast iron several centuries before the technology was known to Europeans. At the end of the 19th century China's main iron and steel-producing areas, and therefore its main iron ore mining areas, were Dabieshan, Guangdong, Shanxi and Sichuan. Dabieshan was an expanding iron province then but the other three were past their peak with China coming under pressure from western nations such as Great Britain and France, and of course Japan.

Mining in China in the late 19th century was not an activity that seemed to make either Chinese miners or mine owners in Shanxi, in the central part of the country, particularly rich, as was observed by German traveller Ferdinand von Richtofen following an extended visit to China in the 1870s. Part of the reason for this was the availability of cheap iron products from Europe, which could be transported more cheaply several thousands of miles than could products from the interior where China's iron ore mines were to be found.

In Dabieshan in central China the iron ore was found in ironsands where the sand was treated in large sluices using river water and a rich iron residue was obtained; this was then transported to ironworks and furnaces in the immediate area. In Sichuan in the southwest the ore used by the iron makers was often clay ironstone which was mined from open pits dug into the ore. Fuel to power the furnaces depended on the technology used, with depleting charcoal steadily being replaced with coal in the 20th century as operating and technical problems associated with coal, such as high sulphur content, were overcome.

The European competition issue meant that two of the four main iron ore mining and iron making areas, Guandong and Shanxi, were in decline as the 19th century waned. Sichuan, perhaps the biggest iron producer, was holding its own and only Dabieshan was expanding, helped by a combination of rich ores and a market that could not be supplied cheaply with imported iron. Although very limited records exist regarding the iron ore workings we can surmise from the relatively small smelting operations in parts of Guandong and Shanxi that the mines were also often small.

With the coming of the First World War the supply of cheap iron into China dried up, providing the local industry with the first good news for many decades. The first half of the 20th century though was difficult for China with world wars swirling around it, the Japanese marching through it and then the arrival of Mao and the Great Revolution. After the end of the Second World War there was a major increase in iron and steel making throughout China and therefore in the output of iron ore, many of the iron and steel-making operations remained small scale but increased markedly in number.

In the last 20 years there has been an explosion in demand for steel as China has begun to modernise, and the iron ore mining and steel industries have undergone massive expansion and re-structuring. Although a major importer of iron ore, buying over 600 million tonnes in 2009, China is also the largest producer as well, with 2009's output of 900 million tonnes over twice as high as the next largest producer, Brazil. Over the last two decades a number of huge state and private iron ore and steel groups have emerged, taking advantage of the country's huge reserves of iron ore. These mines are widely spread around China but often, as we mentioned above, the ore is low grade and needs to be blended with the high-grade ore imported from Brazil and Australia.

THE GROWING AFRICAN INDUSTRY

As a low-value, high-tonnage industry iron ore mining has always been marginal in Africa's huge interior where local demand has been modest and the distances to port are such that only higher value minerals, such as copper, are economic to transport for export.

SOUTH AFRICA

The continent's major force in iron ore is unsurprisingly South Africa and today its Kumba Iron Ore is the world's fourth largest exporter of iron ore. The industry goes back to the early years of the 20th century when it was decided that South Africa could support a steel industry, and the Union Steel Corporation (USCO) was founded in 1911, ten years after pig iron was first produced in the country. Before that South Africa had relied on Britain for steel products to fuel its infant manufacturing and mining sectors.

For the first few years USCO relied on scrap provided by the railways and the mining industry, but in 1916 the country's first iron ore mine at Pretoria West was developed to provide local feed for USCO's furnaces. In the 1930s the state set up South African Iron and Steel (ISCOR), which built a new steelworks at Pretoria and developed its own iron ore mine at Thabazimbi in 1934 as demand climbed during the decade. After the end of the Second World War ISCOR continued to expand its productive capacity and in 1947 opened the Sishen mine in the Northern Cape, another giant open pit iron ore mine, to supply feed for its steel-making operations. At the same time, as we will learn later, ISCOR was also developing coal mining reserves to fuel its furnaces.

In 1976 a rail link with the port at Saldanha Bay on the west coast was completed and ISCOR began to export iron ore. With the international noose tightening around South Africa's apartheid regime the trade was relatively modest, but following the end of apartheid in 1990 iron ore exports began to increase. In 1990 ISCOR was privatised as part of a liberalisation of South Africa's heavily state influenced economy. In 2001 after a difficult period financially for ISCOR the mining interests, primarily iron ore, were floated off separately as Kumba Resources and in 2006 the iron ore operations were transferred to a new listed company, Kumba Iron Ore, controlled by Anglo American.

MAURITANIA, SIERRA LEONE AND GUINEA

Another important African iron ore exporter is SNIM, the Mauritanian state-controlled iron ore miner. The original mine was established in 1952 by Miferma, a private company, at Kedia d'Idjil in the northern part of the country. A 450-mile railway was built to bring the ore to Nouadhibou on the Atlantic coast. Over the years production and therefore exports have run at around 11 million tonnes per year from open pit deposits which are high grade and allow the ore to be directly shipped to overseas customers, primarily in Europe. In the longer term there are plans to double output and exports. For a while in the 1970s, as the high-grade ore was depleted, beneficiation of lower-grade ore to direct shipping quality was started, but more recently in 1991 new high-grade deposits were brought into production. Processing of the ore is split between plants at the mines and plants at the port.

Two iron ore projects in West Africa, the closed Marampa mine in Sierra Leone and the Kalia deposit in Guinea, have attracted the attention of the Chinese, who are providing assistance in the form of sales contracts and finance for the development of operations at both sites, where production at Marampa is expected to start in 2013/14. Marampa was mined as an open pit operation between 1933 and 1975 by the Development Company of Sierra Leone. Over that period production reached an annual 2.5 million tonnes (an overall aggregate 60 million tonnes) of relatively low-grade iron ore, which was shipped by a purpose-built railway to the deep-water port of Pepel and then on to customers in Europe and the UK. Interestingly Sierra Leone, although primarily an agricultural country/colony, also had developed gold, bauxite, rutile and platinum mines during this period, mostly rather small scale.

The Kalia deposit in Guinea complements the CBG Bauxite mine, which has been in production since 1973. Kalia is expected to be in full production by 2018 and output of 50 million tonnes per year is envisaged. The deposits at Kalia were extensively examined in the 1930s by Soviet geologists, the area having a record of small-scale iron mining going back centuries, and evaluations were periodically carried out, subsequently between the 1960s and the 1990s. However it has taken the rise of China to finally stimulate a development plan for these huge relatively low-grade resources in what was for many years one of Africa's most backward countries.

ALGERIA

Iron ore mining in Algeria has a long history dating back to the period of occupation by the Romans. In 1830 the French levered control over Algeria from the hands of the Turks and began to distribute economic favours to French settlers. Amongst the beneficiaries was Henri Fournel, who rediscovered the ancient iron ore mines near Annaba at Ain Mokra, Bou Hamra, Meboudja and Kareza, where mining re-started in the 1840s. Whilst an attempt was made to utilise this iron ore in making iron in Algeria, the attempt failed and the ore was exported. The arrival of French prospectors from Lorraine, which had been annexed by Prussia in 1870, gave a further boost to mining in Algeria.

The country was thought promising for new mineral deposits following the discovery of the Mokta el Hadid iron ore deposit near Ain Mokra in 1857, which began producing in 1860, and the new settlers were much encouraged in the late 19th century when high-grade iron ore was discovered at Ouenza, another area known to the Romans, in the north of the country. In the early years of the 20th century Ouenza became the subject of ownership disputes where leases were granted to two different companies, one for the surface minerals and the other for underground minerals. The dispute over mineral rights between French, German and Dutch interests was not resolved until 1908 and after that there were complications centring on the railway which had been built to ship Ouenza's ore to the coast. These were not finally cleared up until 1913.

Little happened during the First World War but in the 1920s substantial investment was made in railway lines to bring the ore to coastal ports where steel mills in the UK and Belgium were key customers. During this period the controversial Swedish industrialist Ivar Kreuger acquired control of Ouenza through his Grangesberg associate and for a while Algeria was one of the two biggest forces, along with Sweden, in the sea-borne global iron ore trade. In 1966 the Ouenza, along with other iron ore producers, was nationalised.

Later in the century private capital again was encouraged to invest; global steel giant Mittal took a stake in 2001, with the state holding company, Ferphos, retaining an interest in both the iron ore and phosphate mines in the country. Ouenza continues to mine iron ore today, controlled by Arcelor Mittal, as is the Bou Khadra mine.

EUROPE'S COAL AND STEEL COMMUNITY

The 20th century was dominated by European political problems and the two world wars which wreaked chaos across the continent and brought misery to tens of millions around the world. One of the ruses that a group of European governments, including Germany and France, used to try and wean the continent off its habit of regular and violent political explosions was the establishment of a trading cartel covering steel.

The first cartel was set up in 1926 with a membership of Germany, France, Belgium and Luxembourg, with Poland, Austria, Hungary and Czechoslovakia joining subsequently. Much was made, at the time, of the healing properties of the cartel in terms of trying to reconcile Germany and France by binding them into a treaty that might reduce the chances of future conflict. An empty hope as it turned out.

The dominant western steel producers at the beginning of the 20th century in order of production were the USA, Germany, the UK, and fourth France, as shown in the table.

WORLD STEEL PRODUCTION BY COUNTRY IN 1905

Country	Production (m tons)
USA	20.3
Germany	10.0
UK	6.0
France	2.2

No more than 20 years previously the UK had been the largest producer ahead of the US – which in 1885 produced a mere 1.7 million tonnes of steel. By the beginning of the First World War in 1914 the US was an industrial giant out-punching the UK, although as a political force it did not yet match Germany or indeed the British Empire, a situation exacerbated by its neutrality in 1914.

The situation regarding Germany and France and their sources of iron ore at the start of the war was complicated by the fact that both countries normally drew their iron ore from the historically disputed territory of Lorraine – Germany mining 29 million tonnes from German Lorraine, and France mining around 20 million tonnes from French Lorraine. With the outbreak of war in 1914 Germany seized the iron ore mines of French Lorraine and held them, basically unmolested by Allied forces throughout the war, even though they were only a few miles from the front where most of the fighting took place in the first four years of the conflict.

Similarly France's main coalfield in the Pas-de-Calais was only a few miles from the front, but it was never the target of an aggressive push by the Germans. With the end of the war the idea began to germinate that since the iron ore deposits of Lorraine were geologically linked and exploited by both Germany and France, it made sense to set up an organisation to recognise that fact and draw Germany and France together in a commercial alliance that might also help to settle the two countries' political differences at the same time.

As mentioned above, the cartel was set up in 1926 but unfortunately it failed as a political body and in 1939 France and Germany once more fell upon each other. During the period between the two wars the UK's position as an iron ore producer continued to deteriorate rapidly and its steel output even fell behind that of France. At that time the UK's iron ore primarily came from Sweden and Spain, its own iron ore reserves being too low grade by then to be economic to mine anymore.

Following the end of the Second World War in 1945 the issue of trying to make sure that conflict never broke out again across Europe led to the US paying close attention to European affairs through the strategic alliance, the North Atlantic Treaty Organisation (NATO) founded in 1949, and the occupation of Germany by US, British and French forces immediately after the end of the war. However, the long-term desire to bind Germany and France together in a commercial alliance, which would consign conflict to the dustbin of history, as it was supposed to do in 1926, once more led back to the issue of coal, iron ore and steel, and in 1951 the European Coal and Steel Community (ECSC) came into being. The intention of the ECSC was to pool all the coal, iron ore and steel assets of its members (West Germany, France, Italy, Holland, Belgium and Luxembourg) together for the common good and for common exploitation.

However by the 1950s Europe, as a coal and iron ore mining force, was beginning to waver, a trend that accelerated during the following decades. Today Europe's biggest iron ore producer is Sweden and its 21 million tonnes of annual output is just 1.7% of world production. As far as steel is concerned, according to the World Steel Association in 2009 the European Union produced around 11.4% of world output and the biggest EU producer was Germany with 2.7% of world production.

In today's world of massive bulk carriers and huge mechanised, high-grade, low-cost, open pit iron ore mines in Australia and Brazil, the need for high-value-added steel users in Europe to access domestic supplies of iron ore has disappeared. Indeed, although still a large steel producer, Germany gets around one-third of its steel requirements from imports, a clear indication of the value that German steel users can add when manufacturing their products, many of which, like cars, are for export. The Coal and Steel Community has gone as have the coal and iron ore mining industries that formed the base of the organisation. In its place we today we have something quite different; the EU.

12. CHILE: KING OF COPPER

AMERICAN INVOLVEMENT

If Chile entered the 20th century as king of the global nitrate industry it ended it as king of copper, with output of the red metal three times higher than its nearest rival, neighbouring Peru. For Chile it was the expertise of US companies like Braden Copper, which in 1904 started exploitation of the El Teniente copper deposit, and the Guggenheims' Chile Exploration Company (Chilex), which began development of Chuquicmata in 1910, who were able to harness new mining and treatment technology to economically mine large, low-grade orebodies. El Teniente quickly came under the control of US giant Kennecott and in 1923 the Guggenheims sold Chilex and Chuquicamata to another US mining giant, Anaconda, which already controlled the Potrerillos copper project through its Andes Copper subsidiary.

The extent of US interests in the revival of Chile's copper industry was as a result of concerns back in Washington that British commercial ties with Chile, and indeed other Latin American countries, made the UK uncomfortably dominant in the region. With US government encouragement American copper companies backed by US capital carved themselves the leading position in Chile's mining industry. Later, nationalisation and then more recently the expansion of the operations of the UK giants Rio Tinto, BHP Billiton, Anglo American and Antofagasta in Chile, have reversed the position, pushing US mining interests back down the ladder.

El Teniente was located just 50 miles north of Santiago on the western side of the Andes and surrounded a dormant volcano. The copper ore was mined by driving a huge tunnel system into the outer sides of the volcano where the ore had settled. The mastermind behind the modern El Teniente was American mining engineer William Braden. His foresight in involving American mining peers such as Barton Sewell in funding El Teniente led to the eventual participation of Kennecott in building the giant underground copper mine. Braden also discovered other huge low-grade copper deposits like Los Pelambres, but it was decades before they were developed. In the meantime Braden had also begun to explore the Potrerillos copper deposit in the north of Chile.

POTRERILLOS

Potrerillos was an enormous technical undertaking requiring considerable investment in not only the mine and plant situated in the remote Atacama Desert but also in road and rail links to take copper to the coast for shipping. Construction of the mine started in 1916 but it was not until 1927 that copper metal was finally produced. Potrerillos cost $45 million when completed in 1927, or $550 million in today's money. It was a massive and difficult project, and because of the depth of the orebody and the relatively low grade at 1.5% copper, the mine was developed as an underground operation using block caving which meant that large tonnages could be mined. At its peak the mine produced over 100,000 tonnes of copper a year and was worked until closure in 1959 when the nearby El Salvador copper mine was brought into production.

Since Potrerillos was a very isolated deposit it, like El Teniente, required the building of what was in effect a company town to house both Chilean and foreign staff. Although this kind of set up was very common in the mining industry it nonetheless was a source of tension between unions, workforce and management for decades. The foreign area largely housed highly-skilled technical, engineering and management personnel and their families, and the accommodation and services available to them were far superior to those built for the Chilean miners – they even had their own golf course and country club.

By contrast, Chilean housing was cramped, sanitation fairly basic, recreational facilities were poor and the miners also had to use the company store to buy goods and food. Choice was poor and although necessities like food and clothing were often free they were rationed and coupons were distributed to workers to be used in the store where quality, service and organisation left much to be desired. In 1959, a few years before nationalisation, Anaconda opened the nearby El Salvador mine, a small producer in comparison with Chile's other copper mines, with an initial reserve of 78 million tonnes grading 1.6% copper. This mine's current output at just over 40,000 tonnes of copper puts it in the minnow class as far as Codelco mines go, but it has enabled the life of the Potrerillos complex to be extended.

CHUQUICAMATA

The development of Chuquicamata as a large, low-grade integrated mining operation was also a rather drawn out affair. In 1910 the American industrialist Albert Burrage played a crucial early part in re-structuring and financing the project but he did not have the capital to build the mine, which was based on a reserve of almost 700 million tonnes of ore grading 2.6% copper. He therefore approached the Guggenheims who controlled ASARCO and who agreed to incorporate Chilex and buy out Burrage's interests for $25 million of Chilex shares in 1912.

Extensive metallurgical testing of the ore was carried out in the US and in 1914 Chilex began to develop a mine based on the eastern section of Chuquicamata.

Production started the following year and rose steadily from around 4,000 tonnes of copper to 135,000 tonnes in 1929, by which time it had been acquired by Anaconda. For almost 50 years the massive open pit mine, the largest in the world, was based on exploiting the oxide cap to the Chuquicamata reserves and in that form it produced around 5 million tonnes of copper. But in 1951, with the oxide cap exhausted, a new treatment complex was built to handle the huge sulphide ore reserves that lay below the oxide cap. In recent years a new initiative has been launched at Chuquicamata to develop the massive reserves that lie beneath the current open pit as a new underground mine and production is expected to start by the mid-2010s.

DANIEL GUGGENHEIM (1856-1930)

These days the name Guggenheim is synonymous with the world of modern art and the Guggenheim Museum in New York. However in the 19th and the early 20th centuries the Guggenheims were better known as the most powerful force in US mining, having built a mining empire in both North and South America.

A Jewish immigrant from Switzerland, Meyer Guggenheim started his business career in the US in manufacturing and then in importing fine lace from Switzerland. He invested some of the gains from his importing business in silver and lead mines at Leadville, Colorado. In 1884, encouraged by the success of his Leadville investment, and of the rapid industrialisation of the US, he closed the lace business to concentrate on mining interests and founded Philadelphia Smelting and Refining to treat his Leadville mining output. At that time he was almost 60 and his eldest son, Daniel, one of eleven children, who had worked for the Swiss end of the family's importing business, took up the reins, and progressively became the driving force behind the family's mining strategy. Daniel was educated at a Catholic high school in Philadelphia and by-passed university to enter the family business and was supported by five of his six brothers.

In 1890, following the imposition of new duties on imported silver and lead concentrates used by the Guggenheims' Colorado smelters, Daniel acquired Mexican mines and built a smelter there to circumvent the US restrictions. He also developed the Guggenheim Exploration Company which operated worldwide, financing and developing new mines. He gained control of American Smelting and Refining

(ASARCO) in 1901 after a short but bitter battle with the Rockefellers and Henry Rogers who had tried unsuccessfully to tempt the Guggenheims into joining his Smelter Trust, one of the new monopolistic industrial trusts being established by powerful industrialists in the US at that time. Daniel used ASARCO to build a global mining giant operating Bolivian tin mines, African diamond mines, Alaskan gold mines and copper mines in the US and Chile. ASARCO was a dominant economic force in a large number of developing countries where Daniel Guggenheim's influence was subsequently considerable. As a mining entrepreneur he was also very innovative, profitably mining low grade ores through the technique of treating huge quantities of ore efficiently, and also undaunted by logistics, developing mines such as Kennecott Creek in Alaska and the remote high altitude Chuquicamata in Chile.

Daniel Guggenheim was an autocratic man, which helped him become one of the most powerful late 19th and early 20th century mining figures. He was ruthless in his treatment of labour unions and was not much interested in the environmental damage caused by some of the Group's mines. The Guggenheim family fortune reached $250 million by the end of the First World War, and the war itself, which the Guggenheims thought the US should participate in at an early stage, provided rich pickings. Demand for copper in particular soared, in fact so much so that President Woodrow Wilson threatened to nationalise Guggenheim's mines if he did not lower the copper price in support of the US's latecomer war effort.

The adage 'those who live by the sword die by the sword' has some resonance with the decline of Daniel Guggenheim's influence and an increase in internal family disputes over the running of the family company after the war. One of the younger brothers, Will, was excluded from a Chilean copper deal in the early 1920s and took Daniel and other family members to court; an out-of-court settlement to hide the family's wealth was the outcome, but decline had inexorably set in. In 1922 the Guggenheims were thrown off the ASARCO board, accused of mismanaging company funds. A family dispute arose over the sale of a family-controlled copper mine in Chile – the mine was sold but Daniel re-invested in a nitrates operation that proved a dud. In 1923 Daniel retired to his estate on Long Island, his mining days were over but he had left a huge footprint on the industry, and perhaps also a smouldering resentment overseas of profiteering and exploitation that finally surfaced in the Latin American expropriation of US mining assets in the 1960s.

The latter years of Daniel's life were spent in philanthropic work and the Daniel and Florence Guggenheim Foundation made huge endowments to Mount Sinai Hospital, the New York Botanical Gardens and the Guggenheim Museum, amongst others. Importantly the Foundation also supported and financed the development of the US aviation industry and aerospace research, which laid the foundations for the space age. Daniel died in 1930 a hugely respected figure as a supporter of good causes, but the ruthless accumulation of the riches that enabled him to be so generous was less well publicised.

ANDINA

The other major Chilean copper mine is Andina which is essentially two operations, one owned and mined by Codelco, Rio Blanco/Sur Sur and the other, Los Bronces, owned by Anglo American. Mineralisation was first discovered at Andina in 1920 but it took 50 years for the deposits to be developed. At the time of nationalisation Codelco took control of Rio Blanco, then owned by Cerro de Pasco, and Chile's new state company for smaller mines and projects, Enami, took control of the Los Bronces part of Andina, buying it from French miner, Penarroya.

With Allende's fall non-Codelco mining assets could be sold back to the private sector and in 1978 Enami sold Los Bronces to Exxon Minerals, who in turn sold it to Anglo American in 2002; Los Bronces is now the subject of a dispute between Anglo and Codelco over the sale of a 49% stake in it to the state copper giant. Both Codelco and Anglo American also have new developments under consideration in the area. Anglo is looking at Los Sulfatos just a few miles from Los Bronces where 1.2 billion tonnes running around 1.5% copper have been outlined with substantial higher-grade material in part of the deposit. Codelco for its part plans to double production from its Andina operations by the mid-2010s.

RETURN OF PRIVATE CAPITAL

Whilst Chile's experience of nationalisation in the 1960s was not, as we saw earlier, a happy one, it enabled the industry to renew itself in the years following the military overthrow of Allende. The state company Codelco, as we mentioned earlier, continued to operate the historic copper giants, El Teniente and Chuquicamata, the new Andina mine and the more modest El Salvador, but the exploration and development of many new mines reverted to the private sector.

Indeed since the establishment of Codelco the parastatal has developed only two mines over the period, the oxide open pit Radomiro Tomic in 1997 and the Gabriela Mistral in 2008. The material reduction in the size of US exposure to Chile's copper sector today is perhaps not overly surprising in the light of the Allende experience. However Freeport McMoran is involved in three joint ventures, including the El Abra with Codelco, a joint venture formed in 1994.

The controversy over whether or not US mining companies like Anaconda and Kennecott raped and pillaged Chile in the early decades of the 20th century is difficult to resolve. Allende certainly thought so as he declined to pay compensation following the 1971 nationalisations. It is the case that during the early years of Chile's copper expansion the US companies raised huge amounts of capital to bring these remote mines to production. Infrastructure spending was also considerable and the mining companies were responsible as well for housing their thousands of mineworkers. The living conditions, though, were poor and pollution from the mines remained heavy for decades, with adverse impact on both the health of miners and their families, and on the surrounding environment.

On the other hand, though union disputes were frequent, many of the miners enjoyed much higher living standards than they would have done without the mines. It is also the case that in the early years the mining companies did not pay much if anything in the way of taxes, but as the century wore on royalties and taxes were steadily raised and in 1952 they had reached 70% of taxable income, and had become a major source of tension between the US mine owners and the government, with the former unwilling to invest in new orebodies. In the 1960s the Chilean government purchased a majority stake in the main mines, Chuquicamata, El Teniente and El Salvador, as a result of the disagreement over these tax rates. This was followed by full nationalisation later in the decade.

However, foreign-owned copper producers – with UK miners BHP Billiton, Rio Tinto, Anglo American and Antofagasta amongst others having major operations in the country – now dominate the Chilean scene following the flood of privatisations in the 1980s and the encouragement of foreign capital to return to the industry. The current dispute between Codelco and Anglo American over the Los Bronces mine, however, demonstrates that the relationship between foreign companies and the Chilean state can still be a difficult one. Whether or not keen foreign participation will be permanent is quite another issue, particularly as taxation is an ever present area of sensitivity, but the recent history of copper mining in Chile points compellingly to the fact that the industry is at its healthiest when private participation is encouraged. In the case of Antofagasta its expansion owes a great deal to the efforts of Chilean businessman Andronico Luksic.

ANDRONICO LUKSIC (1926-2005)

Andronico Luksic was one of Chile's key players in the rise of the privately-owned section of the copper industry in the years following the fall of Allende. He was born in the port of Antofagasta of a Croatian father and Bolivian mother. There is a perhaps apocryphal story that he was brought up in a house close to the railway tracks of the Antofagasta and Bolivia Railway Company which hauled silver from the Huanchaca mine in Bolivia, and listening to the comings and goings of the trains as a boy, Luksic vowed eventually to acquire the British-owned company, which he did in 1980.

Luksic studied law at the University of Santiago and then economics at the Sorbonne in Paris where he spent a lot of his time trading currencies. On his return to Chile, using money that he had earned from his currency trading he set up a motor spare parts business based on his uncle's Ford dealership in Antofagasta, having found his personal shyness a block to success in practising law. Luksic's father had been involved in nitrates, Chile's biggest mining business in the late 19th and early 20th centuries, and Luksic, following the family mining tradition, bought the Portozuela copper mine in company with a French associate in the early 1950s. He then sold this to Japan's Nippon Mining in 1954. It is said that despite selling the mine for $500,000 Luksic had actually set a much lower price of half a million Chilean pesos but the Japanese had not understood that and had been happy to pay the figure in US dollars!

During the 1960s and the 1970s – difficult years for Chile as politics swung from capitalism to communism and then back to capitalism – Luksic, who was something of an amateur geologist, acquired the Lota Schwager coal mines and the Madeco copper products operation. Over the years he also invested in a wide range of businesses in Chile covering brewing, food, telecommunications, transport and banking, many purchased after the fall of Allende, with whom Luksic had got on quite well, at very advantageous prices. The prices were particularly favourable in the mid-1980s, which was a period of extreme weakness for the Chilean economy.

Luksic had also begun to invest abroad, first in Argentina and then in Colombia and Brazil. In the latter country he had acquired an iron ore mine, further underlining his interest in mining. In the last two decades of the 20th century Luksic began to emphasise this side of his business interests following the acquisition of the Antofagasta company. As an Anglophile Luksic did not want to repatriate the company to Chile but instead wanted to build on UK investor interest in Latin America, particularly Chile, and also in the 1990s in mining. He therefore used UK incorporated Antofagasta as the vehicle for the copper properties in Chile that he had purchased when the Pinochet government allowed private and foreign capital to develop new copper mines in the country. Luksic's strategy paid off handsomely and Antofagasta now has a copper production level fast approaching that of Africa's copper giant Zambia.

In his later years Luksic turned over much of the running of the company to his sons from his marriages to Ena Craig and, after her death, Iris Fontbana. He spent increasing amounts of time managing his charitable foundations, one set up in memory of his first wife, and providing resources for the development of tourism in Croatia, the land of his father. In 2004, a year before he died, Luksic had the immense satisfaction of seeing Antofagasta admitted to the FTSE 100 Index in London.

---

13. THE RISE OF URANIUM

One of the most controversial minerals in the latter half of the 20th century was undoubtedly uranium. The modern uses of uranium in nuclear weapons (atomic and hydrogen bombs) and in nuclear power stations is well known, but it was used in its natural state as pitchblende over 1,000 years before that to make yellow ceramics and later yellow glass.

In these early days German miners were the sole European producers of the mineral. Martin Klaproth, a German chemist, discovered uranium in pitchblende in 1789; French chemist, Eugene Peligot, was the first man to isolate the uranium; and Anton Becquerel, a French physicist, discovered that uranium was radioactive in 1896. The work of these three was vital in preparing the way for uranium's changing role from colorant to power source, but it was the German chemists Otto Hahn and Fritz Strassman in 1939 who completed the journey when they discovered nuclear fission.

The first and only nuclear devices exploded in anger were the two atomic bombs dropped on Hiroshima and Nagasaki in August 1945, which led to the surrender of Japan to the Allies, and the end of the war. So terrible was the destruction these relatively small bombs caused that nuclear bombs were never used again in warfare, although the years since 1945 have often been uneasy ones on the nuclear front. The first nuclear power station to generate electricity for a power grid was built in the Soviet Union in 1954; two years later at Sellafield in England a nuclear power station generated commercial electricity for the first time.

THE BEGINNING

The first uranium mine was developed by Union Miniere at Shinkolobwe in the Katanga Province of the Belgian Congo in 1921. The purpose of the mine lay not in the uranium that was mined but in the tiny amounts of radium that were extracted from the uranium ore, for in the early days it was radium with its ability to emit very strong radioactive rays that scientists coveted. Union Miniere constructed a mill at Oolen in Belgium to extract the radium, which was hugely valuable; the yellow uranium waste was simply dumped. In due course that uranium dump fell under the control of the German Reich during the war, but Germany was some way behind the US in its development of nuclear technology and fortunately for the Allies, German control of the uranium did not threaten their war effort.

Whilst power generation now is the prime influence in the market for uranium, in the 1940s and 1950s it was the development of the atomic bomb that drove demand for the mineral. The US, starting in the 1930s, had gathered together a number of scientists to research and develop a nuclear bomb, although much of the research, particularly in the field of nuclear fission, had application in the area of peaceful use in the potential of nuclear power, and also in areas of medicine such as x-rays and cancer treatment. In these early days only relatively small amounts of uranium were needed but as interest in nuclear power increased exploration budgets began to take account of this new mineral.

CANADA AND AUSTRALIA

The two countries with historically the largest reserves of uranium are Canada and Australia. The first uranium discovery in Australia was made in 1906 by Douglas Mawson, who was eventually knighted for his exploits as an Antarctic explorer. Mawson found Radium Hill in the northern part of South Australia, north of Port Augusta, but it took almost 50 years for a mine to be developed, stimulated by the 1950s boom in uranium demand. The largest Australian uranium discovery in the 1950s was Mary Kathleen in Queensland, not far from the huge base metal mines of Mount Isa.

Mary Kathleen was found in 1954 but it was not the first Australian uranium mine to be developed – that was the Rum Jungle mine found by a prospector, Jack White, who had to hand it over to the Australian government. The government contracted Consolidated Zinc, which eventually became part of the Rio Tinto group, to develop Rum Jungle. The uranium product was then shipped from Darwin to the UK and the US. No financial accounts were ever published. Mary Kathleen was controlled by Rio Tinto which had purchased it from the discoverers, the Walton Syndicate, for £250,000 plus a royalty. With Canadian production coming on stream in the 1960s and with the market squeezed between reduced military demand and only slowly-growing nuclear power demand, both Rum Jungle and Mary Kathleen were closed.

With demand for uranium beginning to increase again in the 1970s Australian uranium exploration stepped up a gear. The nickel boom of the late 1960s and early 1970s was fading fast but the huge increase in the oil price, driven by worrying political instability in the Middle East, saw power utilities and governments shifting attention to trying to generate more electricity using nuclear power.

Historically there had been concerns in Australia that the country was deficient in certain metals and exports were therefore at times not permitted. In fact any shortages were largely as a result of low levels of exploration, and spurred by global economic expansion from the 1950s onwards that situation changed. In any case, since Australia had (and indeed has) no nuclear power stations, uranium for a time could be freely exported. In the 1960s and 1970s the Beverley, Ranger, Narbalek, Koongarra, Yeelirrie and Jabiluka uranium deposits were discovered, neatly filling in behind the fading nickel boom. The discoverer of Yeelirie, Western Mining, also found the huge multi-metal deposit at Roxby Downs (now Olympic Dam) in South Australia, which contained significant amounts of uranium.

OPPOSITION IN AUSTRALIA

Plans for mining these large deposits ran into severe trouble when a Labor government took power in 1972 and instituted an anti-nuclear policy. The Ranger and Narbalek projects were developed before the Labor government put a blanket ban on new uranium mines and an exception also was made for Olympic Dam due to the importance of its other products, particularly copper. Jabiluka and Yeelirrie were not so lucky; test work was done at both projects but no operating mines were developed. The return of Liberal governments led to some easing of the uranium development ban but some projects got bogged down in negotiations with aboriginal rights groups and in the 1990s a very weak uranium market led to further problems.

CANADA

Canada's uranium mining industry was born in the Great Bear Lake region of the Northwest Territories in 1930 where brothers Gilbert and Charles LaBine discovered radium. As in the case of Union Miniere at Oolen, the value of the uranium was not recognised at first. However the National Research Council of Canada established a project to use uranium and heavy water to produce plutonium in an atomic reactor.

The uranium used came from the Port Radium mine near Great Bear Lake, which ran out of ore in 1955, and the Gunnar Mine at Lake Athabaska in Saskatchewan, which took over the contract supplying uranium until 1963. As world demand for uranium to fuel nuclear power stations grew, several new uranium mines were opened in Ontario at Elliot Lake and in Saskatchewan at Lake Athabaska. Many of the operations were quite small, taking advantage of enriched easy-to-get-at ore and supplementing output from the bigger mines in the two areas. The key players were Rio Algom at Elliot Lake and Eldorado at Lake Athabaska.

As the advanced economies began to accelerate their programme of nuclear power station construction the demand for uranium began to increase again in the 1960s and 1970s. In Canada many of the Elliot Lake and Lake Athabaska mines were coming to the end of their lives and exploration began to pick up in response. Cameco was very active in the Athabaska Basin bringing new mines, such as Rabbit Lake, Collins Bay and Key Lake, on stream from the mid-1970s onwards. More recently it discovered the very high grade Cigar Lake at Waterbury Lake in Northern Saskatchewan; Cigar Lake has unfortunately suffered major flooding delaying its opening by several years.

We have seen that uranium mining in Australia met opposition from federal and state governments as well as a wide range of environmental groups and aboriginal interests plus their, not always Australian, lawyers. In Canada political and environmental pressure on uranium mines, though considerable, was not as troublesome in terms of exploration and development. Having said that Australian production continues to run Canada a quite close second in terms of annual uranium output.

THE RISE OF KAZAKHSTAN

Historically the largest uranium producer was the Soviet Union. In the aftermath of the break up of the union, Kazakhstan – whose nuclear power and uranium programmes had been controlled from Moscow until that time – became the largest uranium producer of the former Soviet Bloc countries. In terms of current world output and reserves Kazakhstan is the third largest producer, after Canada and Australia. With global demand on a sharply rising uptrend Kazakhstan is making expansion of its uranium mines a priority. Along with that commitment Kazakhstan is planning to build two nuclear power stations in conjunction with Japanese companies and also as many as 20 small reactors to provide power for outlying towns and cities. In the past the country had one reactor which operated from 1972 to 1999 producing electricity and also powering a desalination plant.

Kazakhstan's uranium exploration programme began in 1948 as the Cold War intensified and the Soviet Union began to develop nuclear weapons. The early uranium mines were hard rock operations but since the 1960s Kazakhstan has concentrated its uranium mining on lower grade sandstone deposits where it has pioneered *in situ leaching* (ISL) of the uranium using sulphuric acid.

The mining method consists of drilling holes into the orebody and then pouring sulphuric acid onto the ore. This dissolves the uranium into a solution which is then pumped to the surface where the uranium is removed in an ion exchanger. The waste solution is then poured back into the sub-surface of the deposit. This method of mining does not agitate the uranium as it is being mined and because the uranium comes out in a solution there is much less ground disturbance during the process. The processing plant is consequently smaller as there is no waste rock to deal with. The driving force behind Kazakhstan's nuclear power and uranium mining programme is Kazatomprom, which is government-owned.

WORLD PRODUCTION FIGURES

Uranium is one of the commonest minerals found and theoretically could even be extracted from the sea if the price was right. This means that uranium can be found in many countries and indeed the global production base is spread very wide, as the table shows.

URANIUM MINE PRODUCTION BY COUNTRY (2003-2011)

	Production by year (tonnes U)								
Country	2003	2004	2005	2006	2007	2008	2009	2010	2011
Kazakhstan	3,300	3,719	4,357	5,279	6,637	8,521	14,020	17,803	19,451
Canada	10,457	11,597	11,628	9,862	9,476	9,000	10,173	9,783	9,145
Australia	7,572	8,982	9,516	7,593	8,611	8,430	7,982	5,900	5,983
Niger	3,143	3,282	3,093	3,434	3,153	3,032	3,243	4,198	4,351
Namibia	2,036	3,038	3,147	3,067	2,879	4,366	4,626	4,496	3,258
Uzbekistan	1,598	2,016	2,300	2,260	2,320	2,283	2,657	2,874	3,000
Russia	3,150	3,200	3,431	3,262	3,413	3,521	3,564	3,562	2,993
USA	779	878	1,039	1,672	1,654	1,430	1,453	1,660	1,537
China (est)	750	750	750	750	715	769	1,200	1,350	1,500
Ukraine (est)	800	800	800	800	846	800	840	850	890
Malawi	-	-	-	-	-	-	104	670	846
South Africa	758	755	674	534	539	655	563	583	582
India (est)	230	230	230	177	270	271	290	400	400
Brazil	310	300	110	190	299	330	345	148	265
Czech Repub.	452	412	408	359	306	263	258	254	229
Romania (est)	90	90	90	90	77	77	75	77	77
Pakistan (est)	45	45	45	45	45	45	50	45	45
Germany	104	77	94	65	41	-	-	-	52
France	0	7	7	5	4	5	8	7	6
Total world	35,574	40,178	41,719	39,444	41,282	43,798	51,450	54,660	54,661
Tonnes U3O8	41,944	47,382	49,199	46,516	48,683	51,651	60,675	64,461	64,402
% of world demand	n/a	n/a	65	63	64	68	78	78	85

Source: World Nuclear Association data

Many countries that no longer produce or that have never produced uranium are involved in exploration following the sharp rise in the uranium price since 2005 and the bullish outlook for demand as nuclear power station construction plans are formulated. The high natural occurrence of uranium also encourages this. Amongst previous producers that have currently stopped mining uranium are Gabon, Hungary, Argentina and Spain, and they are all assessing their position and have exploration programmes currently in progress. However, the controversial nature of nuclear power and uranium does mean that many uranium exploration programmes are the targets of often widespread protest.

The table shows the eight largest uranium mining companies in the world. These companies dominate uranium mining today, responsible as they are for 85% of global annual production.

URANIUM COMPANY OUTPUT (2011)

Company	Country of operation	Production (tonnes)
Kazatomprom	Kazakhstan	8,884
Areva	Canada/Kazakhstan/Niger	8,790
Cameco	Canada/USA	8,630
Armz-Uranium One	Russia/Kazakhstan	7,088
Rio Tinto	Namibia/Australia	4,061
BHP Billiton	Australia	3,353
Navoi	Uzbekistan	3,000
Paladin Energy	Namibia/Malawi/Australia	2,282
Other companies		8,521
Total		54,610

Source: World Nuclear Association

The table on the next page shows the 12 largest operating uranium mines in 2010, with just four being responsible for 30% of all uranium mined annually.

LARGEST PRODUCING URANIUM MINES (2010)

Mine	Country	Main owner	Type	Production (tU)	Proportion of world production (%)
McArthur River	Canada	Cameco	Underground	7,686	14
Olympic Dam	Australia	BHP Billiton	Underground	3,353	6
Arlit	Niger	Areva	Open pit	2,726	5
Tortkuduk	Kazakhstan	Areva	In situ leach	2,608	5
Ranger	Australia	Rio Tinto/ERA	Open pit	2,240	4
Kraznokamensk	Russia	ARMZ-Uranium One	Underground	2,191	4
Budenovskoye 2	Kazakhstan	Kazaktomprom	In situ leach	2,175	4
Rossing	Namibia	Rio Tinto	Open pit	1,822	3
Inkai	Kazakhstan	Cameco	In situ leach	1,602	3
South Inkai	Kazakhstan	Uranium One	In situ leach	1,548	3
Akouta	Niger	Areva	Underground	1,548	3
Rabbit Lake	Canada	Cameco	Underground	1,463	3
			Total	27,951	54

Source: World Nuclear Association

THE FUTURE OF URANIUM

Whilst this book is essentially a history of mining we are also interested in the future of the industry and no mineral excites so much controversy and discussion as uranium. For some it will be eternally wedded to the bombing of Japan in 1945 and the Cold War, which lasted from 1945 until 1990, when nuclear arms were built to deter both sides of the ideological divide. Others see nuclear power as the green option for generating electricity without emitting carbon dioxide. Then again others see nuclear power as too dangerous to contemplate due to the long active life of toxic nuclear waste. They also fear that such waste may be easy for terrorists to hijack with frightening consequences if the terrorists build a 'dirty' bomb. There is also the problem of the extreme nervousness over the safety of nuclear power engendered by events like the March 2011 Japanese earthquake and the damage done to a number of nuclear reactors in the aftermath of the shock.

Apart from the lack of carbon dioxide emissions another benefit of nuclear power is the fact that fuel, unlike in conventional power stations using coal, oil or natural gas, is a very small percentage of the total cost of producing electricity at around 8%. That explains the recent situation when the uranium price rose almost 500% in the space of three years between 2005 and 2007 without having the slightest effect on rising interest in new nuclear power stations. Another plus is that undeveloped uranium deposits are widely spread around the world and while these will take time to bring to production there is no shortage of uranium at the right price. The right price is perhaps not as high as some of uranium's most fervent supporters think, but clearly the early century price level below $10/lb is completely inadequate.

In the last decade nuclear power stations got their uranium from a combination of mine output and stockpiles built up for the manufacture of nuclear weapons – stockpiles that are no longer needed. The latter source is finite and, therefore, must be replaced by new mine supply over the medium term. In the shape of Canada, Australia and Kazakhstan the market should be well supplied over the longer term. In the shorter term the position is less clear with a number of new nuclear power stations being built and requiring considerable quantities of front-end starter nuclear fuel.

It is important, however, not to lose sight of the fact that nuclear power stations historically have come on stream late and well over budget, sometimes due to technical problems, sometimes due to environmental issues. It is also the case that, as we move into the second decade of the 2000s, the greatest number of nuclear power stations are in the category of planned/proposed and many of those will not see the light of day until the end of the decade. A considerable number of them indeed have yet to start the permitting process.

China, India, Russia and Ukraine have the largest nuclear power expansion programmes. It is also interesting that the Middle East, in the form of the UAE, has a substantial development programme planned as does the US, Italy and even little Vietnam, determined as ever to be considered a serious economy in a region dominated by China. Germany and Japan are at the other end of the scale with little or no extra capacity likely to be built and both have major question marks hovering over their longer term commitment to nuclear power.

ENVIRONMENTAL HEAT

The issue of environmental opposition to nuclear power cannot be shrugged off and this equally applies to uranium mining. Coal-fired power stations also face significant opposition of course, even if they utilise the latest technology for cleaning generating emissions. The problem with nuclear power and uranium, as far as environmentalists go, is that the issue of nuclear waste disposal has not been satisfactorily dealt with. The waste can, of course, be securely sealed in concrete and buried but significant earth movements and eroding water ingresses can threaten the stability of the containers, almost wherever they are buried. The risks are very low that this will happen if the burial site has been carefully chosen but cannot be completely ruled out. At the same time anti-nuclear groups think that renewable energy, wind and wave power, for example, are the green way to go and think that nuclear power, with its nil carbon emissions, gets in their way. They also believe that civilisation needs to re-examine the way it organises itself and cut down on its use of energy in any case.

It is a philosophical point but as time has progressed it is often the case that when technology, where metals and therefore mining often play a key part, takes a backward step so does civilisation. So it can be seen that after the Roman Empire collapsed Europe experienced a dark age when Roman advances such as running water, central heating and sewage disposal broadly disappeared for quite a number of centuries.

More recently after a period of massive technological advance we have seen the ending of supersonic air passenger flight, the first backward step (technologically) in aviation since air flight began a century ago. Whatever one's view of supersonic flight, civilisation undoubtedly advances in step with technology and it is risky to break the link. Uranium miners are certain that they can provide the raw material necessary for the new generation of nuclear power stations and governments increasingly believe that nuclear is the only way to square the circle of the economic ambitions of the BRIC countries and other developing economies and the need to lower carbon emissions.

It is not necessarily a majority view and anti-nuclear interests have seemingly inexhaustible reserves of energy with which to fight nuclear power expansion. Events like the 2011 Japanese earthquake only strengthen their hand.

14. COAL IN THE WESTERN ECONOMIES

Whilst uranium may well be the future for electricity generation, the Industrial Revolution was built, as we have already seen, on *king coal*, and coal remained the prime source of power well beyond the end of the Second World War. When the 20th century opened the US was the largest coal producer in the world at 350 million tons, with Great Britain in second place at 236 million tonnes and Germany in third place at 173 million tons. By the end of the century the US had fallen into second place behind China although its output had trebled over the period, and Germany had slipped to seventh place with output largely in line with 100 years previously. In contrast Great Britain had fallen off the map, producing around 40 million tonnes in 1999, a figure that has since halved, although interestingly it remains about the only advanced economy with a reasonable market for coal for domestic (open fire) heating.

THE US

The main coalfields of the US as the 20th century opened were to be found in the east, stretching from Alabama in the south to the Canadian border. At this time the main coal mining states were Pennsylvania, Illinois, Ohio, West Virginia and Iowa, with Pennsylvania the largest producer by some margin. By the end of the century Wyoming's huge surface mines made it far and away the largest producer in the US with West Virginia and Kentucky the next two largest producers.

As in the 19th century most US mines before the Second World War were underground operations and even today if Wyoming is removed from the equation slightly more US coal comes from underground sources than from the surface. However, since Wyoming's Powder River Basin mines produce 40% of the nation's coal today, this means that 70% of US coal is now surface mined.

One of the curiosities of coal mining in the early part of the century related to the isolation of some of the coal mines and their communities. This isolation meant also that there was a shortage of banking facilities and therefore of US legal tender. To overcome this and enable coal mining families to buy essential supplies, the coal mines paid their workers in company scrip, money that carried a dollar value but was not legal and could, therefore, only be exchanged in a limited number of (usually) company-owned outlets. These outlets, general stores in the main, discounted the scrip; so for example a one-dollar coal scrip coin might only purchase 50c worth of goods. However, the store then sold the scrip back to the company at a higher discount, say 80c. This meant that the coal company effectively got its labour at a discount. This shoddy practice eventually died out after 1945.

WYOMING THE GIANT

For most of its history the most important economic activity of the state of Wyoming has been farming and ranching, although coal was first discovered in the early 1800s and the first coal mined in 1859. Anthracite was the main coal product for many years. The coal seams of Wyoming, including those of the Powder River Basin, were formed from huge peat bogs that over millions of years have been compressed and altered to become coal. The first commercial mines in the state established in 1868 were at Carbon near Medicine Bow and nearby Rock Springs. These were owned by Wyoming Coal and Mining which was taken over in 1874 by Union Pacific Railroad which already controlled the company as well as transporting the coal. By the turn of the century the mines had closed but not before severe labour disputes had led, as seems always the way in the coal mining industry, to tragedy. This came in the form of the 1885 massacre of low-wage Chinese miners by white miners at Rock Springs following a wages dispute with Union Pacific.

As elsewhere coal mining in Wyoming was a hard business with safety never much of an issue with the mine owners. In 1903 the worst accident in the state's history took place at Union Pacific's Ludlow mine when 169 miners were killed by an explosion. As the years passed further accidents took place at Union Pacific mines and other mines in the state, and union problems led to countless disputes, sometimes as a result of company greed, sometimes as a result of union militancy.

After the end of the First World War Wyoming found demand for its coal weakened as the US economy entered a long period of instability which only ended with the Second World War 20 years later. As a harbinger of what was to come in the 20th century, Wyoming coal mining also became increasingly mechanised, causing considerable hardship in terms of employment amongst the mining community. Wyoming output revived as the US economy recovered rapidly due to the start of the Second World War in 1939, with the US initially providing wartime supplies and equipment to Great Britain and its Commonwealth and Empire allies, and then its own armed forces after it entered the war in 1942. By the end of the war Wyoming's coal output had reached almost 10 million tonnes but, as happened at the end of the First World War, after 1945 production collapsed again, falling to just over 2 million tonnes in the following year. Depression in the industry was to last another 20 years until the mid-1960s, with Wyoming's anthracite becoming increasingly shunned due to its main and declining role as a domestic fuel for heating and the advance of oil and gas as domestic heating fuels.

However in the mid-1970s in the wake of the huge increase in the oil price following the Seven Days War in the Middle East, mining companies began to look at the huge coal deposits of the Powder River Basin (PRB) which lay on the border between Montana and Wyoming. The main field in the Basin from which most of the coal is extracted is the Gillette field which lies mainly in Wyoming. There are two other significant coal blocks, the Sheridan-Birney and the Birney-Custer-Recluse fields, the former is roughly split between the two states and the latter lays two-thirds in Montana.

Gillette is the largest of the producing fields, providing over 90% of Wyoming's output. Its resources are enormous but much of Gillette is buried too deep to be economic at current or even forecast future prices; current stripping ratios, dirt to coal, are around one-to-one, but much of the Gillette resource would require stripping of ten-to-one to be developed. The original resource was estimated at 800 billion tons, enough to last centuries at current mining rates, but the number in terms of economic mining is nearer 10 billion tonnes or little more than 20 years production.

The coal itself is ideal for power station burning, as it is low in sulphur and is easy to mine using massive draglines. It is also easy to load, an important factor when the coal has to be transported on long trains over many hundreds of miles to customers as far away as Texas, Missouri, Arkansas and Georgia, as well as to nearer power stations in Colorado, Nebraska and Minnesota. The first railroad consortium to transport coal from the PRB was Western Railroad in 1982. In due course the whole network was amalgamated under the Union Pacific line. Originally the route serving the Basin had been a single track built by Burlington Northern to link Chicago with Denver.

In due course as volumes from the PRB increased, the line had to be extended and also an additional track added, and this process was not without further problems as contracted coal volumes threatened to outrun line capacity. In the mid-1990s and then again ten years later, breakdowns and capacity problems gave miners and railroads considerable headaches. In 1996 a major investment programme allowed for an extension of the dedicated coal line and the addition of a third track. The volumes on the line now can run to 15,000 tonnes of coal per train with around 80 trains daily and these trains can be up to 150 trucks long.

Wyoming, and its Gillette field, remains the dominant producer in the US today. Unlike in the dark days of 19th century Wyoming coal mining, the Gillette workforce is compact, very well paid and highly productive. The main three producers in the field are Peabody, Arch and Rio Tinto through its Kennecott subsidiary. The Powder River Basin is likely to dominate US coal mining for the foreseeable future and in an energy-scarce world is one of the US's most valuable natural resource assets.

GREAT BRITAIN

When the 20th century dawned Great Britain was arguably the most powerful economy in the world, and its Empire made it probably the most politically influential country as well. However, rivals were gathering in the form of the US, Germany and Russia, and in the Far East, Japan, and as the century progressed Britain's power, undermined fatally by two world wars, slowly faded. This steady decline was mirrored by its once mighty coal mining industry. But we should not get ahead of ourselves.

LEADING THE WORLD... JUST

As the new century dawned Great Britain was the largest coal producer in the world feeding the fast growing appetite of its economy for capital goods and technologically advanced infrastructure. Its peak year of production was symbolically 1913 when almost 300 million tonnes of coal were mined – British coal output never exceeded that peak again and neither did Britain's political power.

However Britain's coal mining industry had been climbing a wall of worry for decades even as it expanded and demand for its output grew. The concerns as to whether its structure, an amalgam of many small producers and a few very large ones, would be able to sustain this demand were a permanent issue right through the 20th century. There were fears that the smaller mines would mine inefficiently leaving, in aggregate, substantial quantities of coal behind that might become unrecoverable. The inability to attract sufficient capital to make long-term investment in the smaller mines was also a worry, and the spectre of war that hung over Great Britain for most of the 20th century added a political dimension to the industry's situation, and also worried British governments of all persuasions.

Britain was also beginning to lose its way as a technical innovator with mine productivity poor as mechanisation, particularly the introduction of coal cutting equipment, ran well behind international rivals such as the US. Counterintuitively the big increase in percentage terms in coal cutting in Britain arose during the 1930s years of depression when the use of cutting equipment expanded from 25% to 55% of coal mined.

THE BRITISH GOVERNMENT INTERVENES AND FINALLY NATIONALISES

With the coming of the First World War, government participation in the running of the British coal mining industry first reared its controversial head. Concerned that labour relations in the industry were poor enough to threaten the country's war effort the government took over control of production, although not of ownership. Whilst this underpinned output during the war, a strategic necessity, the costs were long term as investment in the mines was no longer a priority. In 1919 a Royal Commission recommended that the coalmines be nationalised but in 1921 the government returned control to the owners. The 1920s saw a continuous run of disputes over pay and conditions between the mine owners and the unions.

This followed into the next decade when economic slump led to demands for labour pay cuts at the same time as demand at home and export sales fell sharply. Continuous threats of labour strikes were a feature of the 1930s and though no strikes eventuated the tradition of poor relations between owners and miners carried on into the Second World War. In 1942, to support the war effort the Coal Commission came into being with its task to regulate the industry and draw royalties from the still privately-owned collieries. After the war a new radical and reformist Labour government nationalised the coal mining industry in 1946, which numbered around 1,300 separate mines, and this survived the fairly swift return of a Conservative government in 1951 which did not have the stomach to de-nationalise the industry.

DECLINE BEGINS

By the end of World War One UK coal production was running at around 200 million tonnes per year and stayed at that level until the 1960s when it began a steady decline as natural gas, nuclear power and oil began to push it to one side. This process was exacerbated by the labour strikes of 1972, 1974 and 1984/5 as left wing union leadership put pressure on, particularly, Conservative governments over wage levels and the government's desire to slim down the industry to improve profitability. Between nationalisation in 1946 and the first coal miners' strike of 1972, Britain's National Coal Board had lost money in 12 out of the 25 years, and talk of profits always greatly agitated the miners' leaders who saw the industry more as an extension of Britain's welfare state than as a business. This attitude also had an impact on investment in Britain's mines, with Conservative governments being particularly unwilling to countenance public money going into an unreformed mining industry.

OIL AND GAS, THEN PRIVATISATION

The discovery of first natural gas in the mid-1960s and then oil in the North Sea sealed coal's fate as a widely used fuel outside the power generation sector in the UK. The coal mining industry began to shrink with output falling below 150 million tonnes by 1979 when the election of Margaret Thatcher's hardline Conservative government precipitated a major deterioration in relations between the government and the coal miners. This trend continued through the end of the 20th century, exacerbated by the long strike in 1984/5, which itself precipitated the privatisation of the coal mines in 1994.

The biggest force in the privatised industry in England was RJB Mining, which later became UK Coal. The years following privatisation were not easy as energy prices remained low worldwide and the UK, in particular, turned to building natural-gas-fired power stations to the detriment of coal demand. By 2008 UK production had fallen to 17 million tonnes from just 16 underground mines and 35 open cast operations. In order to meet annual UK power station demand today a further 40 million tonnes of coal has to be imported. It is indeed a case of how are the mighty fallen but it should not be assumed that Britain has simply run out of coal – current recoverable reserves are believed to run to many billions of tonnes, with in-situ resources (not economically recoverable with current mining technology) significantly higher. Recoverable reserves are equivalent to over 100 years of demand at current levels.

The biggest UK mine is Daw Mill in the Warwickshire coalfield which itself has been worked since the early 1900s. Daw Mill, which was first mined in 1965, currently produces around 3 million tonnes of coal annually or 17% of total UK output. The underground operation has workings, taking in two historic operations, extending over several miles originally served by individual shafts for both lowering workers and raising coal to the surface. In 1983 a drift was developed to allow coal to be continuously hauled to the surface, greatly increasing productivity potential. Although a large operation Daw Mill is one-tenth of the size of Colombia's Cerrejon open pit mine, probably the world's largest, and one-fifth of the size of the giant Chinese underground mine at Daliuta in Shaanxi Province. It is also the subject of doubts about its economic future.

In keeping with many coal mining countries in the West, the future of the UK industry relates to the speed with which carbon capture technology can be developed in order to mitigate and even eliminate carbon emissions from coal-fired power stations. At the moment a great deal more work and investment is required to make carbon capture an effective and viable process. In the interim Britain's coal fired power stations are ageing and new ones are facing stiff resistance from environmental campaigners. It is ironic though that after several decades, indigenous natural gas and oil production and reserves are now falling rapidly leaving *old king coal* as Britain's most abundant energy resource, although shale gas excites some interest.

SOUTH AFRICA

As we learnt earlier, the second half of the 19th century saw the establishment of a South African coal mining industry based on three main centres: the Orange Free State/Transvaal, Natal, and the East Rand/Eastern Transvaal. These centres served the diamond fields of Kimberley, the gold fields of the Witwatersrand and the burgeoning railway system associated with these massive mining developments. This history has been divided into eras throughout and when looking at South African coal the Anglo-Boer War neatly separates the era of the Industrial Revolution from the modern age. This is thus where we pick up the story of South African coal once again.

After some disruption caused by the Anglo-Boer War, normal coal production resumed in 1903. Just before the war in 1898 production from all Transvaal coal mines had reached 2 million tons, earlier in the decade output had been a quarter of that figure. By 1900 as the war dragged on output had slumped to 500,000 tonnes but with the end of the war the mines soon recovered and production climbed to new highs. A number of issues arose as the industry expanded, some of which continue to this day. The main ones related to the cost of carrying the coal on the rapidly-growing rail network, the position of unions and the economic issue of mining costs and market prices for coal.

COAL PRICING, QUALITY AND SAFETY

As regards prices (using old UK money), in 1902 the average coal price at the mine head was 96 pence per ton, but by 1909 this had fallen to 60 pence per ton. Meanwhile rail freight charges were relatively high – the charges themselves varied but the SA Chamber of Mines at that time calculated that on average US railroads shipped coal 180 miles for the cost that SA railways charged to carry it 30 miles. The Netherlands SA Railway Company, which served the Witwatersrand area, also compounded transportation issues as it did not allow gold mines to build sidings so coal and other supplies could be delivered right to the mines. It also insisted that all coal it carried be bagged not shipped in bulk which only added to transport expenses.

Despite these problems the market for coal was growing rapidly as the mines and adjacent towns expanded and their appetite for power grew in line. In the second decade of the new century, power station building accelerated in the wake of industrial and municipal demand, with Johannesburg a particularly substantial customer as city and gold mines expanded. This growth of power generation also helped the coal mines as it provided a serious market for their lower-quality coal that was ideal feed for power stations.

In order to provide a commercially acceptable product in these relatively early days of the SA industry, many mines, particularly in the Transvaal, upgraded their coal by screening and hand picking – a labour intensive process. However, from the early days mines in Natal used washing plants for their upgrading, where in essence coal was passed through fluid allowing the impurities to be detached from the raw coal and be flushed away.

Safety was an issue, with methane at depth and coal dust, both of which caused explosions. Ground instability also was an occasional problem. In the 20th century though there was a marked improvement in mine fatalities – the figure per 1000 miners in the five years to 1905 was 3.2, but this fell to 1.8 in the comparable period to 1955. The industry, which had started out in the 19th century as extremely fragmented with a large number of small, often uneconomic mines, developed larger, more mechanised and efficient mines as the years went on.

The industry also became more organised in terms of labour, management and ownership. Provincial and central government supervision increased too, especially after the formation of the Union of South Africa in 1909. Black labour, however, was not involved in this consolidation process, for even in the days before apartheid natives had little industrial bargaining power. Various industry bodies were set up to support the SA mining industry as a whole and the coal owners in particular. This extensive organisation eventually led to muscle flexing, although the industry itself, when in dispute with the railways, was usually pushed back to seeking government intervention.

INDUSTRIAL PROBLEMS

The labour base of the South African coal mining industry was built on the foundations of British immigrant miners and so it is perhaps inevitable that the 20th century witnessed regular, and often violent, industrial disputes.

In 1912 a short but bitter dispute on the gold mines over working hours left 11 miners dead. The following year a major industrial dispute launched by Indian workers in Natal and supported by Mahatma Gandhi sucked in the coal mining industry. In 1914 a major general strike in the Transvaal again sucked Natal coal miners into a dispute about railway retrenchments. The start of the First World War and South African participation led to a cessation of strike activity in the country and the coal industry also fell quiet.

Wages of white miners were increased during the war years to encourage productivity for the war effort, although shortage of railway wagons and bunkering space, both on the mines and at ports, was a problem as were the rapacious railways themselves, as their carriage charges represented almost half the all-in production costs. The war years also saw the incorporation of Anglo American Corporation of South Africa, and the Transvaal Coal Trust became the new group's coal division, changing its name to Rand Selection. Johannesburg Consolidated, eventually to become an associate of Anglo American, also entered the coal sector after the war.

Both Transvaal and Natal coal producers had built up a strong export business for their coal and in the leaner post-war years both sought to position themselves at the quality end of the export market by grading their output. The coal was broadly of a high quality and acceptable as an export commodity, but like coal elsewhere South African coal did on occasions spontaneously combust – this added a further hazard to exporting if the coal combusted whilst on the sea. There were additional problems associated with exporting where a more cooperative approach was suggested in terms of shipping coal. However the individual collieries tended to land their own supply contracts and it took many years before a more integrated industry approach was adopted.

Unfortunately as the world slipped into a post-war economic slump warfare broke out between the gold mines and the collieries on one side and white labour on the other. Fuelling the dispute was the huge increase in the cost of living in South Africa during the war years and the collapse of international coal prices after the war, causing the collieries to seek cost cuts, particularly in wages. There were also problems relating to the conciliatory approach taken by Evelyn Wallers, who headed the Transvaal Chamber of Mines, towards the aspirations of the gold and coal miners both white and black.

Many white miners and labour leaders took the Chamber to be weak and became increasingly fractious. In 1922 events came to a head on the Rand as both gold and coal prices fell sharply and the mine owners proposed wage reductions. A general strike ensued which lasted almost three months. The eastern Transvaal mines were relatively peaceful, as were those of the Orange Free State, but on the Rand a large number of guns fell into strikers' hands and bloodshed ensued, leading the government to declare marshal law. The coal mines, however, managed to keep working during the strike with mine officials and black miners keeping the coal flowing, and the strike collapsed when government forces finally subdued the strikers' armed militia.

COMPETITION AND MODERNISATION

The early 1930s were a very difficult period for SA coal. The British pound came off the gold standard in 1931 but the SA pound stayed on it until the following year and strengthened as a consequence. This led to British coal being able to undercut SA coal in what were hardly very buoyant export markets due to the worldwide economic slump.

When South Africa came off the gold standard in 1932 the SA pound weakened, easing the coal industry's competitiveness problem. Then in 1934 the US devalued the dollar thereby raising the dollar price of gold by almost half. This led to a surge of prosperity for the SA gold mines and this quickly transferred itself to the coal mines, which benefited from demand from a strong local economy. During this period the Lewis and Marks group reorganised its coal interests into the Amalgamated Collieries of South Africa. The new company operated mines in both the Orange Free State and the Transvaal and had extensive rights over other coal-bearing ground that it developed in due course.

As the 1930s went on coal output reached almost 18 million tonnes in 1938 against 14 million tonnes at the beginning of the decade, although this was still small by world standards. The industry was also making good progress on addressing some old issues, particularly that of productivity, the use of coal cutting machinery and also power hoisting, which was becoming industry wide. Progress was also made on delivery quality, with an increasing use made of coal washing to improve the consistency of coal sold as well as its calorific value. As the decade came to an end the SA coal industry was beginning to consider the possibility of using coal to produce petroleum products as Germany was doing. The onset of the Second World War temporarily brought that research to a close.

EXPORTING FOR THE WAR EFFORT

The war years brought opportunities for the SA coal industry to supply the wartime economy and the export trade as well, where the US and the UK in particular were no longer active as they needed to direct coal output to their own war effort. The SA coal industry also had to integrate its marketing effort more closely due to local price controls, which were not applied to exports; the rule was that coal groups had to pool the profits from local and export sales. During the war coal output rose significantly, by 50% in the Transvaal fields and 20% in Natal, and prices, partly as a result of cheap black labour, remained low relative to the rest of the world – a tonne of Natal coal sold for 95 pence, against 150 pence for US coal and over 500 pence for a tonne in the UK. Wage rates for white coal miners in 1946 were 265 pence per shift against 35 pence per shift for a black miner. This disparity was largely based on the fact that the higher paid mining jobs were not open to black miners, an issue that was slowly confronted and reformed in the post-war years.

South African coal output in 1946 was just over 25 million tons, with the Transvaal and Orange Free State responsible for 20 million tons. The major producing area was the Transvaal with the Witbank/Middleberg field the largest and the Eastern Transvaal's Ermelo/Breyten the next largest. The pioneering East Rand mines were in decline and down to two producing collieries. The large

Vereeniging field in the OFS was a smaller producer than the Transvaal, but still ahead of Natal which remained important as the main source of coking coal for the manufacture of town gas.

There were also important corporate moves around that time, in particular the acquisition by Sir Ernest Oppenheimer's Anglo American group of the mining interests of Lewis and Marks, which made Anglo the largest coal group in South Africa. The man who Sir Ernest hired to head the coal side was Tom Coulter, a Scotsman who arrived in South Africa in 1912 and spent his working life in the coal industry, firstly as a mine surveyor and then as a manager and executive. He encouraged the use of officials to keep coal mines running during the strike prone years of the early 20th century and foresaw the need for a major expansion in coking coal production in Natal after 1945 on the back of growth in the steel industry. He became Chairman of the Collieries Committee on the Chamber of Mines in 1948 in recognition of his position as Head of Anglo American's coal division.

SUPPLYING THE STATE

In the 1950s two major influences on the development and expansion of the coal industry were the requirements of Iscor and Escom, the state steel and power generation companies respectively; these were closely followed by Sasol, the oil from coal operation which ultimately became the second largest coal consumer behind Escom. This huge expansion in demand for coal meant that mechanisation took another step forward, and the industry also began to look at the possibility of developing opencast mines which, though capital intensive, threw up the possibility of a significant improvement in productivity. Security of supply was of increasing importance for these huge consumers as their own operations expanded, and Escom power stations were deliberately located near to major urban centres and potential coal deposits to capture development economies.

DISASTER AT COALBROOK

One of the giant coal providers to Escom was the Coalbrook North colliery near Vereeniging, owned by Clydesdale Collieries. On 20 January 1960 the worst mine accident in South Africa's history and the fifth worst in recorded mining history occurred at the colliery, when 437 miners lost their lives. The massive and sudden collapse of underground workings meant that there was no hope of finding any survivors. The accident shook the industry to its bones, leading to extensive investigations into underground mining methods, particularly in regard to support for working areas.

In due course a number of recommendations regarding underground development and safety were implemented, and a huge investment was made in purchasing and maintaining a machine capable of drilling at the rate of 12 feet an hour through the hardest rock. The hole drilled was able to take a rescue module and it was eventually used by the Mines Rescue Brigade in 1991 to free miners trapped after a major roof fall at the Emaswati mine in Swaziland.

CHEAP COAL, CHEAP POWER

One of the issues that has most exercised South Africa's coal industry over its history has been the price of its product. Price controls for domestically sold coal were tightened during the Second World War and retained after the war. These controls enabled Escom to offer cheap power, in terms of world prices, to local electricity customers and to expand its network rapidly with consequent benefits to the South African economy.

The greater good may have been well served but the mines were less happy, operating on the basis of mining profits of 25c per tonne (South Africa by then had abandoned sterling and decimalised) against a coal price of R1.52 per tonne, a mine margin of 16% before central and financial charges. Escom bought its coal on the basis of cost plus contracts, so returns were fixed which gave the mines some stability but it restricted growth opportunities in operating margins. But the 1970s oil price crisis was just around the corner and this would change everything.

EXPANSION AND EXPORTS

The coal industry's corporate structure, as it unknowingly prepared to enter a new, almost golden, age, was robust and dominated by financially-powerful mining houses. The 1960s had seen the rapid expansion of the Afrikaner mining house General Mining controlled by Federale Mynbou; from this Trans Natal Coal was born. Alongside General Mining in terms of size in coal mining was Anglo American, which in the early 1970s operated as Amalgamated Collieries and SA Coal Estates. These two companies produced around 75% of the country's coal. The balance was spread around a number of smaller producers including Rand Mines, JCI/Tavistock, Gold Fields and Anglo Vaal. The largest colliery in the country was the infamous Coalbrook.

Led by Anglo American the SA coal industry began to think more about profitability and the possibilities of increasing exports where prices were much higher, with the oil crisis of 1973 around the corner its timing could not have been better. Its efforts on the technical front were also well timed, South African coal had always had quality problems and Anglo decided that investment in coal beneficiation was warranted. With the Japanese always keen to expand their supplier network for strategic reasons, the Transvaal Coal Owners Association (TCOA) began talks in Tokyo in 1970. As a consequence of these talks the industry signed a 13-year contract to deliver 30 million tonnes of coal to the Japanese steel industry; it also signed a contract with SA Railways to help develop a new port for coal exports at Richards Bay, Natal, which opened in 1976. During the construction process, which occurred during a period of high inflation, there were moments when the deal looked in danger of collapsing but the Japanese, anxious for diversified energy sources, renegotiated parts of the contract to take account of changing economic circumstances.

THE OIL CRISIS SPURS COAL

With the 1973 oil crisis leading to a major increase in energy prices worldwide the issue of SA coal exports caused some concern on the part of both the government and the public. The Petrick Commission was set up in 1970 to examine the state of the coal industry and an important part of its investigations concerned the ability of the country to be both an exporter of coal and a major user, with particular reference to the requirements of Escom. The fear was that South Africa's coal resources were not as substantial as had been thought in the past and that coal mining had historically been wasteful, with much coal left un-mined due to inefficient mining methods.

The commission worked throughout the 1970s energy crisis and reported in 1976, when it basically supported the coal export trade but called for more realistic pricing to encourage better and more economic mining methods and an increased exploration effort. Two years later the SA government granted coal exporters another allocation of 12 million tonnes annually over 30 years. This allocation also included three global oil companies as well as the SA mining houses. In 1970 industry coal sales had been R110 million with exports of just R10 million, 9% of the total; by 1978 sales had risen to R870 million with exports of R325 million, or 37% of the total.

By the end of the 1970s coal, spurred by the second oil crisis which followed the fall of the Shah of Iran as well as the Three Mile Island nuclear plant accident in the US, reinforced its position as a key energy source for the future. South Africa had become the third largest exporter of coal globally but as the 1980s progressed political factors began to make themselves felt, as large importers like Denmark banned South African coal in protest over apartheid. The industry, however, in terms of its corporate structure and financial strength, was well able to withstand the weakening in demand for SA coal.

Anglo American's Amcoal had been re-structured by Graham Boustred, Anglo's coal czar, and was the country's largest producer. Gencor's Trans-Natal Coal, which eventually merged with Randcoal, and more recently was renamed Ingwe Coal, was the next biggest producer and all the other active mining houses including Gold Fields, JCI and Lonrho had well-financed coal divisions. With the coming of F.W. de Klerk and political reform in 1989 the pressure eased and SA coal producers were once more able to plan for long-term export growth.

THE RISE OF THE OPEN CUT

During this post-oil crisis period the industry also began to change technically. Mechanisation with coal cutting machinery was not widely used, as the underground mines were essentially labour-intensive traditional drill, blast and haul operations – low coal prices meant that profitability was not sufficient to support the level of investment needed for mechanisation. However there was a move during the 1970s towards opencast mining of shallow coal seams using draglines, as was the practice increasingly in the US. The first of these large-scale opencast mines was Trans-Natal's Optimum Colliery in the Eastern Transvaal, which had been originally developed as an underground operation in 1970, but by 1982 had become an opencast mine with a maximum capacity of 12 million tonnes per year. In 1979 Randcoal and Shell opened the Rietspruit opencast mine in Natal, which had an annual capacity of 9.5 million tonnes. Around the same time Amcoal's Kleinkopje and Kriel opencast mines near Witbank were commissioned. Both had started life as conventional underground operations.

RAPID EXPANSION IN THE 1980S

The politically difficult 1980s saw the coal mining industry forging ahead with new colliery developments to feed Escom and Sasol and steel maker Iscor, all key strategic industries in an increasingly hostile world. A further allocation of export permissions were granted by the government in 1982 taking the total annual allowed export figure to 80 million tonnes, although actual export volumes at the time were around half that figure. The long-term requirements of all these interests stimulated an accelerated exploration programme which led to a number of potential new coal fields being identified – namely Grootegeluk near the Botswana border, the nearby Waterberg field and Soutpansberg to the north, and further east the Tshikondeni mine as well as the huge Springbok field. Such is the long lead-time for major coalmine projects that these areas remain very much development areas today. They also help South Africa in terms of expanding its coal product range, with large quantities of coking coal having been identified at Soutpansberg and Tshikondeni.

The industry was also reasonably profitable despite continuing cost inflation problems, with rail haulage on the key export route to Richards Bay a particular issue, where by the second half of the 1980s rail charges equated to half the boat-loaded cost of export coal. The arrival of Indonesia and Colombia on the coal export scene also brought two lower-cost producers to compete with South Africa, a matter of concern but helpful in persuading the government and the railways to rein in haulage rates. Coal prices too had performed satisfactorily, with domestic and export prices having risen from R2 and R7.50 per tonne respectively in 1970, just as the great inflation began to bite, to R27 and R78 per tonne respectively as the apartheid era came to an end in 1989. Today South African export prices have hit R500 per tonne and domestic prices, if the costs of transportation are taken into account, were very close to export price levels in 2008.

Currently South Africa produces around 250 million tonnes of coal annually (30% exported) and has reserves of about 53 billion tonnes, an inventory that should last 200 years at present production rates. In terms of the environment many of the major mines and future development projects are located well away from areas of material population, which makes it much easier to plan for expansion and development. Although better known for its gold and diamond interests in the past, Anglo American, through Amcoal, is South Africa's largest coal group, as it has been for decades, and today coal is often its largest profit contributor. The other current major producers are BHP Billiton, Sasol and Xstrata.

INDIA

We have already seen the modest beginnings in the 18th century of the now mighty Indian coal mining industry – these ended with the expansion of the Jharia coalfield at the end of the 19th century. As the 20th century dawned India's coal mining industry was beginning to boom and during the First World War its growth was rapid. However, the end of the war saw Indian coal production decline sharply, something seen in other parts of the world where economies experienced a material fall off in post-war economic activity and weakening demand. Coal production, which was running at an annual 6 million tonnes in 1900, increased to 18 million tonnes by the end of the war before falling back. As the Second World War loomed output began to rise again, reaching 29 million tonnes in the opening years of the war, a level it maintained through the war years.

THE PRIVATE SECTOR GETS A LOOK-IN...

The coal mining industry continued to expand in the post-war years and with substantial private investment encouraged by the State's National Coal Development Corporation (NCDC) – which also took over the state-owned railway coal producers – production grew to 56 million tonnes by the mid-1950s. Oddly for one of today's most dynamic emerging economies, India in the 1970s became a highly centralised socialist economy with left winger Indira Gandhi as prime minister – evidence of this was seen in 1972 when the coking coal industry was nationalised under the name of Bharat Coking Coal (BCC). A year later non-coking coal mines were also nationalised under the control of the Coal Mine Authority, which also took over the NCDC. In 1975 Coal Mine India Ltd (CMI) was formed to hold the country's various coal mining undertakings, including BCC, in a five-division structure. The industry began to expand quickly and two new coal companies were carved out of CMI in 1985 – Northern Coalfields and South Eastern Coalfields – and in 1992 the latter spawned Mahanadi Coalfields.

...BUT STILL STATE CONTROLLED

Today coal mining is predominantly under state control but a private mining sector has rather hesitantly been born out of government relaxation of rules regarding mining for restrictive or captive use. That means that companies like Tata Iron & Steel are allowed to develop coal mines to provide them with coal for their own use. No coal can be sold to outside customers, who still have to buy coal mined by the nationalised CMI, although the distribution of coal was de-regulated in 2000.

One of the problems that Indian coal mining faces in the new millennium is the fact that demand for coal is beginning to rise above the ability of local supply to meet that demand. This has led to a steady rise in the amount of imported coal coming into India, which now stands at more than 140 million tonnes and rising. In response to this the government had a production target of 540 million tonnes by 2012 doubling to around 1 billion tonnes per year in the 2020s.

FOLLOWING THE CHINESE OVERSEAS

Another recent development, perhaps influenced by the expanding appetite of the Chinese for overseas sources of coal, is the search by CMI for overseas coal reserves. Its aim mirrors China, whose own coal industry we deal with in the China section below, in looking at Australia, Indonesia, the US and South Africa, but it is also looking for strong overseas partners for these ventures, a recognition, perhaps, that its nationalised structure may prove a problem both in political and entrepreneurial terms.

Underpinning this structure is the Central Mine Planning and Design Institute responsible for exploration and mine development in particular. Around 80% of India's coal comes from large open pit mines, the much smaller underground sector operates traditional bord (roadway/tunnel) and pillar techniques. It has been expected that in due course the Indian coal sector would be further opened up to private capital, and the proposed sale of 10% of Coal India in 2010 was a first step in that process, although the government will remain the controlling shareholder. At the same time the private captive coal producers seek a more liberal operational structure as regards leases, distribution and sales.

The table shows the major Indian coal producers in 2011.

INDIAN COAL PRODUCERS (2011)

Coal company	State of operations	Coal company	State of operations
Bharat Coking Coal	Jharkhand	South-Eastern Coalfields	Chhattisgarh, Madhya Pradesh
Central Coalfields	Jharkhand	Western Coalfields	Maharashtra, Madhya Pradesh
Eastern Coalfields	Jharkhand, West Bengal	Singareni Collieries	Andhra Pradesh
Mahanadi Coalfields	Orissa	Neyveli Lignite	Tamilnadu
Northern Coalfields	Madhya Pradesh, Uttar Pradesh		

15. CHINA

One of the most critical trends influencing the mining industry in the modern world is the rise of China to economic super power status. It has become a major importer of a whole range of raw materials, which has had a profound effect on the international mining industry and its development plans. At the same time China is itself a major exporter of a range of minor but strategic metals such as antimony, tungsten and rare earths. One of the bigger issues surrounding Chinese demand for metals is how it will respond to the more difficult economic times that the world faces over the next few years. In the recent past it has often stockpiled metals in the face of rising consumption to avoid shortages, and as it has built up its foreign currency reserves it has been able to invest in mines abroad as another way of improving security of supply.

THE FOREIGN INFLUENCE

China is often seen as a closed, opaque country. As we observed much earlier, its civilisation is indeed an ancient one but records, especially covering ancient times, are quite thin as regards the mining industry and this remained so until the end of the 19th century. At the turn of the century the important Xikuangshan antimony mine in Hunan province and the Wanshan mercury mine in Guangxi province were operated by European owners hungry for supplies of these strategic metals. Despite this, for many years contact with the marauding Western imperial nations undermined the economics of China's often small-scale, high-cost mining industry, embedding a deep suspicion of Western economic motives in the Chinese psyche. Since the 1990s we have begun to see foreign participation in the mining sector but it has been low key and slow. The main foreign companies have been small, exploration-driven groups primarily aiming at gold, but Griffin Mining, a small UK AIM listed company, brought a small zinc mine to production in 2005. Eldorado, Sino Gold and Jinshan have all brought gold mines to production in the last ten years as well.

JAPANESE IMPERIALISM

The Geological Survey of China was established in 1922, unfortunately just after Japan began its incursions into China, which led to full-scale war in 1937. These incursions and the subsequent war were centrally aimed at first creating a political situation where a number of Chinese provinces became friendly to Japan and so provided buffers against anti-Japanese sentiment, particularly from Russia. But perhaps the most important aim of Japanese imperialism was to secure privileged access to China's huge raw material wealth for Japan's own industrial development and political expansion. For Japan this proved a disastrous ambition, pitting itself against the US, the British Empire, the Soviet Union and China. But the ensuing conflict saw China suffer huge damage to its economy and when the Communists assumed power after the revolutionary struggle that followed the Second World War there were only around 200 qualified geologists in the whole country. This paucity of expertise underlined the backwardness of China's mining industry in technological terms; perhaps not surprising considering the continuous upheavals the economy went through as a result of the Japanese occupation.

COAL

The renewal and expansion of China's coal mining industry had proceeded very slowly in the 19th century, as we have already seen, but in the 20th century the pace of advance picked up. China had once been the largest coal producer in the world but that was before the Industrial Revolution and related to a time when economic activity utilising coal was very minor.

Companies from Russia, Japan, Great Britain and China were at the forefront of foreign investment in Chinese coal mining in the late 19th century but the terms surrounding their participation changed early in the 20th century, and 50/50 joint ventures with Chinese companies using local management became the required norm. The situation became more difficult after the fall of the Qing Dynasty in 1911 and a number of foreign mining groups lost their rights; this excluded the Japanese and so began a period of expansion for Japanese mining interests.

In the 1920s virtually all foreign-produced coal was from Japanese-owned mines with the Fushun mine in the province of Liaoning in the northeast, once controlled by the Russians, dominant – it represented 70% of foreign controlled output. The split between Chinese and foreign controlled coal output was around 50/50, and this lasted into the 1930s. With the invasion of China by Japanese forces in 1937, Japan gained almost total control of China's coalmines with consequent benefit to Japan's war effort. With the Allied defeat of Japan in 1945 the coalmines returned to Chinese ownership, which meant the Communists, as Mao's revolution pushed the Nationalists from the mainland onto the island of Formosa (now Taiwan).

Since a large percentage of Chinese coal production was in Manchuria in the northeast of the country, when the Russians invaded the region following Japan's surrender they walked off with huge quantities of industrial equipment including plant and machinery from the coalmines. Any equipment that could not be moved was destroyed. It is believed that the coal mining industry lost almost 90% of its operating plant and machinery this way, and the ensuing civil war between the Communists and the Nationalists finished off much of what was left.

At the end of the civil war in 1949 China had coal production of just 32 million tonnes against 64 million tonnes in 1942; in comparison, in 1949 the US produced 436 million tonnes and the UK 219 million tons. The concentration of coal mining in the north and northeast of China made things lopsided in terms of industrial development in the 1920s and 1930s with mines serving the foreign-run Treaty Ports of Manchuria and Shanghai. At the end of the 1930s this northern region of China produced 93% of all national coal output and attempts to diversify coal mining were initially unsuccessful, often thwarted by bureaucracy.

The mines were primarily underground operations, the Chinese-owned pits being normally very small. Although in the Japanese-owned mines investment in mechanisation had begun, productivity at the end of the 1930s – at 0.6 tonnes per man shift – compared poorly with European and US productivity where mines in the Netherlands, for instance, mined 2.6 tonnes per man shift. In the decades before the Second World War, coal consumption was primarily industrial/transport with domestic demand largely confined to the cities, which due to China's agrarian nature were not the power centres they have become today. In more recent years, as the Chinese economy exploded upwards, coal consumption in electricity generation has generated rapid development of local mines and sucked in huge imports at the same time.

PRONE TO ACCIDENTS

It is one of the facts of mining, particularly coal mining, in China in the modern era that the country has, and continues to have, a very poor record when it comes to mine safety. The disaster in Japanese occupied Manchuria in 1942 remains the worst coal mining accident in recorded history when over 1,550 miners were killed following a coal dust explosion at the Honkeiko mine. More recently in 1991, around 147 miners were killed by an explosion at the Sanjiao River coalmine in northern Shanxi province. In the year 2000, 162 miners were trapped and then died following a gas explosion at the Muchonggou coal mine in Guizhou in the south west of the country. The year 2004 was a bad one for accidents with 148 miners being killed in October by an explosion at the Daping coal mine in Henan province and a month later 166 miners were killed by a gas explosion at the Chenjiashen coal mine owned by the state in Shanxi province. The following year, 2005, was no better, with 203 deaths at the Sanjiawan mine in Liaoning province and 161 miners killed by an explosion at the Dongfend mine in northeastern China.

All these accidents were interspersed with literally dozens of smaller accidents where, nonetheless, fatality numbers were significant, often above 50. For example the total number of mine deaths in China in the first nine months of 2003 exceeded

13,000, the largest percentage of them in coal mines. In the last couple of years the Chinese government has announced a major safety crackdown on several thousand mines, particularly coal mines, with some success as coal mine deaths fell to around 2,400 in 2010.

MINING TODAY

China is one of the largest producers and also consumers of metals in the world today as a result of its economic dash for growth, this powered by a massive transfer of labour from the countryside to the city and the development of large-scale industrial enterprises exporting finished goods to the West. Although China is a substantial importer of a wide swath of metals, it is also a big producer of most of them in its own right.

The table on the next page shows the significance of China as a producer for a variety of metals, giving data for its own production and this as a percentage of world production. It is the largest producer in the world by volume of aluminium, coal, gold, lead, steel, tin and zinc amongst the major metals, and of antimony, magnesium, mercury, rare earths and tungsten of the minor/strategic metals. Of particular note has been China's assumption of the role as the world's largest producer of gold, a position held by South Africa for decades. As we have mentioned before gold is an area where there is considerable foreign interest in getting involved and this can only underpin China's leading position as a producer.

CHINESE MINING OUTPUT (2011)

Metal	China (tonnes)	World (tonnes)	China as percentage of world
Aluminium	18,000,000	44,100,000	41
Antimony	150,000	169,000	89
Coal (1)	3,520,000,000	8,000,000,000	44
Cobalt	6,500	98,000	7
Copper	1,190,000	16,100,000	7
Gold	355	2,700	13
Iron ore	1,200,000,000	2,800,000,000	35
Lead	2,200,000	4,500,000	43
Manganese	2,800,000	14,000,000	20
Molybdenum	94,000	250,000	38
Nickel	80,000	1,800,000	4
Tin	110,000	253,000	43
Tungsten	60,000	72,000	83
Zinc	3,900,000	12,400,000	31

Source: US Geological Survey, 2012. (1) Energy Information Administration: short tons, 2012.

In 2008 the CIA calculated that China's economy as measured by GDP was the third largest in the world just behind Japan and equal to 5% of total world GDP – this was still well behind the US whose GDP was 18% of the world figure. Historically in the early 19th century China's GDP was around 30% of the world total, much the same as it was 200 years before. The main reason for this would have been China's very large population, well in excess of Europe's, and in a predominantly rural world, food and basic materials would have constituted the core of any economy then.

Despite China's historic position as one of the most advanced countries of both the ancient world and the Middle Ages, the industrial age was fashioned by the West and led by Great Britain. Perhaps the Industrial Revolution moved too fast for Chinese political structures. It is also the case that China itself came under pressure from the imperial powers in the second half of the 19th century and the early 20th century. As the 20th century progressed China also had to endure Japanese invasions in the 1920s and 1930s and then after the Second World War the communist revolution led by Mao Tse-Tung held China back economically for almost 50 years.

LOCATION OF CHINESE MINING AREAS

Many of China's main mining areas today are to be found over on the east side of the country (as shown in the table) reflecting its huge size and the establishment of the main population centres (and thus industrial enterprises) where there was clear access to the sea.

LOCATION OF CHINA'S MINERALS

Province	Region	Metals
Anhui	East	Coal, copper, iron
Gansu	Central, north	Copper, lead, nickel, zinc
Guangdong	South east	Lead, uranium, zinc
Guizhou	South central	Coal, gold
Hebei	East, north central	Iron
Henan	East, central	Alumina, coal, gold, silver
Hunan	South east, central	Lead, silver, uranium, zinc
Inner Mongolia	North central east	Coal, copper, iron, lead, zinc
Jianxi	South east	Copper, gold, silver, uranium
Jilin	North east	Nickel
Liaoning	North east	Iron
Qinghai	Central west	Gold, lead, nickel, uranium
Shaanxi	Central	Coal, nickel
Shandong	East, central north	Alumina, iron, gold
Shanxi	Central, east north	Alumina, coal, copper, iron
Sichuan	Central south	Iron, lead, zinc
Tibet	South west	Copper
Xinjiang	North west	Coal, copper
Yunnan	South central	Copper, lead, nickel, silver, zinc

THE CHINESE LOOK ABROAD

Among the larger Chinese mining investments recently made abroad are Chinalco's in Rio Tinto, Wuhan Group in Australian and Brazilian iron ore projects, Shenhua Energy in Mongolian and Australian coal, China Investment in Fortescue (iron ore), Minmetals in OZ Minerals (zinc), Shenzhen Zhongjign in Perilya (lead/zinc), Yanzhou Coal in Murchison (coal) and Sinosteel in Midwest (iron ore), the latter five targets being Australian.

It is likely that over time this list will expand but even now the direction of China's interest is clear – it has shifted towards Australia. This is intriguing as China has also being tying up deals and investments in Africa but has found its motives and practices being questioned by sceptical African governments, some of them perhaps mindful of the Chinese mining industry's very poor historic record in the area of mine safety, as we outlined above.

16. THE SOVIET UNION

As one of the two so-called superpowers for a large slice of the 20th century the Soviet Union, consisting of Russia and the now independent republics to its west and south, relied largely on its own raw materials to fuel its economy. The Soviet Union's military power and heavy industry were huge consumers of metals and energy minerals, and its competition with the US in the areas of outer space and advanced weaponry meant that it also required material quantities of specialist metals to be used in high-technology applications.

FOREIGN PARTICIPATION AS THE NEW CENTURY DAWNS

Of course in the early years of the century Russia was still under the rule of the Tsar and the country's economic structure was still largely capitalist with foreign involvement – led as always by the British – a key part of the structure. One of the most successful of the foreign mining entrepreneurs was Leslie John Urquhart, a Scotsman born in Turkey, whose Australian interests we mentioned earlier. His first steps in business were in the importation of chewing tobacco into the UK and the operation of oil wells in the Caspian Sea in Baku where he was bounced out of his interests there by threats of assassination.

In 1906 he turned his attention to Russian mining and acquired a number of prospects in Russia and Siberia, which he financed by offering shares to British investors in London. The companies included Anglo-Siberian, Kyshtym, Tanalyk and Irtysh, covering coal, gold, copper and lead, and represented a massive investment in Russian mining and processing. All remained fine until the Russian Revolution led to expropriation of foreign assets, including Urquart's operations. He placed his Russian assets into Russo-Asiatic Consolidated and fought unsuccessfully for years to have his concessions re-instated. He was also involved in Lena Goldfields – the biggest gold mine in the Soviet Union (USSR). This mine was the subject of a long international court case in 1930 over compensation from the Soviet government following its nationalisation in 1917.

THE RISE OF THE GULAGS

The Soviet years of central control and direction saw a major push to develop the vast country into an economic powerhouse to match the West. These were the Stalin years and the expansion of the mining industry was often achieved by the use of labour transported to the Gulags of the eastern USSR. In these transportations dissident professional and manual workers alike were settled in camps, often for decades, until the death of Stalin in 1953 led to most of them being closed by 1960.

The Gulags had a number of key political functions, but economically they played an important role in the establishment of heavy industrial complexes for steel, manufacturing and mining, including mining of coal, iron ore and base metals. Gold production was also an important activity given that the rouble was unconvertible and the USSR was not a major manufacturing exporter like Germany or the UK, but was from time to time a heavy importer of food stuffs and advanced machinery, and therefore in need of convertible assets.

One of the largest gold areas discovered then was in the Kolyma region in the far northeast of the USSR. The initial finds were in the early years of the 20th century and sparked a mini rush to the region by prospectors. It was not until 1928 that mining in the area began in earnest, with geologists and miners provided by the start of Stalin's mass transportations. By 1937 official figures for output showed annual gold production from the Kolyma region of around 1.6 million ozs and during the Second World War production rose further to 2.9 million ozs or about half of the output of the USSR.

Living conditions for all Gulag workers were very tough and few survived the camps. The climate of the region also made mining a very difficult activity as the winters were long and extremely cold. The mines were primarily surface placer operations in nature and were extensive and many were high grade. The gold-bearing earth had to be dug up and then melted before the process of washing, panning and separation could begin. The introduction of expensive mechanical technology was often trumped by the fact of an almost limitless supply of slave labour. The mines, whilst many in number, were often very small, although high grade, and did not lend themselves to mechanisation in any case. The gold was transported through the port of Magadan, which was used to bring supplies in and transport minerals out, and had been established as the administrative hub for the region. It is estimated that as much as 1,000 tonnes of gold (32 million ozs) was produced in the whole region in the peak years of the Gulag, 1930-1953, as well as substantial quantities of tin and the key strategic metals, tungsten and cobalt.

In the years between the two World Wars the push to build up the USSR's mineral production led to further major discoveries and developments and new Gulags were established across the Soviet Union's northern regions to service these discoveries. Additional copper discoveries were also made in the Urals and in Siberia. Bauxite was discovered in the 1930s, in the Urals as well, and aluminium plants were built in Severouralsk – a town with a mining tradition derived from the establishment of iron ore and copper smelters there in the mid-18th century.

In 1934 tin was discovered in the far east of the USSR in Primorye, an area of the country where coal, lead/ zinc, tungsten and gold mines were also developed. This geological push in the 1930s led to the opening up of the huge Upper Kama potash and potash magnesium salt deposits to the west of the Urals, leading to the establishment of processing plants in the city of Solikamsk which had a centuries-old tradition of salt mining.

NORILSK

One of the largest mining complexes discovered and developed during Stalin's era was the Norilsk nickel, copper, platinum and palladium deposit in northern Russia on the Taimyr Peninsular above the Arctic Circle. Like many regions in Tsarist Russia the Peninsular had been known as prospective for minerals for at least three centuries so this provided the USSR with a target list of areas suitable for the establishment of Gulags. Exploration and feasibility planning started in the mid-1920s and like most Soviet mining and metal projects of the time its labour came from the transportation of dissidents and others to the camps of the Gulag.

Rather sinisterly, control of the Norilsk project was passed to the NKVD, the USSR's secret police, in 1935, when the NKVD assumed responsibility for the Gulags. This also gave the secret police primary influence over Norilsk's production targets, important due to the strategic sensitivity of the mines with their output of high-tech metals. Production of nickel and copper began in early 1939. During the Second World War nickel production rose from 4,000 tonnes to 10,000 tonnes in 1945, largely reflecting military weapons demand. At the end of the Stalin era in 1953 Norilsk was producing 90% of the USSR's PGMs, 35% of its nickel, 30% of its cobalt and 12% of its copper. Today annual output from Taimyr includes 120,000 tonnes of nickel and 340,000 tonnes of copper.

Norilsk was developed as an underground mine, sensible in such a cold climate, and after 60 years of operation still contains as much as 2 billion tonnes of ore to be mined with present treatment grades of 0.8% nickel and 2.2% copper. Access by decline has been added to increase operating efficiency. The mine, unfortunately, is a dirty operation overall with environmental damage high, something that has been with the mine since inception. Much of the damage comes from the onsite smelter; the surrounding area is significantly polluted with heavy metals as a result of the inefficient smelting process at the site. Around the same time nickel and copper were also found near Murmansk, on the Russian side of the border with Finland in the northwest, and in due course after the war the mines developed became part of the Norilsk group.

In more recent years Norilsk has experienced considerable change, particularly in the corporate area and also in its strategy of acquiring complementary mining operations both in other parts of Russia and overseas. At the start of the 1990s with the fall of the Soviet Union there was a major restructuring of Norilsk Nickel and in due course in 2003 private capital was introduced with the controversial Oleg Deripaska of RUSAL and Vladimir Potanin of Interros International each acquiring a 25% stake. The relationship between the two oligarchs has been a difficult one and legal disputes over control have rumbled on for years. One of Norilsk's most high profile diversifications was the acquisition of a 55% control stake in the only significant PGMs producer in the US, Stillwater Mining, in 2003.

THE SEARCH FOR INDIGENOUS DIAMONDS

Mineral exploration programmes in the Soviet Union were very much driven by the central requirement for industrial raw materials but this did not mean that geologists could be stopped from academic speculation. So it was that in the 1930s Vladimir Sobolev, an academic, observed that Siberian geology had a number of similarities with the geology of Southern Africa, a major source of diamonds. Interestingly, as a communist society the Soviet Union did not regard luxuries such as diamond jewellery as important, although no doubt out of the view of the toiling masses the Party apparatchiks and their families, like the Tsarist aristocracy before, enjoyed owning them.

In addition, as the Industrial Revolution began to give way to more technologically-based industrial development, particularly in the area of precision engineering, the merits of using industrial diamonds for accurate cutting of materials, drilling and polishing became increasingly appreciated. At that time major economies such as the US, the UK, Germany and the Soviet Union had to import all their diamonds from Africa, particularly British-influenced countries like South Africa. We addressed this issue and its political ramifications during the war earlier, but suffice it to say just as the US was anxious to secure reliable supplies of industrial diamonds so was the Soviet Union.

In the years after the end of the Second World War Russian geologists began the search for diamond indicator minerals in the frozen wastes of Siberia. Their search, which began in 1947, was successful, and follow-up work to uncover diamondiferous kimberlite pipes was eventually even more successful. In 1954 in Yakutia (now called Sakha) the Zarnista pipe was discovered and this was followed the next year by the discovery of the Mir and Udachnaya. The Mir mine was in production just two years after its discovery and over the years another 500 plus kimberlites were discovered in the Yakutia craton in eastern Siberia.

Of course few of these pipes were diamondiferous but additional mines were established in the region over the years, including the Interationalnaya. Other diamond pipes in Siberia include Anabar, which is really a surface alluvial operation, Aikhal and Jubilee. In due course exploration pushed west into Arkhangel'sk where further diamondiferous pipes were found at Lomonosov and Verkhotina, the latter being a project under the control of Canadian mining junior Archangel Diamond. Since the liberalisation of the Russian economy following the break up of the Soviet Union, most of Russia's diamond mining industry has come under the control of Alrosa. As the marketing agreement between Russia and De Beers was abandoned some years ago this allowed the Russians to build their own marketing effort.

THE GROWING GOLD INDUSTRY

In the decades before the fall of the Soviet Union gold mine output was a state secret and outside estimates proved to be unreliable. Over the last twenty years more reliable historic figures have been released and it is believed that gold production in 1970, on the eve of the great 1970s gold bull market, ran at around 218 tonnes per year. Today Russia on its own produces almost 200 tonnes of gold annually from mines largely located in the eastern part of this vast country.

Historically Russian gold production has been dominated by alluvial mining and it remains a major alluvial producer into the 21st century. However, since the break up of the USSR, hardrock mining of gold lodes/seams has expanded considerably going from around 20% of output in 1997 to nearer 75% in 2010. The hardrock operations are largely big open pit operations although there is some underground mining. The problem with alluvial mining in the remote regions of eastern Russia is the very low winter temperatures which can lead to operational shutdowns; this is less of a problem with the big open pit mines in the region where ore is usually dynamited.

Independent Uzbekistan is the largest producer of the former USSR republics (excluding Russia) and output currently is almost half that of Russia. The biggest operation is the Muruntau mine which was discovered in 1958 and began production in 1967. The mine produces around 60 tonnes of gold per year and also has one of the largest gold resources in the world, with over 5,000 tonnes in situ. In addition other republics including Kazakhstan, Kyrgyzstan, and Tajikistan have been substantial gold producers in the past and in recent years have been raising their gold production partly, as in Russia's case, as a result of foreign participation. Amongst the foreign gold groups operating in Russia are Kinross Gold, Highland Gold, High River Gold and the UK-incorporated Petropavlovsk (formerly Peter Hambro Mining). The largest Russian producer is Polyus Gold, incorporated in Russia but with ambitions to move its registration to London.

Russia's two biggest gold deposits are the state-owned Sukhoi Log, once owned by Australian group Star Mining in the 1990s, and Natalka being developed by Polyus Gold. The Natalka mine started producing gold in 1945 and continued to do so until 2004 when an exploration programme was carried out to test whether a much larger operation was feasible. This was positive and first production from the new expanded pit is expected in 2014 when output should be around 500,000 ozs, rising to 1.5 million ozs in 2020.

The Sukhoi Log deposit could, when developed, be potentially a bigger gold producer that Natalka, with annual output of around 1.75 million ozs. The problem is that the project has a very murky past going back to the days when Star Mining tried to develop a mine with Russian partner Lenzoloto (now controlled by Polyus Gold) and with South African group JCI. The Russians moved the goalposts on several occasions and any legal success that came Star's way as it battled to develop Sukhoi Log was overturned. In the end Star, after sinking millions in the project, had to admit defeat and withdraw. The deposit was then acquired by the Russian government who then began the search for a new group to develop the mine, a process which continues in 2012.

Amongst the major mines in Russia are Polyus Gold's Olympiada mine, which produces around 600,000 ozs annually, and its new Blagodatnoye mine, which produces about half that figure; both mines are located in the Krasnoyarsk region. It is also a substantial alluvial miner with a number of operations in the Irkutsk region. The UK miner Petropavlovsk operates the Pioneer mine in the historic gold region of Amur in the south east of the country not far from the border with China. Annual production at the mine is 360,000 ozs and the British-owned miner has other projects in the Amur region including the Pokrovsky mine with annual output of 145,000 ozs.

It is expected that Russian gold production will continue to increase as projects such as Sukhoi Log and Natalka are finally developed and exploration successes are turned into mines. Part of this expansion is politically driven by the interest of the Russian government in increasing the percentage of gold it holds in its foreign reserves, still heavily weighted towards the US dollar as a result of soaring receipts from the energy price boom since the early 2000s. In 2010, for instance, it added 140 tonnes of gold to its reserves, a 17% increase. Whilst Russian gold production is still well short of Chinese output the gap may well close over the 2010s as the financial importance to Russia of the yellow metal increases and new production comes on stream.

17. THE MIGHTY USA

In both economic and political terms the 20th century clearly belonged to the USA. It entered the century as the largest producer of coal and steel in the world, underlining its position as the world's leading economy. Its political primacy was less established and it took two World Wars and a shrewdly timed entry into the conflicts on its part to see off the old global political powers, particularly the UK as it steadily abandoned its Empire to build up a welfare state and improve relations with Europe. After the UK the only rival to US hegemony around the world was the Soviet Union, whose rise and fall lasted but 45 years from 1945 to 1990.

During this century of rapid US economic expansion the mining industry, in terms of growth and development, did not in general terms match that of economically less powerful countries like neighbour Canada, Australia or South Africa. Many of the US's productive mining areas were already well established by 1900 and the direction of the economy was also firmly towards technological innovation and value added. Mining and raw materials, including energy, remained important, but growth in the modern world was increasingly a function of brainpower rather than muscle power.

ALUMINIUM

One of the new signs of industrial power in a high-tech age was a growing aluminium industry. Aluminium in the form of unrefined silicates may have been used in the making of pots and dyes millennia ago, and certainly had been the subject of much research during the 19th century in both Europe and the US. As a result of this, in 1886 American Charles Hall patented the process for a new, much cheaper method of producing aluminium metal, and thereby laid the base for the Aluminum Company of America (Alcoa). Following this development the

US began to develop aluminium capacity in the early 20th century and by the start of the First World War in 1914 it produced over 60% of the world's output.

However the raw material for the manufacture of alumina, bauxite, was unfortunately not found in abundance in the US. For some years bauxite was mined by Alcoa and American Cyanamid in Georgia, Alabama and Arkansas, but in due course these rather small deposits were inadequate for Alcoa's needs and the company began to look overseas for its bauxite supplies. Today Jamaica and Guinea are the two major suppliers of bauxite to US smelters, and Australia and Suriname the two major suppliers of alumina.

Before the development of aluminium, bauxite had been used for a number of other purposes, particularly in building, where bauxite blocks were manufactured in the 19th century. The use of bauxite waste, red mud, to make building bricks and blocks continues today, and is much encouraged in poorer producing countries like Jamaica where the benefits of environmental clean up go hand in hand with the economies of using local raw materials. Eventually aluminium, as a result of its great strength and lightness of weight, made possible the manufacture of large aircraft where the high strength-to-weight ratio of aluminium when alloyed to magnesium was critical. In the early days of powered flight planes were made out of wood and canvas, but the use of planes during the First World War underlined the need to move on to using a more robust construction material like aluminium sheet.

The role of aluminium in the advance of technology in the 20th century has not been confined to just aerospace. The properties of the metal make it ideal for a range of new products and technical advances. It is almost as strong as steel yet is just one-third as dense; it is resistant to corrosion and has useful flexibility under pressure; it has excellent thermal and electrical conductivity; it is non-toxic and extremely ductile and malleable; and it is easy to increase the metal's positive attributes by alloying.

The abundance of bauxite allied with its many attributes made aluminium a key metal in the growth of mass household consumer goods, construction and transportation, and some would argue that our whole modern way of life is built on the metal. Certainly in combination with energy minerals aluminium has wrought huge changes to the modern economy and has played no small part in the rise of the US to economic superpower status in the 20th century.

According to the US Geological Survey primary US aluminium production rose from 2,296 tonnes in 1900 to 1,190,000 tonnes in 2011 (the peak year being 1980 with 4,654,000 tonnes). The unadjusted price of metal rose over the period from $721 to $2,570 per tonne, a massive fall in real terms of over 80%, implying huge gains in productivity and unit cost economies.

GOLD

US gold production rose far less spectacularly than that of aluminium over the period between 1900 and 2011 – from 122 tonnes to 237 tonnes – but these figures do not give a true reflection of the path of output over the 100 or so years because output was spectacularly volatile. Thus a closer inspection of the figures is warranted.

We have earlier looked at the great California gold rush of 1849 and the huge surge in US production brought about by that event. At the peak of the California rush the US produced around 150 tonnes of gold, a large percentage coming from the west. Output stayed above 100 tonnes until the end of the First World War but then slipped back to 64 tonnes in the year of the Great Crash, 1929. The increase in the official price, engineered by President Roosevelt as part of a strategy to kick-start the supine US economy, saw production rise back to 150 tonnes by 1945. When the US entered the war in 1942 gold mining was restricted and production slumped to 30 tonnes by 1945.

Over the next 20 years US production yawed between 30 and 60 tonnes. Despite the rise in the price of gold from $35/oz in 1971 to $850 in 1980, production came in at an unimpressive 30 tonnes, so the huge price rise did not immediately stimulate output. However, over the next two decades US gold production revived, spurred finally by the 1970s price increase. Output peaked at 366 tonnes in 1998 before succumbing to a combination of sharply rising development and mining costs. There was also a loss of confidence in the yellow metal as it plunged down through $300 as the new millennium beckoned, before beginning a steady recovery which took it eventually to over $1,900 in 2011.

US GOLD MINING IN THE 20TH CENTURY

The story of gold mining in the US in the 20th century is an interesting one, for its revival in the latter part of the century was, unusually, down to the efforts of the US's largely ignored neighbour to its north. But before we examine the role of Canadians in the revival of US gold mining, we should first go back and look at the state of the industry as the century dawned. At that time most of the country's gold was mined in California and South Dakota, although there were promising discoveries in Nevada and Utah that would be material later.

NEVADA AND UTAH

Although there was no way of knowing it then, one of the most significant gold mining areas for the long term was the Carlin area of Nevada which in the early

1900s was host to a number of very small placer gold deposits. For many years the largest source of gold in the state was the Comstock Lode, which as we have seen before was primarily a silver mine. For most of the period up until the 1960s US gold production also came from old mines such as the Homestake discovered in 1876, and as a by-product from copper mining in Utah and Arizona in particular.

For 50 years no significant new gold mines were unearthed in the US and then in 1961 Newmont Mining discovered the Carlin in central Nevada. Although the grade was low for the time at less than 10gms/tonne, the deposit was close to the surface and a large open pit was developed. The gold was also extremely disseminated throughout the ore and was not visible, which led to the development of heap leaching techniques using acid to separate out the gold during the treatment process. Over the years other deposits were found and additional pits were developed including Gold Quarry, Post and Genesis. In the 1990s some underground workings were developed at Carlin, Post and Genesis. Following the Carlin discovery a number of gold companies descended on the state and began to explore in the region around Carlin. It is at this stage that the Canadians enter the picture.

The main company in question was Barrick Gold, which had been formed in Toronto in 1983. The following year it bought Camflo Mines which had a small operation in Nevada, thus bringing Barrick south of the border. In the next couple of years it expanded its interests into Utah as well, but the company-making acquisition came in 1987 when Barrick bought the then modest Goldstrike property on the Carlin trend, discovered in 1982, from two small companies who had debt problems, Pancana and Western States. Fortuitously one of Barrick's directors was also a director of Pancana.

Goldstrike now consists of the Betze-Post open pit operation and the underground mines of Meikle and Rodeo. Betze was discovered in 1987, cheek by jowl with Newmont's Post deposit, just after Barrick acquired the property and started its own drilling programme at Goldstrike; two years later the deeper level Meikle and Rodeo were discovered. Overall annual production from the Goldstrike complex is currently around 1.4 million ozs which compares with the 1987 operation it bought where output was a tiny 40,000 ozs per year. Barrick's gamble on Goldstrike had paid off handsomely and its Nevada operations laid the base for the building of a worldwide Canadian gold giant.

Other companies involved in Nevada included Battle Mountain, which was eventually acquired by Newmont in 2000, and Placer Dome, which was acquired by Barrick in 2006. Battle Mountain's Nevada mine was the Fortitude developed in 1967 by Duval, the mining division of Pennzoil which was re-named Battle Mountain Gold and floated off in 1985 as an independent company. A number of other companies were also active in the Battle Mountain region, including Placer Dome, which opened the Pipeline mine on the Cortez Hills/Battle Mountain trend in 1969 in the central part of the state. With Barrick now in control of Placer Dome, the Cortez Hill and South Pipeline projects have also been developed and doubled the company's Battle Mountain output to over 1 million ozs, with full production reached in 2011. Barrick also has a number of other Nevada mines including Bald Mountain, Turquoise Ridge and Round Mountain.

Nevada has always been a relatively easy state to operate in as it has had a pro-mining state government since the gold industry began to revive in the 1960s. Whilst this has created a lot of environmental complaints with regard to pollution and the rights of indigenous Indian tribes it has made the state the largest gold producer in the US with around 80% of total US gold output.

IRON ORE

We alluded earlier to the way that the US's assumption of the position of number one economic power in the world was in some important part due to its self sufficiency in key minerals like coal, oil and many industrial metals. As the years have gone by this self-sufficiency has been lost in certain areas, oil in particular. Output of some metals and minerals has also fallen as the shape of the US economy has changed and new high tech industries that are low on raw material consumption have grown up.

Above, the section on iron ore concentrated on the huge exporters and consumers – Australia, Brazil and China amongst others – and largely ignored the US. In one sense this is unfair on the US as for the first half of the 20th century it was responsible for over 50% of the world's iron ore output, largely from the iron mines of Michigan and Minnesota. Indeed after the end of the Second World War output soared to over 150 million tonnes per year as the economy boomed and the Korean War loomed. Then heavy industry including car manufacturing was booming but in recent years these industries have either disappeared or been severely downsized as the US economy has moved on.

Despite this the US remains largely self sufficient in iron ore with any import requirements largely met by Canada. In recent decades the grade of US iron ore has fallen and this has meant that instead of being able to provide direct shipping ore to US steel mills the mines have had to install treatment plants to upgrade ore by producing pellets from the low-grade ore. In terms of output, present day US iron ore production – at little more than 50 million tonnes – is little different from South Africa's and a third of levels achieved in the 1940s and 1950s. Needless to say it trails a long way behind China, Australia, Brazil and others.

In many respects its decline as iron ore miner and steel maker matches that of the country it supplanted as the world's number one economy a century ago, Great Britain. The pattern of ownership of the iron ore mines has changed over the last century with individual mines being acquired by steel companies as the industry grew in the first half of the century. In the 1980s the trend reversed itself with Cleveland Cliffs and other miners expanding their iron ore mining divisions, but today US iron ore mines are again often owned by steel makers such as US Steel and Mittal, who are anxious for security of raw material supply.

COPPER

For many decades in the 20th century the US was the largest copper producer in the world based on mines located in areas first developed in the previous century, which we have looked at earlier. At the end of the late 1920s economic boom in 1929 the US produced 1.2 million tonnes of copper – very much in line with current output levels – and this level was also reached during the Second World War. By the end of the 1960s economic boom, and part stimulated by Vietnam War spending, US copper production reached 1.5 million tonnes making the US by far the largest copper producer in the world, 50% greater than the next largest producer the USSR. In the current age the US remains the second largest producer at around a stagnant 1.2 million tonnes, although very much on a par with Peru; both trail well behind Chile where output is now over 5 million tonnes and still rising.

18. EUROPEAN MINING

The modern age for Europe has been a time of war, political crisis and the attempted rise of the European superstate but it has also been a time of economic growth and, despite everything, a time of rising prosperity for its often beleaguered citizens. As the 20th century wore on the industrial bias of the continent began to change as technological advances encouraged the emergence of new industries.

The continent's mining industry could not but be affected by these changes, and the decline of the coal mining industry was one manifestation of this. Indeed, like the US, European production of metals and minerals as a percentage of world output has been sliding since peaking in 1860 at over 60%. During the 20th century the slide continued down through 10% in the 1970s to reach 5% in the new century.

Today continental Europe produces around 5% of the world's copper (850,000 tonnes), with Poland being responsible for more than half that figure. In terms of consumption Europe uses around 16% of copper produced from primary and secondary (scrap) sources. The differential in other metals such as iron ore, nickel and zinc is similar and this has been a trend for a material part of the 20th century; for instance in the late 1960s Europe produced just 6% of the world's copper with Poland being a negligible producer then.

So what happened in Poland to make it the only European copper producer where production grew strongly in the second half of the century?

POLAND

In the early 1950s Poland, then signed up to the Warsaw Pact, at the bequest of Soviet military and strategic needs began exploration in the southwest of the country near Legnica (this area had been part of Germany until the dispensation after the Second World War incorporated it in the newly drawn borders of post-war Poland).

This work started in 1952 but it was not until 1957 that the geologists came upon huge potential deposits of copper and silver in what is known as the Fore-Sudetic Monocline. Poland had been a modest copper producer before this but the new discoveries made the country a major prospective producer. Two underground mines were initially developed, the Lubin and Polkowice mines, which came into production in the late 1960s, and following that a smelter and refinery were built. The Polish government established KGHM to run the mining and treatment operation, by far the biggest in continental Europe. Following the fall of the Soviet Union KGHM was successfully privatised in 1977 and became a fully listed public company.

The mines have been in production for 40 years already and working depths are now down to almost 4,000 feet, but the mines still contain over 1 billion tonnes of ore grading 2% copper and 58gms/tonne of silver. KGHM is the eighth largest copper producer in the world and the second largest silver producer. With substantial production of coal, sulphur, lead and zinc, and industrial minerals such as gypsum Poland is arguably the only major European minerals producer outside Scandinavia, although Spain would probably protest at this assertion.

SPAIN

As with the US many of Spain's operating mines in the first half of the 20th century were essentially mines developed in the previous century as the Industrial Revolution gathered pace. Although Spain supplied metals and minerals to the fast-growing industrial economies of northern Europe it remained, itself, a primarily agrarian economy until well into the 20th century.

Today it still mines base metals and gold, but in no metals category is its output above 1% of world output; it does better in industrial minerals where it produces, for instance, almost 11% of the world's gypsum. This, however, represents a material fall from grace in comparison with its position earlier in the century when Spain was an important producer of copper, lead and particularly zinc.

As its own industrialisation gathered pace in the 20th century its coal mining industry also began to expand. However in the last quarter of the century employment in Spain's coal mines collapsed from over 50,000 to less than 10,000, and as the new century dawned the government, which heavily subsidised the mines, was trying to resist pressure from the European Commission to bring the subsidies to an end – this would seal the fate of the industry at a time of general economic crisis for the country.

IRELAND

Ireland has a long history of mining going back many centuries. One world-class mining project of the modern era that made it to production was the Tara zinc and lead mine near Navan in the east of the Irish Republic. Amongst the many mines, most of them small, that operated for much of the 19th and 20th centuries were Silvermines and Avoca, the former producing lead, zinc and silver from its mines in Tipperary in the south of the Republic until closure of the last mine in the area in 1993. Avoca's mines in County Wicklow in the east closed in 1982 after centuries of mining first iron ore, and then copper and lead. There were many other small base metal and coalmines in Ireland, many of which survived well into the 20th century – Tara was the biggest of them all, the largest zinc mine in Europe and the largest mine ever developed in Ireland.

Mineralisation had been known about in the Navan area for many years and Rio Tinto Zinc at one stage had worked unsuccessfully on what was to become known as the Tara lease. It was in 1969 that Tara Exploration, a subsidiary of Northgate Exploration, a Canadian company formed in 1953 by four Irish Canadians led by Pat Hughes, began to explore the Navan area. Hughes had already had some success with zinc and lead projects in Ireland having developed the Gortdrum mine in County Tipperary and the Tynagh mine in County Galway (both mines have now been closed for many years and the latter is acknowledged as one of the most polluted mine sites in Europe). In 1970 the Navan exploration bore fruit with the discovery of high-grade zinc and associated lead; an extensive drilling programme outlined 70 million tonnes of ore with a grade of 10.1% zinc and 2.6% lead. A relatively shallow underground mine was developed and as part of the financing of the mine Tara Exploration was floated on the Toronto Stock Exchange. Mine development began in 1973 and the mine opened in 1977.

During this period a corporate struggle ensued between Canadian giant Cominco and Northgate over control of Tara Exploration, where the Anglo American/Charter Consolidated group also had a stake. In 1975 Northgate enlisted the help of Noranda, another Canadian base metal group, and together they saw off Cominco's challenge. Despite the high grade of the deposit and extensive investment in mechanisation and advanced mining technology, low zinc prices in the 1980s and ensuing losses saw Noranda selling out to Finland's Outokumpu in 1986; three years later the Irish government, which owned a residual 25% stake in the mine, sold its interest to the Finns.

These problems were not the only ones faced by Tara and its owners for throughout this period, and for many years after, there was a dispute over additional ore on the lease, this ore being owned by Bula Mines, which Bula wanted to develop as a separate mine. Permission was never granted for this even though the Irish government had a stake in Bula, and a long and expensive legal action ensued which bankrupted Bula.

In 2004 the Tara mine was sold to the Scandinavian mining house, Boliden, as the long period of low metal prices, which had dogged Tara almost from the beginning, came to an end. Today Tara remains a major world producer of zinc with an annual output of 200,000 tonnes of metal; despite having mined around 75 million tonnes of ore since 1977, its current reserves are a healthy 35 million tonnes and exploration in and around the mine continues.

OTHER PARTS OF EUROPE

Although the emphasis of global mineral exploration has in recent decades tended to focus on almost anywhere but Europe, a number of significant programmes have been instituted, particularly in Scandinavia, which has a long tradition of base metal and iron ore mining. Indeed the iron ore operations at Kiruna in northern Sweden which were taken over by LKAB in the late 19th century, and which had been going since the early 1700s, today – as the world's largest underground iron ore mine – continue to provide LKAB with ore for both local and foreign steel producers.

In eastern Europe outside Poland there has also been interest in historic mineral deposits. A number that have been looked at over the last five years, and in some cases much longer, include European Goldfields's Certej gold project in Romania and its Skouries and Olympias gold projects in Greece; European Nickel's Caldag nickel project in Turkey; and Cape Lambert's Sappes gold prospect in Greece, which has proved a dead weight to previous owners such as Greenwich Resources and Mineral Securities. None of these projects are easy and they have come up against local opposition and government foot dragging, and some of them, Sappes for instance, have been the subject of feasibility work for over 20 years.

19. LONDON RISES FROM THE ASHES

By the end of the 1980s London, once the giant of the mining world in terms of control and finance, hosted just one major world class mining company, RTZ, which was a founder constituent of the FTSE 100 when it was formed in 1984. Twenty years later London was awash with large mining groups seeking money and primary listings in an atmosphere reminiscent of the City's heyday as the centre of world mining finance at the height of the British Empire's power in the Victorian era.

The reasons for London's fall from grace as a mining centre, before this astonishing revival, were multiple. Perhaps the key reasons, related to Britain's much reduced political and economic position in the world following the Second World War, were the imposition of foreign currency controls and the abandonment of the Sterling area. It therefore followed that mining companies had to look outside the UK for finance in the 1960s and increasingly in the 1970s when metal prices, particularly gold, were strong. London, of course, retained management of a lot of foreign money and therefore continued to have knowledgeable corporate advisers and mining industry analysts which kept the door open for fundraising, but these were nationalistic days and mining companies found it politically prudent to incorporate where their primary assets were located.

THE OLD 'COLONIAL' COMPANIES LEAVE

So it was that South African and Australian miners with London incorporation like Johannesburg Consolidated, Union Corporation and New Broken Hill upped sticks in the 1960s and 1970s and went home. The advent of exchange controls and the ending of the Sterling area, as well as the institutionalisation of equity investment, meant that private British investors, who for instance played a major part historically in financing the South African gold mining industry, no longer had the resources to fund new mining investment.

London's impact on other mining countries was also severely undermined, with its role in South America reduced to very modest levels, and independence for countries in Asia and Africa, like Malaysia and Zambia, leading to both re-incorporation and nationalisation. In the 1980s takeovers and mergers in the UK saw the disappearance of Selection Trust (to BP), Consolidated Gold Fields (to Hanson), CAST (to Selection Trust), Tanks (to Soc Gen Belgique) and the restructuring of Anglo American's international mining arm, Charter Consolidated, as a UK-orientated industrial group.

THE END OF EXCHANGE CONTROLS

The ending of UK exchange controls with the advent of the Conservative Thatcher government in 1979, which ultimately was so critical to the rise again of London as a mining centre, did not immediately arrest the decline of London's mining influence, and many of the corporate actions listed above happened in the 1980s. Perhaps this was due to the fact that long-term trends often have long tails, perhaps because mining was seen as a sunset industry and attention was increasingly drawn to the technology boom of the 1990s.

Whatever the reason the 1990s opened with London's mining hierarchy consisting of RTZ and a handful of minor players including Antofagasta, Anglesey Mining, Greenwich Resources and Cluff Resources. It was a sobering fall from grace and for a number of years Canadian companies took up the running, scouring the world for money and projects, becoming dominant in Latin America and increasingly important in the Far East and Africa. The parochial Australians also broke out of their shell, helped by the abandonment of their exchange controls and the increasingly fractious problem of native land rights in Australia.

London's revival as the global centre of mining finance was spread over a number of years later in the 1990s and into the new millennium, and was due to a number of influences. Important to the trend was the pool of corporate and broking expertise which lived on in the City despite the reduced image of the mining sector. Julian Baring, who we meet below, was a particularly distinguished example of this expertise. When foreign banks and brokers began to acquire City firms in the wake of Big Bang they inherited considerable experience in the mining sector.

Of material help in creating a platform for the listing of junior mining companies mostly involved in exploration was the establishment of the Alternative Investment Market (AIM) in London in 1995. This light touch, relatively cheap market attached to the London Stock Exchange enabled smaller mining companies, particularly with interests in Africa, to incorporate in the UK, raise money and then obtain a quotation for their shares in London on AIM. In due course a number of mining companies in both Australia and Canada obtained a secondary listing on AIM, often combined with a London fund raising. In the new millennium interest in the sector increased as metal prices, driven by exploding demand from China, started to rise sharply.

JULIAN BARING (1936-2000)

Although Julian Baring, part of the famous Barings banking family, was not a mining man in the sense of being a geologist or an engineer, he arguably had more influence on the development of the mining industry, particularly in South Africa, than most of the industry's practitioners in the second half of the 20th century.

There are many facets to the mining industry. For years financing was in the hands of colourful entrepreneurs who were usually rich men in their own right. They and their friends – often London merchant banks and private investors – provided money for both exploration and development. However as industrial demand for metals exploded in the 20th century a more structured approach to finance became necessary. Increasingly detailed analysis of projects was needed and it was into this burgeoning new world that Julian Baring, hotfoot from the public relations department of Anglo American in Rhodesia, arrived in 1967 when he joined London stockbrokers, James Capel. He persuaded the firm to let him set up a specialist mining equities team, probably the first integrated equities team seen in London.

Over the next two decades he established James Capel's mining department as the leading mining team in the City. A comprehensive computer valuation program for gold shares, GoldVaal, was developed which trumped anything that even the gold mining industry had. Whilst the team was very active in the Australian nickel boom, both in terms of fund raising and share trading, its sector dominance really came to the fore in the long gold boom which stretched for almost ten years from 1972. Baring spearheaded other innovative initiatives over the period. These included the re-purchase for the Zambian government in the stock market of some of the nationalised copper industry's compensation bonds then selling at a huge discount to redemption value. Subsequently as metal prices soared in the 1970s the Zambians borrowed to buy back the bonds at par direct from the bondholders at great long-term cost to the country.

As the 1980s developed and the long gold bull market came to an end Baring decided that it was time to step down from broking equities and he set up a specialist mining share unit trust within James Capel. This did very well and in 1991 he left James Capel with the trust to join Mercury Asset Management (MAM). At MAM Baring floated a new trust in 1993, Mercury World Mining, which raised over £500 million making it the largest investment trust fund-raising achieved in London at that time. Also at MAM Baring launched his campaign to reform the clique ridden South African mining industry. He was one of the key backers of the groundbreaking restructuring of Rand Mines and slowly but surely his arguments against the *heads we win, tails you lose* system of SA mining house fees, where in bad times the mining houses still collected huge management fees from their gold mines even if shareholder dividends collapsed, forced the South Africans to change their operating structures.

When that battle had been won Baring had another issue up his sleeve – the growing tendency in the 1990s for gold mines to hedge. As a gold fund manager he berated the industry for its presumption in trying to hedge the gold price risk that he, as an investor, wished to shoulder. When the Central Banks in 1999 decided to curtail their market sales of gold, the gold price soared but a number of gold miners almost went bust due to the complexity and toxicity of their hedges, fully justifying Baring's view that hedging was seldom in the interests of shareholders. Today there is little hedging done any more in the gold mining industry and many gold giants like Barrick and AngloGold have bought back their hedges.

In 1998 Baring left MAM to concentrate on his chateau in Provence in France where he had established a vineyard, and also fulfilled a lifetime passion by opening a pottery in the local village. He sadly died of a heart attack at the age of 64 in 2000.

THE GIANTS GATHER AND HEAD FOR THE FOOTSIE

At the same time – the late 1990s – things were also stirring at the top end of the sector. In 1997 RTZ and its listed Australian subsidiary CRA reorganised themselves into a product-based related group divided into a UK and an Australian registered company with identical global mining interests, and the name of the group was changed to Rio Tinto. In 2001 BHP and Billiton followed the same route, merging and then dividing into a UK and an Australian company with identical mining interests. A few years before, Anglo American, anxious to break out from its South African base to build up its global mining interests, and stymied by SA exchange controls and the high cost of capital, applied to the SA government to relocate to the UK and permission was finally granted in 1999.

Interestingly gold miner Randgold Resources had tried to go this route through parent Randgold and Exploration (originally the venerable Rand Mines) around the same time. Permission was refused by the SA authorities so the group floated Randgold Resources independently in London in 1997; ten years later on the back of a rising tide of gold discoveries and mine developments in West Africa, particularly Mali, it made it to the Footsie.

The next mining company to enter the top 100 UK shares was Xstrata, a South African-orientated coal and chrome miner floated out of the private Swiss metals group Glencore. It had global ambitions for its mining arm and chose London as the most likely location and market to help fulfil this strategy. Soon after Antofagasta, which despite its perennial UK listing was a small Chilean railway company, saw its stock rise as its plan to emphasise its growing copper mining side began to pay off with rising copper prices and rising group copper output. *Fags* was followed by a raft of companies from the old Soviet Union and Asia including Kazakhmys, Vedanta and Lonmin, the latter the mining arm of the old Lonrho that had bounced in and out of the Footsie in the late 1980s.

In the late 2000s another group of companies from a range of different countries seeking new investor capital appeared in London and often went straight into the Index. Some – like Ferrexpo – did not survive the tanking of the resources sector as the credit crunch of 2008 bit deep. However, Randgold, with its growing suite of West African gold mines, profited as battered investors sought protection in gold and its market capitalisation exploded to Footsie size.

As the gold price continued to rise throughout 2009 and 2010 the Canadian giant Barrick decided to float off its east African assets in a company, African Barrick Gold, which was big enough for an early Footsie entrance and an enhanced valuation, in a market more comfortable with Africa than was Toronto.

The table on the next page gives a record of UK incorporated mining companies that have been listed on the FTSE 100.

FTSE 100 UK INCORPORATED MINERS

Company	Year in	Year out	Main operations
RTZ/Rio Tinto	1984		Global
Consolidated Gold Fields	1984	1989	Global
Lonrho	1984	1985	Africa
	1986	1987	
	1988	1992	
Billiton	1996	2001	South Africa
Anglo American	1999		South Africa/Global
BHP Billiton	2001		Global
Xstrata	2002		Global
Antofagasta	2004		Chile
Kazakhmys	2005		Kazakhstan
Vedanta	2006		India/Zambia
Lonmin	2006	2008	South Africa
	2009	2011	
ENRC	2008		Kazakhstan
Ferrexpo	2008	2008	Ukraine
Fresnillo	2008	2008	Mexico
	2009		
Randgold Resources	2008		West Africa
African Barrick	2010	2011	East Africa
Glencore	2011		Global
Polymetal	2011		Russia

One of the interesting things about this rush to London was the fact that the first part of it, before and just after the millennium, was in many respects a welcoming home of old Commonwealth friends, from Africa and Australia in particular. Later the surge was led by companies from countries with minimal familial links with Britain who simply wanted a listing in the premier global mining market. With AIM, at the other end of the industry, providing a platform for juniors to raise money to pursue development projects and exploration programmes it could have been argued that London's mining prowess had been fully rehabilitated.

The credit crunch did dent some of that optimism but it is the case that many of the world's largest diversified mining groups are now incorporated in the UK, listed in London and part of the FTSE 100 Index.

THE LOSERS

The rehabilitation of London as a major mining finance centre has meant that there have been losers among other world mining centres. One of the hardest hit has been Canada where an important number of mining companies have been absorbed by foreign groups. From the UK Rio Tinto acquired Alcan, Canada's aluminium giant, and Xstrata took over Falconbridge, which itself had earlier merged with Noranda to form a very large diversified Canadian/Global resources group. Canada also lost nickel giant Inco to Brazil's Vale. Although Canada continues to host many of the world's largest gold groups such as Barrack and Goldcorp, and retains a large and lively junior mining sector, it has lost out in the base metal field.

Australia has also seen major corporate activity and again Xstrata has been in the thick of it with the acquisition of MIM in 2003. Western Mining lost its independence in 2005 when BHP Billiton beat off Xstrata, who had opened the bidding for the Nickel Boom legend. However, because of the dual nature of the bidder at least a part of WMC remained Australian owned. Australia has also found that a number of its medium-sized miners, such as Oz Minerals and Fortescue, have attracted the attention of Chinese interests anxious to secure long-term mineral sources.

South Africa has also been a sizeable loser, with Anglo American and Billiton relocating to the UK. It also suffered during that period from the near implosion of its gold sector with production collapsing and mass mergers shrinking the gold mines sector from over 40 listed companies to less than ten. At the same time Brett Kebble of JCI tried to amalgamate a number of group gold companies into a new vehicle and march it off to Toronto. The attempt failed, which led to Randgold

Resources, one of the Kebble group companies at the time, cutting itself adrift and departing for London as we saw above.

Sometimes the losers were less easy to spot. In welcoming Anglo to London, De Beers came under effective UK corporate control for the first time since Cecil Rhodes's time. This came about as a result of De Beers's wholesale re-organisation but in due course many of the functions within De Beers on the sales and marketing side, which had traditionally been carried out in London, were relocated to Africa. Following the passing of Anglo's final 2008 dividend, as a result of De Beers credit crunch woes, it was easy to say that London had in fact lost out big time as the group's debts headed into the City and the jobs headed out.

As to whether London's current pre-eminence can be maintained, or whether it will slip away as it did before, is a moot point. The possible problem lies in the UK's serious financial situation at the end of the 21st century's first decade, where the government is on the prowl for any conceivable source of tax revenue it can find, and therefore also crawling over every tax benefit it might have handed out in the past to make London attractive as a centre for global business.

Some of us are old enough to remember the post war years of austerity when sky high taxes and fierce exchange controls pushed many businesses traditionally located in the UK offshore as we have related earlier. We need to be careful that these days do not return, but already rising taxes in the UK have encouraged some companies (although not yet mining ones) to relocate head offices outside the UK.

20. THE ENVIRONMENT

As we have seen, the business of mining can be both dangerous and environmentally despoiling. Whilst it would be wrong to assume that in the distant past issues of the environment were totally ignored it would also be true to say that the legal framework under which a mine is developed, and under which it is run, has been tightened up considerably over the last few decades. This was inevitable particularly as the 20th century saw the advance of the welfare state in most western countries. Governments came under popular pressure to increase the regulation of industrial activities, both in terms of the workplace and the impact of industry on both the local and the wider environment. Whilst the global mining industry ended the century a far safer place to work in than it had started it, disasters still occurred and one seared into the consciousness of the British is Aberfan.

ABERFAN

One of the greatest human disasters in post-war Britain was the Aberfan disaster in South Wales in 1966. Here an artificial hill comprised of dumped coal mine waste, which loomed over the mining town of Aberfan, collapsed on a school built at its foot killing over 140 people, most of them school children in class at the time. The collapse came after very heavy rain which had destabilised the tip, causing its structure to weaken and then disintegrate into an avalanche which crashed down onto the town below.

The tip, of course, came before the school, indicating that planners had failed in their duty to assess risk effectively, especially as it was known that the tip was criss-crossed by meandering underground streams. However, the tip, to avoid the need to haul the waste uneconomic distances, was always positioned in modern hindsight too near the town, which in the style of British coal mining communities was built right on top of the coalmine itself. Following the disaster the public enquiry discovered that only in parts of Germany and in South Africa were there any laws governing mine tips and their structure.

LOW-GRADE MINES LEAD TO HIGH VOLUME WASTE

One of the trends in modern mining has been the increasing development of low-grade orebodies generating a high ore throughput and therefore requiring ever-larger waste dumps. However, large or small, high or low grade, the problem of mine waste is an increasing headache to be factored into mine development plans. One recent example in Alaska illustrates the problem of mine development in remote areas, let alone in areas with some population density.

In Berners Bay in Alaska's Tongass National Forest, US precious metals group Coeur d'Alene has a medium-grade gold deposit which it has developed. The deposit, called Kensington, lies at height overlooking the bay and the original plan was to use wetlands lying behind the deposit to build the mine waste dump, an arrangement that had been agreed with both the state and local environmentalists as well as native Indian tribes in the area who stood to gain jobs from the mine. The company then noticed that there was a deep lake much nearer the proposed mine and therefore much cheaper as a dump site. A proposal was then put forward to dump in the lake, which contained pure water and consequently many varieties of fish that would be poisoned by the waste. The idea was that after the mine was mined out the lake would be drained and the bottom sealed, pure water would be put back into the lake along with new fish stocks. A large dam would secure the pond waste from leaking down the mountain and contaminating the bay below.

An argument broke out over the revised plan between Coeur d'Alene, the state of Alaska and the US Corps of Engineers – who said that the new approach was environmentally sound, giving a better all round result as far as green issues are concerned – and the Sierra Club and other environmental groups who fundamentally disagreed and were also concerned about setting precedents for future developments in the region. It is not the purpose of this book to go into considerable detail about this issue and the arguments on both sides, but it does underline how difficult things are becoming for mining companies in planning new mines and getting the necessary permits, especially as the Kensington development, as is inevitable in the US, ended up in the courts. Permission was eventually given for the development to proceed and mining started in 2010.

DISASTERS

Mining has always been open to criticism with regard to waste disposal. In the 1990s Canada's Galactic Resources abandoned its Summitville gold mine in Colorado leaving the state and local authorities with a major headache as the high-level waste ponds, with poisonous heavy metals and residue treatment fluids, threatened to overflow, discharging into the local environment via mountain streams and moving slurry. The situation was stabilised with only a limited amount of run off but the clean-up cost over $100 million.

South Africa, with its gold mining industry now well over 100 years old, has an abundance of slimes dams holding huge amounts of accumulated waste. Breaches of earth-made dams happen regularly and occasionally serious accidents occur as in 1974 at Impala Platinum and in 1994 at Merriespruit. Among the most tragic tailings dam breaches elsewhere was the one in 1985 in Trento at the Italian fluorite mine of Prealpi Mineraia where 268 people were killed when the dam failed because run off pipes became blocked.

POLLUTION

Environmental issues do not have to be current to cause today's miners severe headaches. The echo of past asbestos mining operations has dogged a number of mining and industrial companies around the world in recent years. One of the worst affected was the UK's Cape which was hit in the 1990s from two directions, its old South African based asbestos mines and its processing and manufacturing operations in the UK. Its South African mines were substantial producers of the highly dangerous blue and brown asbestos (white asbestos is thought by some, but by no means all, to be relatively safe). Its UK factories manufactured insulation products from its South African asbestos output – in due course both South African miners and UK factory workers contracted asbestosis followed by cancer. Long-running court cases, which almost destroyed Cape, led to substantial damages being paid out and these payouts continue today.

In the US WR Grace, which operated a vermiculate mine in Libby, Montana, has had to pay out hundreds of millions of dollars in compensation for asbestos poisoning relating to the fact that Grace's vermiculite unfortunately was naturally laced with asbestos, and the mine dust circulated for years in the windy environment of central Montana. Past executives of the company have been prosecuted for covering up the dangers relating to the vermiculate dust, although in 2009 three executives were cleared of criminal charges.

In Australia James Hardie in New South Wales operated an asbestos mine for several decades until 1979, where many residents eventually contracted asbestos related conditions, and we have already touched on Colonial Sugar's blue asbestos mine near Wittenoom in northern Western Australia, the closure of which destroyed a once 20,000 strong community before revival came as the Pilbara iron ore mines were developed.

It needs to be remembered that taking a green stand on mining developments or operations can have material downsides. One case in point is the shutdown over recent decades of rare earth mining in the US. The US was once the leading producer of these minerals, which include cerium and promethium used in polishing and nuclear batteries respectively (there are 15 other rare earth metals with many different high-tech applications). The pollution caused by the mining and treatment of rare earths and also the collapse of prices around the turn of this century meant that US rare earth production declined to almost nothing. A big surge in Chinese output filled the gap and prices were held down to boost exports. Now the Chinese have reined back their production, leading the US to scramble to re-start its own rare earth mines.

The point here is a clear one – although US communities near rare earth mines may have enjoyed a much enhanced living environment with the closure of US mines, it has come at what the US government considers an unacceptable price – reliance on supplies from a country with which it has a difficult political relationship and which is a major economic competitor. As Washington might say, green does not always mean safe.

21. MINING AND LABOUR

It would be impossible to write a history of mining without referring to and commenting on the subject of labour, and indeed we have looked at the conditions of work and the issues facing mine labour over the centuries in a number of contexts already. We look at some aspects of this topic in more detail now.

THE FIRST FORMAL MINERS' STRIKE

One of the more interesting case studies is that of the Mexican silver miners in 1766 at the rich Real del Monte mine in the central eastern region of the country and their dispute and strike. The dispute concerned the altering of working practices that had afforded the miners extra incentives to encourage their efforts and supplement their income. Mining has always been, and remains today, a tough occupation, and 18th century Mexico was no exception. The miners at Real del Monte worked 12-hour shifts and were organised in small gangs which had mining quotas to achieve. Anything they could mine above the quota rate they could sell for their own benefit. The mining methods were crude and consisted of driving holes by hand into the ore and then setting explosives. The resulting broken ore was then carried in sacks to the surface.

Following a major investment in the dewatering of some of the mine workings the owner of the mine, Pedro de Terreros, decided to change the working agreements with his labour force. Perhaps unsurprisingly the changes were based on trying to retrieve the pre-investment profit margins by squeezing the miners' returns, in essence charging for some of the equipment provided by the mines, cutting wages and increasing the amount of ore that had to be delivered for the owner, so making it very difficult to mine surplus ore for the mining gangs' own benefit. The final nail for the miners was that when they did mine a surplus their ore was mixed with the owner's ore before splitting, and this was done in such a manner that the owner's split contained most of the higher grade material, further reducing the miners' rewards.

This led to one of the first recorded instances of a straight miners' strike. The resolution of the strike was also interesting because the mine was not remote and so the dispute was conducted in a relatively open community where the issues and grievances could be aired widely. The miners drew up a petition which was first presented to officials in the regional capital of Pachuca; when nothing happened it was sent to the Spanish Viceroy in Mexico City. Following this Terreros appeared to back down and agreed to reinstate many of the old working arrangements, and work resumed at the mine. However, he reneged on the deal and this then set in train a period of violence that only ended when the Viceroy's emissary, Francois de Gamboa, arrived to seek a solution. In the end Gamboa sided with the miners and the old agreement was reinstated.

This incident is fascinating because it shows that even in the 18th century the authorities could take the side of the miners. Fast forward to Britain in the 1970s and the first coal miners' strike of that decade, and once more the final outcome in a dispute, this time largely about wages, was decided by the authorities in the form of an independent enquiry and a massive wage increase of 21%. In the case of the UK strike the mine owner was the British government, which is different from the Real del Monte case, but Lord Wilberforce, who headed the UK enquiry, is very much a figure like Gamboa, the Viceroy's emissary.

INDIVIDUALISM AND SOLIDARITY

As we have previously seen the labour force arrangements for mines have varied materially over the ages. We have gone from small scale and basic surface operations carried out by independent miners many millennia ago to bigger projects requiring more organised working arrangements where slave labour, as in the days of the Egyptians and Romans, was often employed. We have then seen the arrival of the era, essentially still with us today, where metal demand is driven by technological developments, and the consequent scale of mining operation, and thus work force structure, is influenced by the growth of this demand. In the modern era we have seen the resurgence of the individual prospector turned miner as in the great gold rushes of the 19th century, and we have seen the huge organised work forces of, for example, the British coal mining and the South African gold mining industries of the 19th and 20th centuries, both of these now much reduced.

These two latter examples also bring us close to a stereotypical picture of the average underground miner as a low-paid worker employed to work long hours in dangerous conditions, loyal to his community and family, and hostile to his employer whether they be private entrepreneur or the state. In more recent decades mining has become increasingly capital intensive and work forces have become smaller, more technically competent and therefore much better paid. Safety has also become a major priority as customers have become more concerned about the ethics of their suppliers, i.e. the mines and the working conditions therein.

THE ENTREPRENEURIAL TRADITION OF PROSPECTING

There is, of course, a world of difference between the attitudes of the entrepreneurially-driven gold rush prospectors of the 19th century, for instance, and the collectivist philosophy that developed, for example, in many of the coal mining communities of North America and Europe in the 19th and 20th centuries. The gold rush prospector was utterly single minded in his search for gold, and thus his fortune, and completely uninterested in anything that could get in the way of him achieving this end. Thus he might rail against being exploited by suppliers of services in the mining camps, but their profiteering was endurable if in the end the prospector made a bonanza find and consequently his fortune. Labour militancy could have no part in this process, particularly as the prospector's individuality was what motivated him to search for his holy grail, often for his whole life and usually unsuccessfully. Such a man, however, does not think of himself as exploited and the lure of wealth may in some cases in the end take second place to the nobility of the search.

Returning to the 1766 strike in Mexico, one of the key issues that arose was related to the profit sharing element of the mine owner's agreement with his workforce – the quota and the surplus accruing to the miners. Since the unionisation of mine work forces during the 19th century, miners' representatives have tended to shy away from profit sharing or bonus arrangements. They believe such arrangements are too easily manipulated by management to the detriment of the work force, as occurred in Mexico over the reduced workers' surplus.

Another aspect of union suspicion revealed itself in the 1980s when some of the South African mining companies decided to implement share savings schemes amongst their largely black work forces. The unions' concerns lay in the possibility that, having shares in the company they worked for, the workers could be compromised when it came to the issue of industrial action. It could be argued that the unions' point was well made, for if one is going to strike it is better to be a clean protagonist without any stakes on the other side of the dispute.

CAMARADERIE OF DANGER

Whilst the interests of miners in many respects are little different from those of workers in other industries, there is one area where they may have additional cementing loyalties – the area of danger. The fact is that underground mining in particular is a hazardous occupation, whilst labouring in transport or retail is generally not. Experiencing and sharing danger several thousand feet below ground gives workers an enhanced understanding of their relationship with their fellow workers and their responsibility to them. When danger becomes disaster miners tend, understandably, to stick together. When mine owners come into dispute with mine unions, miners generally stand behind their union representatives.

However, it would be wrong to say that, despite evidence historically of pretty poor if not downright bad operating practices in the mining industry, management is always at the bottom of disasters. Sometimes management lays down sound operating and safety procedures but carelessness on the part of an operative at the wrong moment can lead to disaster, as happened at the Vaal Reefs gold mine in South Africa in 1995 when a runaway train crashed down a lift shaft onto the top of a cage carrying over 100 miners, killing virtually everyone. So shattered by the disaster was Anglo American's Gold Division Chairman, Englishman Clem Sunter, that he felt that he had to resign and he did not work in frontline mining again. Of course in the case of such a disaster, management has to take responsibility but the human error that led to the accident was primarily down to the underground worker operating the train's braking system.

LOYALTY HAS LIMITS

Mineworker loyalty to unions and comrades can be stretched when miners think they are being taken for granted and taken advantage of. In the 1980s after the labour triumphs of the 1972 and 1974 coal miners' strikes in the UK, Arthur Scargill, one of the most influential foot soldiers of the 1970s battles, who had become leader of the Mineworkers Union subsequently, decided to take on Margaret Thatcher's Conservative government. Scargill was a Yorkshireman and effectively ambushed the government with a spontaneous strike in 1982 over pit closures, which the government at that time was ill prepared to resist. Two years later Scargill launched a full frontal attack again over the threat of pit closures, but Thatcher was waiting and had prepared for a battle to the death, with power stations fully stocked with coal, ironically dug by Scargill's men over the previous two years, and alternative energy supplies secured also.

One of the other key things Thatcher had done was to divide the highly productive Nottinghamshire coal miners from the rest of the industry, exploiting the Nottinghamshire union leadership's rising disgust at the grandstanding antics of Scargill which they deemed increasingly aimed at self promotion rather than the interests of the miners. The Nottinghamshire coalfields worked throughout the 1984/5 strike and this was critical to the UK government successfully facing down the strike, after which neither Scargill nor the mining unions were ever an industrial force again.

Part of Thatcher's success related to the fact that Nottinghamshire miners worked in modern mines, were highly productive and very well paid, and so were reluctant to strike especially on the say-so of Scargill and his difficult Yorkshire followers. The coal miners' historical loyalties had, after centuries, been successfully divided by prosperity, something that their leaders had ceaselessly fought for over that period. But rising prosperity took miners out of their traditional communities as they sought and gained the trappings of modern life – houses, cars, foreign holidays, etc. Politically inspired labour disputes endangered these hard-won economic and financial gains with little prospect of long-term benefit.

THE MINER BEGINS TO PROSPER

At the same time labour everywhere was becoming more productive as mines became increasingly capital intensive. The advance in mining techniques, as we have seen earlier, meant that mines could introduce mechanised methods which dispensed with large workforces and improved operating conditions, leading to safer working practices. These improvements are perhaps more obvious in western mines and have helped to mitigate one of the most potent sources of labour unrest – unsafe working conditions and consequent accidents and fatalities. In major non-Western mining countries like China and South Africa labour remains vulnerable in these areas.

The other catalyst for fomenting labour unrest is the closed mining community where grievances can be easily nurtured by miners in constant social contact with each other. These opportunities are today less obvious with traditional mining communities no longer popular with modern families, and where mine remoteness might lend itself to a purpose built township many mining companies operate a *fly in, fly out* strategy instead where miners work, say, four weeks on the trot living in dormitory accommodation and then have four weeks off with their families who live elsewhere.

Labour has also benefited from the tendency for producing mines increasingly to be run by large mining conglomerates like BHP Billiton rather than the individual mine owner of history. Such an employer often did not have the resources, or indeed the inclination, to offer his workforce anything more than blood, sweat and tears, and a poor wage. The modern mining house usually provides a full employment package with benefits to a workforce of largely skilled miners with high earning power derived from capital-intensive mining methods enhancing productivity and profitability.

Such a benevolent outcome would not have been in the vision of miners 50 years ago, let alone a hundred, and provides hope for those who favour talk rather than conflict in labour relations. As we learnt earlier conflict in mining has rather gone offshore, for whilst mining companies no longer face permanent hostility from their workforces a new protagonist, the environmental lobby, has emerged.

22. MINING AND THE MEDIA

THE MINING PRESS

A great industry like mining requires the oxygen of publicity to plead its case when under attack or to bring it to heel when it strays too far from the narrow road of proper behaviour. It is certainly the case that over the years the mining industry has generated a substantial amount of written matter both in terms of newspapers and other media forms and books, and has even spawned a number of films of varying quality. We will consider all these areas but our main thrust will be towards newspapers and specialist mining publications and the writers who have made the most significant contribution to the wider understanding of what the industry does amongst the financial and investment community.

In the earlier sections of the book we highlighted the contributions of Pliny and Georgius Agricola but the modern age has unsurprisingly seen the greatest output of mining writing. The 17th century could perhaps be denoted as the time when mining writing began to expand beyond the traditional fare dominated by Argicola's 200-year-old *De Re Metallica*. One of the first English language books, and subsequently English tracts have tended to dominate mining writing, was by Sir John Pettus, Deputy Governor of the Royal Mines, who documented and described the main mines in the British Isles. In the following century Abraham Werner, a German of the Frieberg Mining School, and Englishman, John Hutton, conducted a public debate through their writings on their differing views on the geology of rocks.

THE EARLY DAYS

Two of the oldest mining publications still publishing today are the French *Annales des Mines* and the Britsh *Mining Journal*, the former starting life as *Journal des Mines de la Republique* in 1794 and the latter as the *Mining Journal and Commercial Gazette* in 1835. Before these specialist publications a number of European broad scientific journals had occasionally published writings on mining, usually from a technical perspective, the first dedicated mining publication probably being the German *Magazin Bergbaukunde*. *Annales des Mines* was in many respects a child of the French Revolution, published on direction of the Revolutionary Committee to inform the populace of mining activities in France but also to offer advice and direction to the industry – independent thought was absolutely not encouraged. The journal survived the Terror to prosper thereafter, respected throughout the mining industry.

THE UK

The *Mining Journal* was founded by a young Englishman, Harry English, who believed with the Industrial Revolution in full swing and the mining industry providing the key raw materials driving industrialisation that a regular publication recording and commenting on the issues facing mining was long overdue. Exposed to the oxygen of publicity it was clear that all was not well with the UK's mining industry – its size, excepting coal, was inadequate to provide all the needs of rapidly growing industries such as iron, steel and engineering. At the same time coal mining in the UK was dangerous and fatalities unacceptably common in the fast growing industry.

English became something of a one-man protest movement and his commentaries became increasingly influential, culminating in his success in having a commission appointed to look into the Haswell Colliery disaster in Durham in 1844 following his direct appeal to Queen Victoria. The *Mining Journal* continued to campaign for reform of the mining industry and English's efforts played a major part in the establishment of both the Royal School of Mines and the Camborne School of Mines.

As the 19th century wound on the UK's mining industry, ex coal, became increasingly irrelevant in powering the country's rapid industrial development. The UK turned to countries such as the US, where industrialisation was in full swing stimulating exploration and development, and its own dominions such as South Africa, Australia and Canada and colonies such as Malaya, the Rhodesias and Nigeria, for raw materials. This not only opened up the *Mining Journal's* perspective on mining, increasing its international coverage, but also encouraged new publications into the field.

Two publications relevant to the *Mining Journal* were *Mining Magazine* and *Sturzenegger's, the Rand Gold Mines. Mining Magazine* had been started in 1909 by T.A. Rickard, born in Italy of British parents, and we will hear more of him later. It had an early triumph when its owners Mining Publications, which included Herbert Hoover amongst its shareholders, published Hoover's translation of *De Re Metallica*. The *Magazine* was a publication dedicated to the technical and engineering side of the mining industry but after a bright start struggled for many years. It, however, survived and in the 1960s was bought by the *Mining Journal* which had in 1957 added *Sturzenegger's* to its stable, converting it into a quarterly review of the South African gold mines.

Another long-running UK specialist mining publication is *Metal Bulletin* (*MB*) which saw the light of day in 1913. Its founder and first editor was an Irish journalist, Lawrence Quin, who had been a writer for trade paper *The Ironmonger*. By 1920 the success of the *MB* allowed Quin to take on two other journalists, Leslie

Tarring, who was just 16, and Harry Cordero, both of whom dedicated most of their working life to the *MB* and also introduced their sons into the publication. Tarring died in 1964 still working for the *Bulletin* and his son, Trevor, was made managing director in 1978 and Cordero's son, Raymond, became editorial director around the same time.

Quin's idea was that the *MB* should cover the marketing and pricing of metals, a core activity of today's *MB*, although over the years more comment driven articles were introduced and mining news also became an important part of the modern magazine. The *MB*, which had become a stock market listed company in 1981, remained independent until 2006 when it was acquired by *Euromoney*. *MB* for much of its life published two times a week and built up a terrific reputation as a journal of the metals industry and a price leader for minor metals. In recent years it has become a weekly. For all of its life it has been essential reading for those interested in metal market trends, something it shares with the older daily *American Metal Market* (now published only online), which has become a subsidiary of *MB/Euromoney*.

AUSTRALIA

In Western Australia the Kalgoorlie gold rush saw an explosion in construction and the establishment of the services required by a fast growing town. Newspapers were at the forefront with the broad Goldfields region, which included Kalgoorlie, at one stage publishing half a dozen titles, one of which was the *Kalgoorlie Miner* which started life in 1895 and is still going strong today. Although mining news was important to the *Kalgoorlie Miner* it was always primarily a general newspaper. The oldest specialist Australian mining publication, also still in existence today, was the *Queensland Government Mining Journal* which started up in 1900 as a result of a number of discoveries of gold and copper in the state, and the state government's desire to publicise and track the industry's development. Its first editor was William Hodgkinson, a man given occasionally to poetic flights of fancy despite being a thoroughly professional and experienced explorer and surveyor.

However, the first specialist Australian mining paper was the *Australian Mining Standard*, which saw the light of day in 1888 in Victoria as a result of rising interest in the gold discoveries made in the state during that period. The *Standard* lasted until 1960 when it closed following an unsuccessful attempt to turn itself into a general business publication. If it had managed to soldier on for another six years it would no doubt have received a major lift from the Australian nickel boom.

Another pioneer of the Australian mining publications scene was the *Australian*

Mining and Engineering Review founded in 1908 by Peter Tait. The *Review* survived until the nickel boom and, in recognition of rising interest in mining in the later 1960s, it simplified its name to *Australian Mining* in 1966. Today the Australian mining media is dominated by monthly publications such as *Gold and Minerals Gazette*, a monthly started in 1985, *Australia's Paydirt*, another monthly started in 2004, and its sister publication *Gold Mining Journal*, which is a quarterly, and the *Australian Journal of Mining*, a bi-monthly started in 1982. Whilst all these publications cover the Australian mining scene they also keep a close eye on what is going on in the mining world outside Australia.

SOUTH AFRICA

The discovery of diamonds in Kimberley and gold on the Witwatersrand in South Africa created substantial demand for news of these huge new industries. The gold discoveries on the Witwatersrand in 1886 spawned the *Diggers News and Witwatersrand Advertiser* which started in 1887 and was part-financed by President Kruger of the Transvaal. Of all the mining towns which grew up in the late 19th and early 20th centuries on the back of mineral discoveries, it is Johannesburg that has almost uniquely grown into a major city. So almost from inception a vigorous newspaper industry has been part of the Witwatersrand scene. One of the early newspapers, still published today, was *The Star*, a pro-British newspaper which attracted the disapproval of President Kruger on many occasions. It was joined, among others, in 1902 by the *Rand Daily Mail*, which was to be a constant thorn in the side of the later apartheid governments of the period after the Second World War. Its first editor was the famous British thriller writer Edgar Wallace.

Although these were general newspapers, mining activities on the Witwatersrand did provide an important part of their news stories, but the first specialist South African mining publication was the *South African Mining and Engineering Journal* founded in 1892. The *Journal* was a weekly providing a mix of financial and technical news, but it also offered sometimes trenchant and always independent comment on other mining industry related issues, including the vexed question of native labour which blew up into a big issue after the end of the Anglo-Boer War in 1903. The problem lay both with the wage rates offered to black workers and the rates which imported Chinese workers were prepared to accept. The Boer War saw black wages cut from £2 10s a month to £1, and the *Journal* supported the new lower level as adequate reward. Consequently many blacks left the mines and the Chinese took their place; they eventually returned to China when black labour drifted back to the mines.

Other important South African mining publications included the *Chamber of Mines Mining Survey*, started after the Second World War, which had a substantial and worldwide circulation. Today publications like the old *Mining Journal's Quarterly Gold Mines Journal* (Sturzenegger's), published in the UK, which is now owned by GFMS (Gold Fields Mineral Services) has changed its name to *World Gold Analyst* and has gone worldwide in its coverage. Current South Africa-centric publications include the monthly *Mining Mirror*, first published in 1988, and the bi-monthly *African Mining*, started in 1995, which has a broad African spread. There is also *Miner's Choice*, a monthly that similarly has a broad African focus.

CANADA

Canada has a vigorous mining press as one might expect given its long mining history. Like Australia, as new areas opened up in the wake of gold rushes and other mineral excitements, local newspapers were established and although the content of these was general the activity driving the new economy – mining – occupied an important part of the newspaper.

Perhaps the first specialist Canadian mining journal was the *Canadian Mining Review* (*CMR*) founded in 1879 by W.A. Allan who made his fortune in phosphate mines discovered near Ottawa. The slim paper had a particular slant towards its owner's mining activities and it was not until it took on in 1887 Benjamin Bell as editor, a man better known as one of Canada's leading cricketers at a time when the game had an important foothold in the country, that the paper began to expand its coverage beyond Allan's phosphate activities. It also quickly established itself as a fearless opponent of mining scams and promotions. As well as his pioneering editorial work as editor of the *CMR*, Bell was also an energetic proponent of establishing professional bodies for the mining industry, the Canadian Institute of Mining and Metallurgy (CIM) being one example. In 1904 Bell died aged 43 when he accidentally fell down a lift shaft at work. Following this the name of the paper was changed to *Canadian Mining Journal* (*CMJ*), the name it has carried ever since.

Although the *CMJ* was the official journal of the CIM for many years, in the end the institute wanted its own organ and started the *CIM Bulletin*. This started life as a record of technical papers but eventually became, to the concern of the *CMJ*, a competing publication. In recent years it has changed its name to *CIM Magazine*.

In 1915 Canada's most prestigious mining newspaper, the *Northern Miner* (*NM*), was first published and it remains today the leading publication covering the activities of Canadian mining companies at home and abroad. The founders were Ben Hughes, an Englishman and mining journalist, who quickly sold on his half interest to Richard Pearce and Ernie Hand, a printer, when he returned to Europe to sign up with the war effort. The early years were difficult ones and the *NM* struggled financially but it was another very outspoken publication that added considerable value to investor and industry's understanding of the development of Canadian mining. The arrival of Pearce's brother Norman as joint editor provided further edge to the *NM*'s message and this increased in importance as Canada's mining industry expanded. The paper started life being printed in Cobalt, Ontario, centre of the silver rush of the early 20th century, but in 1929 it moved to Toronto as recognition that its role was increasingly national rather than local.

The *NM* built a reputation for being on the inside of a range of issues, sometimes in wartime those were strategic as when it hunted out examples of trading with the enemy. In peacetime there were campaigns against government attempts to purloin gold profits when the US raised the price in the 1930s. And always the *NM* was on the look out for fraud on the part of mining promoters as well as genuine stories of exciting new mineral discoveries like Red Lake and Kidd Creek providing new wealth for miners, investors, provinces and state.

It also expanded its range of publications and in 1974 bought the equally crusading *Western Miner*, which strengthened its coverage of mining in western Canada. Today the *NM* retains its position as one of the leading world mining information and comment providers. However the importance of immediacy means that the online version of the weekly newspaper, giving subscribers immediate access on publication, is increasingly popular for subscribers, particularly foreign, and the newspaper, once subscription only, is also now sold through newsstands.

THE US

The development of a mining press in the US advanced rapidly in the 19th century, driven by the Californian gold rush of the late 1840s. One of the first specialist mining publications was the *Mining and Scientific Press* (*MSP*) which grew out of the *Scientific Press* first published in 1860. The driving force behind the *MSP* was Alfred Dewey and he directed the paper's fortunes for 30 years. However, although the *MSP* was highly prestigious in its own right as the first specialist mining publication in the US, Dewey, backed by his family company, was anxious to expand his stable of papers and magazines.

Unfortunately one of the papers Dewey bought was the *Pacific Rural Express* (*PRE*), a major farming paper, and this coincided with a period of tension between the rural and mining communities over hydraulic mining which washed contaminated mine waste from the Californian mountains on to fertile farmland. This conflict of interests between the core readerships of both journals resulted in neither being able to launch campaigns in favour of their readership's position, in direct opposition as they were. The *MSP* did not really recover its respect until T.A. Rickard acquired it in 1905 (a story which we describe below).

The *MSP* was a western journal published in San Francisco. The first east coast competitor was the *American Journal of Mining* (*AJM*), a weekly, which was started in 1866. The editor was George Dawson who determined that the *AJM* would take a far broader view of the mining scene than did the west-centric *MSP*. The *AJM* also was not shy of taking a political stance when it considered that the future growth of the industry might be damaged by events. An example of this was the threat posed by native Indian tribes as settlers pushed west – the journal called for reprisals against the indigenous occupants of the great Midwest. Two men, Rossiter Raymond and Richard Rothwell, both mining engineers, became key contributors to the *AJM* and Raymond became editor, and in 1868 changed the journal's name to the *Engineering and Mining Journal* (*EMJ*).

In 1875 Rothwell became joint editor with Raymond and although their strong individuality created problems they were both also highly innovative, with the *EMJ* being a major supporter of the establishment of the American Institute of Mining Engineers (AIME) in 1871. As joint editor of the *EMJ*, Rothwell in particular encouraged the expansion of the journal's stable of publications into coal and steel, but perhaps the most important additional product was the annual *Mineral Statistics – Statistics, Technology and Trade*, which had been first published in 1893.

Whilst Rothwell pushed on with the *EMJ*'s development, Raymond, as well as his contributions to the journal, was busy providing expert comment in mining legal actions, working as a consulting engineer, as a university lecturer in economic geology, as a State Commissioner for New York's nascent subway system and as a consultant for the New York Telephone Company. An active churchman, Raymond was also a member of a number of highly prestigious professional bodies including the IMM in the UK and the Canadian Mining Institute. Raymond died in 1918 having resigned as *EMJ* editor in the late 1880s. Rothwell had died in 1902, being replaced for a couple of years by T.A. Rickard.

The *EMJ* remained an influential leader in terms of mining opinion in the US, which led to its absorption into the McGraw-Hill group in 1917, having been a financially fragile business for a number of years, despite the quality of its content. As the decades went by, the *EMJ* further expanded its stable of publications and with *Engineering and Mining World* it took its coverage on to the world stage. It also consolidated its position as the most influential mining publication in the US and its 1945 campaign to alert the industry to the potential of uranium in the Cold War era and its championing of the development of a US uranium mining sector was an example of this. In due course, though the fall from favour of the mining industry in the latter part of the 20th century caused problems for all mining publications, the *EMJ* had to pull in its horns. It finally closed in 2003. However that was not the end; the journal was bought from the liquidator by Mining Media International in the same year and survives today.

T.A. RICKARD (1864-1953)

Thomas Arthur (T.A.) Rickard was born in northern Italy of British parents with, one might say, mining already in his blood. His father, also called Thomas, was a mining engineer and metallurgist from Cornwall. In his early years the family were constantly on the move from Italy to Switzerland and then to Russia. T.A. was eventually sent to England for secondary education at what is now Queen's College, Taunton.

He was sufficiently bright despite his somewhat nomadic early life to gain a scholarship to Cambridge, but at the urging of family members he chose to follow his father and uncles into mining, and instead went to the Royal School of Mines (RSM) in London from which he graduated in 1885. There one of his instructors was T.H. Huxley of the remarkable Huxley family, a biologist and philosopher and also very public supporter of the then controversial views of Charles Darwin. One of T.A.'s fellow students at the RSM was the future science fiction legend, H.G. Wells.

Following graduation T.A. was invited to take up a job as assayer by one of his uncles in the US who managed a number of British-owned mines in Colorado. He ended up managing the mines and then was hired away to manage a new gold mine in California. The mine, which had been sold to British buyers by an American promoter, was a failure but it led to Rickard striking up a friendship with an Australian mining engineer and in 1889 he went to work in Australia. His job was to assess mining prospects and whilst doing that he also wrote a few mining articles for the Melbourne press.

His next move, at the behest of his father, was to France to manage the development of a marginal cobalt mine. Lacking confidence in the viability of the project he returned to the US where he helped assess projects for an American promoter. He became disillusioned by over promotion of some of the projects and Rickard set himself up as an engineering consultant in Denver. He became Chief Geologist for the state of Colorado but in 1897 he was back in Australia as the Kalgoorlie gold rush broke, again in the role of project assessor where he struck up a friendship with Herbert Hoover. The following year he returned to the US and married his cousin, Marguerite; they had one son, John. Rickard's Denver consultancy resumed and he employed the young Chester Beatty, who was to achieve fame and fortune first on the Northern Rhodesian (Zambian) copperbelt and then in diamonds in West Africa, as his assistant.

As a recently married man Rickard found that he was spending too much time away from home and in 1903 he switched to mining journalism becoming the editor of the weekly *Engineering and Mining Journal* (*EMJ*) in New York. Rickard was suited to this role as he had extensive knowledge of mining from both the technical and the financial angles, and he picked up the skills of journalism swiftly. Unfortunately the *EMJ*, although extremely well regarded, was a financially fragile paper with a constantly changing ownership and Rickard tired of the lack of continuity and left in 1905 to buy and edit the *Mining and Scientific Press* (*MSP*) in San Francisco, with family backing. The following year the San Francisco earthquake struck but the *MSP* survived because Rickard was not only a good journalist and engineer but also a well-organised businessman. Under Rickard's direction the *MSP* flourished using some of the most highly regarded names in mining as contributors, including Herbert Hoover. It was also during this time that Rickard began to write technical mining books and over the years had well over a dozen published.

In 1909 Hoover lured Rickard to London to edit the newly published *Mining Magazine* (*MM*). The magazine was an editorial success from the start but its financial position took some years to secure; when that had been achieved Rickard returned to San Francisco in 1915 where the *MSP* was in poor shape. He had exchanged his shares in *MM* for shares in *MSP* with his cousin Edgar, and set about reviving *MSP* which he did both editorially and financially. In 1926 McGraw-Hill paid Rickard $250,000 for *MSP* but blocked his taking up the secretaryship of the American Institute of Mining Engineers having secured his agreement not to compete as a technical mining journalist in North America. From that time Rickard began to concentrate on broader writing on mining and in 1932 published two substantial histories – *A History of American Mining* and a massive two-volume work *Man and Metals*. For the last 20 years of his life Rickard moved to Canada and lived in Victoria, BC, where he continued to write and lecture on mining and related subjects.

An Englishman granted American citizenship in the 1920s, Rickard's words describing his London period echo down the years to today's mining packed FTSE 100: "the mining finance of London has a geographical amplitude like no other city in the world". Even in our current tightly regulated world Rickard's high ethics, as a mining journalist, would have been above reproach. He believed that advertising needed to be carefully screened to protect editorial integrity, that comment should never be compromised for financial reasons, and that journalists should not own mining shares, sit on mining company boards or advise on any mining promotion. His writing was both elegant and punchy and his *Guide to Technical Writing* remained a core work for aspiring mining journalists for over half a century.

MUZZLED BY REGULATION?

Over the years the nature of media coverage of the mining industry has changed and today the news-chasing journalists of Canada's *Northern Miner*, for example, for regulatory reasons must be very careful in how they handle material that might not be in the public domain and might be of value to insiders. This dilemma often arises because companies can no longer provide information on, say, a new mineral discovery until they have made a public announcement through the relevant stock exchange.

One example of this tightening of news dissemination was witnessed at the Indaba Mining Conference in Cape Town in 1997. In those days mining companies still felt able to make major announcements at such high-profile events where they would have the maximum effect on delegates and on the market when their words were reported – mostly the news would be positive but occasionally negative news would be announced. In 1997 at the Indaba, as an example of the latter, there was an important revelation made by Ashanti Gold, then still an independent company, that it would fail to meet its end-decade target of 1 million annual ozs of production from its Obuasi mine in Ghana. Indeed it said that such a target would never be met.

This announcement, made as part of a presentation on the company by the CEO, came completely out of the blue and was a major shock, especially to those who had only a couple of years before visited Obuasi and as part of their visitors pack had been given a t-shirt by Ashanti proclaiming the 1 million oz Obuasi target. The target was dropped and Ashanti eventually fell, after a disastrous hedging experience, into the hands of AngloGold; those people who attended the 1997 Indaba were the first witnesses of the beginning of the company's fall from grace.

Today, with regulators breathing over companies' shoulders, announcements of significance have to be made within the confines of structures laid down by stock exchanges and other regulators of the securities industry. Loose tongues are no longer tolerated and this makes it very difficult for journalists to tease scoops out of nervous management. Legal and compliance teams within news media have also become nervous about financial scoops, and in recent years this seems to have driven some high-profile organs like the *Financial Times* to write analysis pieces on mining shares using information in the public domain. Mining comment is therefore becoming increasingly bland.

Perhaps in recent times the most influential mining journalists were the *Financial Times* (*FT*) duo of the post-war period, Leslie Parker and Ken Marston. Leslie Parker started his career in financial journalism with the old *Financial News* in 1929 and also edited the annual mining publication, *Sturzenegger's, the Rand Gold Mines*. Parker joined the *FT* in London in 1939 and after the war was appointed the newspaper's mining specialist and immediately started a Monday column under the pseudonym of Lodestar. That weekly column continued until Parker's full retirement in 1978 and for the last 20 years of his career he was also the *FT*'s mining editor. As well as penning the Lodestar column Leslie Parker also wrote extensively on the world mining industry explaining and commenting on the industry's activities and evaluating both company results and strategic corporate developments.

In the 1960s and the 1970s Parker's right-hand man was the able Ken Marston who succeeded him as mining editor in 1975 having been responsible for many years for the *FT*'s Saturday mining review column as well as the daily mining news section. At that stage the *FT* had a premier position in the media world in its coverage of mining, something that sold papers in the age of the 1960s Australian nickel boom and the following South African gold share boom in the 1970s. But when Ken Marston retired in 1987 the *FT* decided, with its two old soldiers gone, that coverage of the industry would be wound down (although for a couple of years George Milling-Stanley continued the Saturday mining review column).

Today the *FT* still has a mining editor but the mining pieces written are very often review pieces on major companies such as Rio Tinto and Anglo American. The paper shows very little interest in charting the progress of smaller mining companies and even the results of major overseas miners often pass without comment. Its approach to metals is scattergun with a daily commodities column which is little more than a sawn-off *London Metals Exchange* daily review. The paper, whatever the evidence, maintains a fundamental anti-gold bias and has an obsession with iron ore and a vague passing interest in other base metals such as copper and nickel. Despite rising interest in alternative, green energy sources the *FT*'s coverage of uranium is weak. The decades of solid quality coverage of mining built up by Leslie Parker and Ken Marston has been squandered by today's newspaper.

THE INTERNET

In the last ten years or so the internet has become a major force in the dissemination of information about mining. A significant number of specialist sites have been developed to follow the industry on a daily basis and this daily coverage does give such sites an advantage over hard copy mining output. One of the earlier sites was the free service from Minesite (www.minesite.com), a UK site which started in mid-1999, hardly a particularly auspicious time for mining, as the tech boom which fired the internet had, in investors' minds, rendered raw material businesses sunset industries.

Another widely followed site is Mineweb (www.mineweb.com), which first saw the light of day as Miningweb, also in mid-1999, and like Minesite is free to view. Mineweb is a South African-owned site but has a worldwide mining news stretch. These internet services are largely financed by industry advertising and there are links between the information sites and the mining company sites that works to the advantage of both sides as a conduit through which information can flow.

The mining websites also seem to be rather more adventurous in terms of their critical analysis of mining industry events than does the conventional printed media. They are, of course, largely subject to the same legal restraints as the print media as far as comment is concerned, but they are far more theme and issue driven than the conventional mining press, and can be outspoken in a way that attracts very loyal readers. There is also a slightly unhinged streak in some of the mining sites where statements and analysis can be challenging and controversial.

There are also a growing number of specialist sites which publish mining news and attach that to a particular metal where they have expertise and where they offer trading services. One example is precious metals site Kitco (www.kitco.com), which first appeared in 1995, although Kitco's actual metals trading business had been established in 1977. Another gold-focused site is Gold Eagle (www.gold-eagle.com) which started up in 1997. The dissemination of gold-based information is similar to other metal sites in that Gold Eagle obtains a lot of its information from third-party providers and commentators who have their own sites which are often subscription based – they are happy to have articles displayed on free sites provided subscribers have seen them first. The use of the site by outsiders is free.

One of the strengths of the internet as a communications network is that it is relatively cheap to set up a website, in sharp contrast to setting up a newspaper or magazine where the costs are very high indeed. The downside, however, is that competition can be intense and the plethora of, in particular, gold and gold share sites can be very confusing. This confusion also, as one would expect, begets contrasting and opposing views and can make it difficult for strong messages to get out into the world. Nonetheless if it can be argued that campaigns and crusades, as led by legendary mining news editors like T.A. Rickard, are less effective now than in the past, it can also be argued that the depth of the internet provides a rich mix indeed of current and historic mining information.

MINING LITERATURE AND FILM

It is arguable whether any industry has generated such a wealth of literature (both history and fiction) and film, as has mining. Perhaps oil would want to stick up its hand but oil has a short history compared with mining and lacks mining's variety. As an indication of the extent of mining industry literature the bibliography of this book, which merely scratches the surface of what was available when I began research, provides some light. However, it is true that if one includes technical books there are many industries that would claim extensive literary coverage.

BOOKS

We have already become familiar with Georgius's *De Re Metallica* which remained an industry textbook for some centuries. Pliny's writings have also been mentioned and Statius, Livy and Vitruvius were other Romans who were familiar with mining and recorded their impressions, the latter's *De Architectura* being an extensive work on all things to do with building and engineering. Diodorus and Theophrastus were two Greek sages familiar with mining and Theophrastus had an interest in natural history, which led him to write about minerals and mining in books such as *Concerning Stones* and *Concerning Mines* (these are not, of course, the ancient Greek titles).

The British Library, which contains one of the largest collections of books anywhere in the world, shows around 17,000 items relating to mining on one of its website databases. Unfortunately there are a large number of items relating to data mining which has nothing to do with mining but is simply data analysis, and it is impossible for outside users to delete this confusing Americanism from the Library's list of mining publications. The Library's list also contains a large number of reports, legal documents and academic and technical pamphlets relating to mining.

Peter McCarthy of AMC Consultants in Melbourne has a large growing private library of mining books and his calculation for published books across the mining sector is upwards of 100,000, perhaps nearer 200,000, which includes tens of thousands of books and reports from local and central governmental sources around the world. Although it is impossible to calculate with accuracy the number of books on mining that have been published over the last couple of centuries, it is clearly a huge number.

Over the last 20 years or so there has been a large stream of technical books published on various aspects of the mining industry. Whilst these publications provide detailed and extensive information for academic study and coursework for graduate studies in geology and engineering, they are not intended to provide the colour and stories that so often fascinate. However there is a very large body of literature covering the history of the industry, which more often than not concentrates on quite focussed issues and subjects. There are also the personalities that make up the mining industry and here we find biographies stretching back into the 19th century and coming right up to the present day. The styles, and indeed depth, vary widely, Anthony Hocking's *Oppenheimer and Son*, published in 1973, being a comprehensive but slightly dull biography of the great South African mining dynasty, whilst Geoffrey Wheatcroft's *The Randlords*, published in 1985, in some contrast, covers the earlier South African mining giants succinctly but with great pace and panache.

Some mining books contain considerable quantities of social comment like Pierre Barton's *Klondike* published in 1972, and some have a strong travelogue feel like Fred Cornell's *The Glamour of Prospecting*, published in 1920, and covering his exploration activities in the north west Cape and South West Africa (Namibia). Another area of mining publishing is local history, sometimes supported by local businesses, written often by local writers and published privately. There are many cases of this in Australia and Canada, the enormous *Golden Destiny* by Martyn and Audrey Webb published by the City of Kalgoorlie-Boulder in Western Australia in 1993 to mark the 100th anniversary of Paddy Hannan's gold find at Mount Charlotte near Kalgoorlie is a particularly sumptuous example.

There is also the genre of the mining or mining-related novel. Many become films like Richard Llewellyn's *How Green was my Valley* and Wilbur Smith's *Gold Mine*, both mentioned below, and Hammond Innes's *Golden Soak*. One of the earliest was Rider Haggard's *King Solomon's Mines*, published in 1885, which has inspired no less than five films. The minerals usually associated with the cinema are coal, gold and diamonds – coal provides the background for social clashes as mine owner grinds the face of his workforce into the coalface, gold and diamonds represent fabulous wealth, encouraging heroic searches and dramatic and often fatal betrayal.

FILMS

One of the earliest mining themed films was Charlie Chaplin's *The Gold Rush* made in 1925 and set in the time of the Alaskan gold rush where Chaplin revives his famous Little Tramp role as a gold prospector.

Gold has always had a key role to play in films with mining themes. The classic *Treasure of the Sierra Madre* directed by John Huston in 1948 and starring Humphrey Bogart and John's father Walter Huston was one of the finest of the genre of prospectors searching for gold to secure them financially for life and falling out with disastrous consequences. In the following year Glenn Ford and Ida Lupino starred in *Lust for Gold*, a film about the legendary Lost Dutchman's gold mine in the Superstition Mountains of Arizona. The theme of rumour, search, betrayal and murder common to many gold mining films is particularly to the fore in this film based on possible fact.

Mackenna's Gold made by British director J. Lee Thompson in 1969 with Gregory Peck and Omar Sharif represents the search theme again as a group of neer-do-wells follow the trail to a fabulous store of gold. In 1974, right in the middle of a boom in gold and South African gold shares, Peter Hunt's adaptation of Wilbur Smith's book *Gold Mine*, shortened to *Gold*, starring Roger Moore and Susannah York, was released. Slick and exciting if rather implausible, the film was shot at the West Rand Consolidated mine in a South Africa then in the grip of apartheid. The location made the late Susannah York's participation a curiosity bearing in mind her strong left wing political views. The gold prospectors versus rapacious landowner theme provided the background story for the 1985 Clint Eastwood film *Pale Rider*, set at the time of the Californian gold rush, although the slow-burn western was as much a showcase for Eastwood's *man with no name* gun handling skills as it was a statement on mineowner and prospector disputes.

Coming up-to-date, in 2009 the James Cameron animated 3D film *Avatar* was released, with its story of a dastardly mining company from Earth attempting to damage the pristine forest of the distant planet Pandora for the mineral,

Unobtanium. This ruthlessness is often a trait of business related films, and oil, like mining, has often been portrayed as a despoiler of the environment. But miners have not always been tagged as bad, as the depiction of happy little diamond miners in *Snow White and the Seven Dwarfs* (made in 1937) attests.

Labour problems have provided many plotlines for mining films, often associated with coal mining which historically is one of the most volatile parts of the mining industry. *Matewan* – released in 1987 and directed by John Sayles with Chris Cooper and James Earl Jones in lead roles – recounted the true story of the 1920 coal miners' strike in West Virginia where unions fought for recognition and also struggled to bring together the mine's Italian and black workforce whilst simultaneously combating violence from the owner. Another strike movie was Herbert Biberman's *Salt of the Earth*, a cult American film made in 1954 by blacklisted Hollywood filmmakers and based on the 1951 New Jersey Zinc strike in New Mexico.

The theme of environmental despoiling and tough pre-union coal mining is portrayed in the John Ford film *How Green was my Valley*, set in Victorian Wales as the arrival of the coalmines begins to radically change the pastoral way of life. The picture, which starred Walter Pidgeon and Maureen O'Hara, is one of the greatest mining-related films ever made and won five Oscars in 1942.

Mining has always struggled to earn a good press and in the world of arts and media it is often seen as an environmental menace, allied with a thoroughly dubious morality, as exemplified by the 2006 Leonardo DiCaprio film *Blood Diamond*, which is concerned with the traffic of conflict diamonds from West Africa. Another example of the greed theme is the 1978 film of Paul Erdman's novel *Silver Bears* starring Michael Caine.

In an historical context mining entrepreneurs of controversial reputation such as Cecil Rhodes were obvious subjects for film makers from *Rhodes of Africa* starring Walter Huston in a sympathetic reading of Rhodes life in 1936, to the somewhat more recent and hostile BBC series *Rhodes*, with Martin Shaw, made in 1996.

More often than not minerals are shown as stimulating greed and giving power to men who cannot but abuse that power in their pursuit of wealth. This can make for exciting film making but these themes hardly improve the industry's image or reputation. On the other hand it is unlikely in the modern world of green issues that mining will be seen as anything other than an unwelcome necessity. The hagiography of past mining entrepreneurs is, therefore, unlikely to be a feature of future output on mining from the broad world of arts and culture. If mining companies want a basically favourable review of their history then they will probably find that they will have to commission the book or film themselves. One example of this is *Gold Fields: A Centenary Portrait* published for Consolidated Gold Fields by Weidenfeld & Nicholson in 1987 and written by Paul Johnson, an interesting choice given his left wing past.

CONCLUSION

A professional mining press began to emerge in the second half of the 19th century in response to the growth in the mining industry and the observed need for information on what was going on, at a time when new mines were being developed to feed the appetite of industrialisation and when exploration, particularly for gold, was accelerating. There were three key areas that publishers identified as needing coverage: exploration, production and allied issues such as financial results; technical aspects; and the pricing of metals.

The latter presented early editors with problems, especially as many suppliers of metals were not always keen on having prices publicly displayed where customers could make reliable comparisons, and could put pressure on the owners who sought advertising revenue. The technical advances made by the mining industry and the design and manufacture of the machinery and equipment required created a keen demand for technical articles. But perhaps the strongest demand was for news of operating, exploration and financial activities as the shareholder base of mining companies widened. It is important to remember that until the 1960s capital and securities markets, in the Anglo Saxon world in particular, were dominated by private investors, and information on what were often speculative investments was much sought after.

The mining press also had an important role to play in encouraging the industry to pay attention to issues of safety, the coal mining industries of the US and the UK being particularly accident prone in the past. As to whether the mining industry appreciated having an increasingly professional specialist press keeping an eye on it is a moot point, especially in the 19th and earlier 20th centuries when mining fortunes could be made by stretching the truth, and when to that end inside knowledge was often critical. We also should not forget the long periods when economic cycles and fashion move against the mining industry – the period from the Korean War until the mid-1960s and then the longer period of the 1980s and the 1990s are two examples when interest in mining was low and mining publications had to fight for their lives and many had to close or merge.

It has to be said that publishing specialist journals on mining is not a guaranteed route to great wealth despite the fact that subscriptions are usually not particularly cheap, certainly today. Even so, the industry has generated huge quantities of news, articles and even entertainment over the years. It does, however, face a challenge from the rise of online information suppliers and in particular specialist mining sites where staffing costs can be kept low. The face of mining media has changed over the years, sometimes as a result of regulatory changes, sometimes as a result of needed economies, and this trend is certain to continue. What is also certain is that without a vigorous and honest media the industry will suffer, as doubts about the industry's ethics undermine investment confidence.

Source: BHP Billiton

The huge Broken Hill mining and treatment complex in New South Wales. Circa 1915-20

Source: Consolidated Gold Fields

Mount Lyell copper mine in Tasmania at its zenith in the 1920s

Sir Ernest Oppenheimer (1880-1957), founder of Anglo American

Source: Mining in Northern Rhodesia

The African workers compound at the Nkana Mine in Northern Rhodesia (Zambia) in 1929

Source: Mining and Metal

Traditional panning for tin in the central region of modern India

Source: Mining

Blasting at the vast Chuquicamata copper mine in Chile in the late 1960s

The Long nickel mine, Kambalda, Western Australia, currently operated by Independence Group but originally discovered by Western Mining in the 1970s

Source: Randgold Resources

Workers harvesting rice, one of the agribusinesses established by Randgold Resources following the cessation of mining at the Morila gold mine in Mali in 2009

Source: Lisa Coulson

The Cerro Ricco silver mines in Potosi, Bolivia, still in production in 2011 after four centuries

The ancient Rio Tinto copper mine in Andalucia, southern Spain, once worked by RTZ (Rio Tinto) and now being re-opened by EMED Mining

Source: Rafael Ibanez Fernandez

The site today of the Las Medulas gold mine in Leon, Spain which was mined in Roman times from the first century AD

Nautilus Minerals lowers its remote control seafloor vehicle in far eastern waters

Source: BHP Billiton

A late winter view of BHP Billiton's Ekati diamond mine site in Canada's Northwest Territories, the first diamond mine developed in Canada

De Beers offshore mining ship, Peace of Africa, operating in Namibian waters

Source: Hilary Coulson

Kalgoorlie's Big Pit gold mine, 2010

The fabulously rich Mir diamond mine in Russia, today

THE FUTURE FOR MINING

1. WHAT LIES AHEAD?

Over the last decade or so there have been moments when the gloom over the future of the global mining industry has been close to impenetrable. Along with agriculture, hunting and forestry, mining is arguably the oldest economic activity known to mankind, and as we have demonstrated in the pages of this book its history spans many millennia. However, from time to time it gets written off as a sunset industry, with a future where financial returns will be modest indeed and where profitability will be driven by cutting costs to improve operating margins. To use the modern jargon, mining has no *pricing power*.

Of course over the last few years this view has been turned on its head and commentators suddenly see a golden future driven by almost inexhaustible demand for metals from China and India (*Chindia*). Does this mean that these two economic giants from East of Suez have changed mining's fortunes forever or will the pendulum swing back against mining in due course?

History is one of those subjects which irritates and frustrates in equal measure. As we look to the future we can easily believe that there are few if any lessons to be learned from past events. Such an arrogant attitude has many times got investors into wealth-reducing trouble. There can be little doubt that in due course the remarkable bull market in metals that has developed since 2001 will come to a decisive end as the equally strong market of the 1970s did. The reason will most likely be the same as then – very high metal prices encouraging a huge growth in new output with a consequent long-term weakening in prices. Of course the supporters of the new paradigm will, as they did during the telecommunications, media and technology (TMT) boom that spanned the years before and just after the new millennium, claim that the rise of China and India changes the situation forever.

One thing that I believe can be said about the future is that infrastructure spending, which has a high propensity to consume raw materials, is not just the preserve of *Chindia*, although as we can see in the table on the next page that the figure for Asia/Oceana is enormous at almost $16 trillion. There are many more large and poor countries where infrastructure spending and demand for raw materials could increase rapidly – Indonesia, Mexico and Brazil come to mind immediately, with spending plans in South America, for instance, set to surpass North America, and many in Russia are living in conditions that no advanced economy would tolerate. At the same time the advanced economies have problems with rotting infrastructure of great age – bridges and power in the US, and water and transport in the UK are just some examples. None of this is particularly new, it is just that for a long time financial markets did not bother looking at these issues.

PROJECTED GLOBAL LONG-TERM CUMULATIVE INFRASTRUCTURE SPENDING (2005-2030)

Region	Potential Spending ($trillion)
US/Canada	6.5
South/Latin America	7.4
Africa	1.1
Europe	9.1
Middle East	0.9
Asia/Oceana	15.8

Source: Booz, Allen, Hamilton 2007

So what might be the biggest influences on the mining sector over the next few decades?

The process of lifting up the standard of living of the poorest will almost certainly continue and that will underpin basic demand for metals and other bulk raw materials. That process in fact has been with us since the end of the Second World War when governments took the view that the Industrial Revolution needed a makeover in terms of who benefited from technical and physical advances.

Then it was the West which led the way as the advanced nations sought to spread prosperity further down the social scale and in the process expand the middle class. This is the pattern being followed in the developing world today and we can expect this to continue for many decades. Although it is economically simplistic to talk about a refrigerator or a car or a television in every home in China, or wherever, such concepts are important and would have a marked effect on consumption of raw materials.

One thing that characterised the view of the metals and mining industries during the TMT boom of the late 1990s and early 2000s was the broad belief that industries needed to have *new paradigm* characteristics to be taken seriously. New metals with economic leading edge uses, it was said, might command respect, but the old *smokestack* metals such as steel and copper were fighting a losing battle as economic growth was increasingly driven by ideas, innovation and services. For those who thought like that, the events that so closely followed the TMT boom came as a complete surprise – massive Eastern economic growth has demanded huge quantities of basic raw materials.

2. NATIONALISM AND NATIONALISATION

One issue which we have already looked at and which we may well have to consider again in the future is nationalisation. The enormous improvement in the outlook for and prices of metals has led to such a boom in mining company profits that an increasing number of governments are looking for an enhanced share in these profits. Some have already begun to change the basis on which foreign and local miners operate in their country – Venezuela, Russia, the Congo (DRC) and Peru are examples of this, as is Australia, a rather more unlikely source of rule change. It is quite probable that this trend will develop further, as it did in the 1960s, although the driving force is unlikely to be post-colonial pride as it was then, but a need to find additional sources of government revenue in an age when globalisation makes it difficult to increase taxes.

Of course pride will play a part in any process leading to nationalisation; the tone of pronouncements in 2006 from the radical Venezuelan government underlined that. The main justification of such action, though, is likely to be that the nation is not getting a fair share of the bounty that the new age of strong metal prices has generated for private and international mining companies. Royalties and higher taxes can be one way for a government to receive more of this bounty. However, as we have already said, globalisation makes it risky, certainly over the longer term, to increase tax rates, because international miners may choose to go elsewhere. Of course if every mining country has a high tax regime mining companies would find little advantage in looking elsewhere, but at the moment this is not the case.

There is also no doubt that national feelings about the ownership of resources are very easy to rouse when the exploitation of these resources enters one of its profitable phases. Many of the countries who seem very relaxed on this point have few natural resources of their own to exploit, the UK would be a case in point, though the seventies did see an outbreak of somewhat contrived nationalism over North Sea oil with the setting up of the government-owned BNOC. But national sensitivities are never far from the surface in poorer countries, as what prosperity they have often depends on the production of raw materials.

One approach in seeking a greater share of the mining profit would be for governments to re-negotiate development agreements and project shares. This has happened in a big way in the energy sector in Russia where government-controlled companies and favoured oligarchs have, in the process, been rewarded with re-negotiated stakes in huge projects that were of little interest when oil was $20/bbl. In some of the *Stans* this trend has reached epidemic proportions in the mining sector, although the foreign victims are mostly small companies. There have also been recent cases in the Congo.

Changing the rules is also easier for a government dealing with local private companies, although care would have to be taken over who owned or was behind a company. In Russia and Indonesia it is clear that some very powerful coves inhabit the corporate world and even governments might be wise to be careful how they act. In the short term, changing the rules for foreign operators may not lead to a mass evacuation, as nationalisation would, but over the longer term the effect would probably be the same. Nationalisation would also be difficult without having experienced personnel to run the mines. Whilst this might be a problem today, with all sorts of technical and equipment constraints due to mining's descent into the sunset sector in the late 20th century, it is likely in the future that many producing countries will materially increase expenditure to produce home-trained management, mining engineers, geologists and miners.

Nationalisation, thought dead as an issue with the fall of the Soviet Union, may have more life in it than some think. In South Africa the ruling ANC is coming under pressure from its youth wing to nationalise the mines, although the government is resisting the call. The problem here is that the idealism of youth means the young ANC members believe the shortcomings of the government in tackling poverty and exclusion can be dealt with by owning the means of production, and thus using the profits not for shareholder dividends but for increased government spending. This is not a new idea in Africa but it is one that has, when tried, singularly failed to deliver in the 50 or so years since the colonialists packed their bags. The main problem is that production and then profits collapse due to operating inefficiencies, both financial and mining.

One other consideration that countries with significant mining sectors have to take account of is the growth of Chinese demand for minerals, and in due course probably Indian demand as well. Venezuela is a good example of the problem. The country has decided to make it very difficult for private raw material companies to operate in the country and has nationalised a number of operations. It is, along with a lot of other countries, now trying to woo China and offer it participation in some of Venezuela's bigger development possibilities. This is fine in the immediate term but China is a hard protagonist, and if a future Venezuelan government wanted to apply pressure for whatever reason on Chinese commercial interests it would find a far more robust response than would be forthcoming from a foreign private company.

3. OTHER POLITICAL ISSUES

Nationalisation of natural resources, though usually billed as a process to obtain proper reward for the 'people' from the people's assets, too often has been a route to enriching a government elite rather than raising national living standards. President Mobutu of Zaire (now the DRC) was a classic example of the autocrat with his hand permanently in the till, as his overseas assets increased spectacularly and the nation's mines and finances collapsed as he pillaged them. Mobutu was of course merely following in the footsteps of the country's Belgian colonial rulers, particularly King Leopold II who had himself economically raped the hapless country in the 19th century. Mobutu's near neighbour, President Bokassa of the Central African Republic, in the 1960s and 1970s was another who enriched himself by plunder. Perhaps more controversially one might point the finger at many Middle East rulers who have built up huge wealth from the proceeds of oil revenues. Most of them admittedly have not ignored the needs of their subjects and have spent heavily on public works, but as we saw recently in Bahrain dissent is an increasing problem for autocrats on the make.

As an adjunct of this we have seen more active involvement by, particularly, western governments in trying to tackle resource theft. An example of this would be the Kimberley Process established in 2003 to stamp out diamond smuggling, an activity which has raised substantial funds for autocrats and anti-government groups, both public and private. As well as western consumers of diamonds, members of the Process include diamond producers committed to the sale of diamonds from carefully vetted sources through closely controlled sales channels, or that's the theory anyway. A number of procedures are laid down to make sure that diamonds coming on to the world market are not from suspect sources whose prime aim is violence and human rights atrocities. The Kimberley Process has proved very difficult to implement and though it still is in place in the 2010s there are an increasing number of questions asked by groups such as Global Witness about the robustness of the Process with regard to countries like Zimbabwe and Venezuela.

4. ENVIRONMENTAL ISSUES

Another area where government influence is increasingly being seen is on environmental issues, where even Russia, a major polluter historically, has embraced green issues, although not always for altruistic reasons.

One thing that can put mining companies directly in the centre of the environmental debate, as we mentioned earlier, is the problem relating to the need to mine increasingly lower grade mineral deposits. In order for these to be mined economically they have to be very large operations and so they generate increasing amounts of residual waste which have to be dealt with; the result is often thoroughly unsightly.

Also the mines are more often than not open pit operations that leave huge scars which also must be dealt with. When open pit mines or quarries are relatively small – the aggregate quarries along England's Thames Valley corridor would be one example – they can be readily turned into leisure lakes and wildlife areas when worked out. A mine like the Kalgoorlie super pit is a different proposition and these sorts of low-grade *mega* mines are what we can expect for the future.

The challenge of mining in an environmentally conscious manner has been an issue for some years and mining companies have been active in trying to find solutions to various problems. The subject of increasingly large and unsightly open pit mines has exercised the minds of a number of mining companies and led to considerable research into other mining techniques. Block caving has its supporters, with its ability to mine up to 100,000 tonnes of ore per day underground using the method of collapsing massive blocks of ore in one go and then carrying the ore away mechanically. Whether this method could be used to mine orebodies very close to the surface is another matter, but where the orebody is very large and of consistent grade, and the ground above stable, it offers options in terms of mine development. There is also the possibility of being able to return the residual waste underground, or at least a proportion of it, rather than heap it into unsightly and sometimes dangerously unstable hills. This kind of mine remediation is likely to become the norm over the next few decades.

Mining companies have over the last 30 years increasingly, if sometimes reluctantly, paid attention to both social and environmental issues and this trend will not reverse in the future – rather the imperative will quicken. This throws up the possibility that mining company profits will retain the potential to yaw unpredictably as they have in the past. The recent surge in metal prices and therefore mining profits has been accompanied by a massive increase in costs, some but by no means all attributable to soaring energy costs. If governments get into the act, instituting or raising royalties and taxes and introducing further social obligations, which are always costly, then the new paradigm will disappear very quickly. An instance of this would be getting public infrastructure funded by the companies rather than themselves, something in which the industry is already involved but where there is ample scope for expansion.

5. NEW TECHNOLOGY, NEW OPPORTUNITIES

Technology will surely play its part in future developments in mining and the revival of interest in undersea mining is an example of this. Although undersea mining is seen as a rather futuristic concept, undersea diamond mining has been carried out off the Namibian coast for over 40 years. To begin with the technology was crude, using large hoses to suck up diamonds which had been washed downriver and out to sea where they settled in pockets and sea floor indentations. In time seabed crawlers, rather like construction diggers, were employed, not always successfully. Dredging techniques were also updated a number of decades ago to mine aggregates in shallow water, a development from the tin dredgers of the Far East which dredged up material from lakes and rivers. And further back mechanisation was applied to mining alluvial gold from river beds in North America.

All of the approaches mentioned here relate to mining at relatively modest depths, whereas the future of undersea mining is rather more ambitious and envisages operations in several thousand feet of water. One of the first really deep sea mining tests was done by Germany metal company Preussag, which collected 15,000 metres of mineralised sludge from the Red Sea from around 7,000 feet in 1979 as part of a deep sea mining exercise for the Saudi Arabian government. The sludge contained around 6% zinc, 1% copper and low grades of silver and was pumped to the surface through a long hose after the mud had been broken up by a vibrating tool. Much was learned from this programme about commercial production rates and the ability to produce a viable concentrate. The collapse in metal prices during the 1980s meant that the experiment went no further; the technical ability to mine at great depths had been proven but absolutely not the economics.

The present approach is to develop sea mining equipment like crawlers capable of operating at around 5,000 feet and excavating mineralised material from seafloor benches. At the moment the waters around Papua New Guinea have interested a number of entrepreneurial exploration companies, and New Zealand and parts of the south Pacific have also attracted interest. The key concerns remain whether it is possible to mine sufficient quantities of material to make the operations economically viable, and also the issue of residue from the seabed material; is it stored on land or returned to the sea, with both having significant environmental implications.

One of the things we have looked at on a number of previous occasions is the way that, in the development of new mines, grades are falling, creating an environmental quandary, as growing demand for metals and falling mining grades means increasingly that new mines are large in terms of the ore they produce and leave greater amounts of waste behind to be dealt with. The unsightly and sometimes dangerous nature of these structures makes it more than likely that in future large low-grade open pit mines with huge waste disposal requirements will find permits increasingly difficult to obtain if they are situated in areas of significant population or outstanding beauty. This will have a fundamental impact on where mining will be allowed and thus on metal prices. Although the search for and production of raw materials has become exciting again, it has also become far more controversial and difficult than it was even 50 years ago.

There is also the subject of new metals and the assumption of minor and even unwanted metals to the status of major metals, as has happened to nickel and platinum over the last hundred years or so. We have mentioned rare earths earlier but in recent decades we have witnessed runs on tantalum in the wake of the 1990s mobile phone boom and there is support for magnesium as a metal to be used in vehicle manufacturing due to its very low weight-to-strength ratio. Lithium used in long-life battery manufacture has been a favoured choice for years for those gazing into their crystal balls to divine future trends in metal consumption. There are also new metallic compounds being developed all the time, an alloy that is spongy and unbreakable metallic glass are two examples of this. Metals might also be worked in different manners in the future – already an alloy has been created in the laboratory that can be blown into shape like glass and plastic.

Even more esoterically in terms of the industry's future, the spectre of mining in outer space has been raised by American scientists, with the long-term (very) possibility that future colonising space travellers from Earth could use bacteria to extract raw materials to use in establishing settlements on other planets.

We started in the Introduction by talking about Danny Boyle's film *Sunshine* and the giant metal space vehicle on its mission to save the Earth by re-igniting the Sun. It is perhaps, therefore, appropriate that we end our journey through the history of the mining industry back in outer space with the concept of intergalactic travel, the establishment of distant settlements and the inevitable search for raw materials to turn such settlements into viable long-term communities.

For as long as the human race survives it will always need raw materials because civilisation and progress are impossible without them. The mining industry then will always be with us – like an Arctic summer the sun will never quite set on it, although cycles of despair will regularly return to test the resiliance of one of the oldest activities on earth and the faith of those who make their living from this often frustrating but vital industry.

APPENDIX

MINING AND MINERALS TIMELINE

Date	Event	Country
BC		
100000	Surface flint mining	England/France
60000	Nile Valley flint mines	Egypt
40000	Haematite mining	Swaziland
9500	Copper pendant fashioned	Iraq
8000	Quartz mining in Washoe Valley, Nevada	North America
5000	Lead used	Egypt
	Gold and emerald mines developed	Egypt
	Gold mined in the Balkans	Bulgaria/Rumania/Serbia
4500	Hammered copper recovered from Chaldean remains	Iraq
4000	Bronze casting	Egypt
	Feinan copper mines operated	Jordan
	Copper mining in the Timna Valley, Sinai	Egypt
	Copper/tin mining/smelting, Khetri	India
	Objects made from surface iron	Egypt
3500	Underground flint mines	England/France
3000	Copper mining, Michigan	North America
	Turquoise mines developed, Serabit el Khadim	Egypt
	Bronze casting	Thailand
	Copper mining started at Copra Hill	Wales
	Lead/silver mining in Almeria	Spain
	Tin mining/smelting, Kestel/Goltepe	Turkey

2800	Development of smelting	China
2700	Copper mining	Cyprus
2500	Silver worked for wire and jewellery, Hissarlik	Turkey
2100	Tin mining in Cornwall	England
2000	Coal used for heating	China
	Separation of silver from lead	Turkey
	Metalworking of gold	Peru
	Laurian silver mine developed	Greece
	Diamond mining started in the Sub Continent	India
1500	Iron smelting and forging started	Thailand
1400	Nubian gold mines developed	Egypt
	Copper mining/smelting, Tongling, Jiangxi	China
1300	Copper mining/smelting at Muhlbach	Austria
1200	First iron objects made from mined ore, Anatolia	Turkey
1000	Copper mined near Agades	Niger
	Copper mining/smelting, Tonglushan, Hubei	China
950	Rio Tinto silver mine developed	Spain
650	First gold and silver coins minted	Turkey
625	First salt works established on the Tiber Estuary	Italy
500	First mining of Chuquicamata copper deposit	Chile
	Underground zinc mining, Rajasthan	India
400	First record of Almaden mercury mine	Spain
250	Platinum jewellery fashioned	Ecuador
104	Italian slave insurrections temporarily halt mining	Italy
100	Gold mines developed, Karnataka	India

AD		
50	Romans revive lead mining in the Mendips	England
125	Hadrian's reorganisation of mining industry	Roman Empire
200	Development of precious metals mines	China
600	Moors reopen gold/silver mines	Spain
	Silver/lead mined for the Crown at Melle	France
700	Gold mines developed in Latin America	Mexico
800	Charlemagne begins Roman mine renovation	Italy
	Copper originally mined at Palabora	South Africa
900	Coal mined by Hopi Indians in Arizona	USA
	Copper mines developed in the Atlas Mountains	Morocco
	Base metal mines at Rammelsburg, Saxony re-opened	Germany
1000	Copper working of tools and jewellery	Botswana/Zimbabwe
1170	Erzgebirge silver deposits discovered	Germany
1238	Newcastle coal mines developed	England
1288	Felun copper mine issues shares, the oldest recorded joint stock company	Sweden
1374	Zinc first identified as a separate element	India
1451	Development of method of separating copper and silver, Johannes Funcken	Germany
1471	Portuguese begin to trade gold in West Africa	Guinea
1494	Copper mines developed at Neustohl (Banska Bystrica) by the Fugger family	Hungary/Slovakia
1500	Diamonds produced from the Golconda mines	India
1516	Silver discovered at Joachimstal	Czechoslovakia
1520	Rapid expansion of Balkan silver output	Bosnia/Serbia
1524	Spaniards mine copper on Cuba	Cuba
	Silver discovered near Acapulco	Mexico
	Gullnes copper mine opened	Norway
1540	First Finnish iron mine, Ojama, developed, Lohja	Finland

1545	Gualpa discovers Potosi silver mountain	Peru
1552	Pizzaro begins to 'collect' gold from the Incas	Peru
1553	Spaniards develop copper/tin/silver mine at Tazco	Mexico
1556	Agricola's *De Re Metallica* published	Germany
1585	First iron ore discovered in the Colonies, North Carolina	USA
1608	First American iron ore shipped to Bristol for smelting	England/USA
1619	Dudley begins experiments to smelt iron with coal	England
1621	Coal mining starts at Villenueva del Rio, Seville	Spain
1627	Explosives first used in mining in Europe	Hungary
1644	Winthrop establishes first iron works in the Colonies, Massachusetts	USA
1652	Copper mining starts at Lokken Verk	Norway
1679	Coal discovered in Illinois	USA
1689	Cornish tin mines introduce drilling and blasting	England
1692	Lead discovered in Mississippi Valley	USA
1693	Minas Gerais gold discovery	Brazil
1709	Darby pioneers coke smelting of iron	England
	Settlers first discover copper, Connecticut	USA
1712	Newcomen's pump installed in tin mine, Cornwall	England
1719	Gold first produced in Russia, Transbaikalia	Russia
1724	Ancient Huelva copper mines re-opened	Spain
1727	Diamonds discovered in Minas Gerais	Brazil
1744	Gold discovered at Ekaterinburg	Russia
1775	Ranigunj coal mine developed, West Bengal	India
1782	Coal first mined in New Brunswick	Canada
1783	Tungsten first isolated at Vergara	Spain
1789	Uranium first identified in pitchblende	Germany
1791	Hunter Valley coal deposits discovered, NSW	Australia

1809	Baan Thungkha tin mine opened	Siam/Thailand
1815	First mine safety lamp, the Davey, produced	England
1820	Huge new tin deposits discovered in Perak	Malaya/Malaysia
1823	First Canadian gold find, Chaudiere River, Quebec	Canada
	First Australian gold find at Bathurst	Australia
1828	Gold discovered in Georgia	USA
1832	Chanarchillo silver deposit discovered, Copiapo	Chile
1834	St John del Rey acquires Morro Vehlo gold mine	Brazil
1840	Copper discovered, Keweenaw Peninsular, Michigan	USA
1843	Burra copper mine South Australia opens	Australia
	Alluvial gold production starts, Lensky district	Russia
1848	Gold discovered at Sutter's Mill California	USA
1850	Compressed air first used instead of steam in Fife coal mine by Thomas Cochrane	UK
	First mechanical rock drill used, Mount Cenis	France
1851	Gold discovered in Ballarat, New South Wales	Australia
1852	Iron ore first produced near Lake Superior, Michgan	USA
1853	Hydraulic mining technique revived, California	USA
1854	Fortuna lead mine developed to fuel English industrialisation, Andalucia	Spain
1857	420 tonne native (pure) copper boulder found, Minnesota	USA
	Mokta el Hadid iron ore deposit found	Algeria
1858	Discovery of the Comstock silver lode, Nevada	USA
1859	Coal mining begins in Wyoming	USA
1860	Pittsburgh coal seam developed	USA
	Gold discovered in Queensland	Australia
1864	Butte gold discovery Montana	USA
1865	Electrolytic copper refining introduced	England
	Alfred Nobel invents dynamite	Sweden

1867	The first diamond, the Eureka, discovered in South Africa, Hopetown Cape Colony	South Africa
1870	Gold mining starts in Amur	Russia
1871	New Rush diamond discovery in the Transvaal	South Africa
	Eastern Transvaal gold field discovered	South Africa
1874	Witwatersrand gold field discovered, Blaauwbank	South Africa
	Homestake gold find, South Dakota	USA
1875	Huelva gold/copper pyrites exports start	Spain
1876	Homestake gold mine discovered South Dakota	USA
1877	Ashio copper mine re-opened	Japan
	Leadville silver mine opened, Colorado	USA
1879	King Leopold begins personal exploitation of Congo natural resources	Belgium/Belgian Congo
1881	Leeuwkuil coal mine opened, Vereeniging	South Africa
	Asbestos mining starts at Thetford, Quebec	Canada
1883	Discovery of Broken Hill, New South Wales	Australia
	Mount Morgan gold discovery, Queensland	Australia
1884	Butte copper mine begins production, Montana	USA
1885	Lead/silver mining starts at Bunker Hill, Idaho	USA
1886	Gold discovered in the Transvaal	South Africa
1887	Stock exchange established in Johannesburg	South Africa
1888	Platinum found in nickel/copper ore, Sudbury	Canada
	BHP floated on stock exchange, Melbourne	Australia
	Rhodes gains exclusive mineral exploitation rights from the Matebele	Zimbabwe/Southern Rhodesia
1889	Coal first discovered in Powder River Basin, Wyoming	USA
	Rhodes's De Beers gains full control of the Kimberley diamond field	South Africa
1890	Formation of the Guggenheim Exploration Co.	USA
	Detection of untreatable pyritic ore on the gold fields causes massive share price crash on the JSE	South Africa

1893	Gold discovered at Kalgoorlie, WA, by Paddy Hannon	Australia
	Rosebery zinc deposit discovered, Tasmania	Australia
1895	Sullivan lead/zinc mine B.C. discovered	Canada
	Burnham reports discovery of extensive copper workings near the Kafue River	Northern Rhodesia/Zambia
1896	Gold discovered in Klondike, the Yukon	Canada
1897	Ashanti gold mine starts production, Obuasi	Ghana
1900	Copper mines developed in Katanga	Belgian Congo
	Iron ore mining started at Kiruna	Sweden
1903	Silver discovered at Cobalt, Ontario	Canada
1904	Mining starts at El Teniente copper mine	Chile
1905	Bingham Canyon copper mine Utah opens	USA
	Cullinan diamond, the largest ever, found at Premier mine	South Africa
	Sunshine silver mine opens, Idaho	USA
1906	Coal dust explosion kills 1,100 miners, Courrieres	France
1908	Luderitz diamond fields discovered	SW Africa/Namibia
1909	Discovery of the Hollinger gold deposit, Timmins	Canada
1910	Expansion of tin mining by the British on the Bauchi plateau	Nigeria
1911	L'Aluminium Francais cartel formed	France
1915	Chuquicamata copper mine opens	Chile
1916	First SA iron ore mine developed at Pretoria West	South Africa
	Patino completes vertical integration of tin interests with acquisition of William Harvey tin smelter	Bolivia/England
1917	Expropriation of Lena Goldfields by Soviet government	Russia
1921	First uranium mine developed, Shinkolobwe	Belgian Congo
1923	Mount Isa lead/zinc mine Queensland discovered	Australia
	Discovery of the Horne copper/gold deposit	Canada

1924	Platinum discovered in the Transvaal	South Africa
1925	Ernest Oppenheimer gains control of diamond selling syndicate, London	England
1927	Hudson Bay base metal mine opens, Flin Flon	Canada
1928	Extensive copper discoveries made on the Copperbelt	Northern Rhodesia/Zambia
1930	Iron ore discovered, Kola	Russia
	Great Bear Lake uranium discovery, NW Territories	Canada
1931	Flooding of the Nchanga copper mine shuts mine for six years	Northern Rhodesia/Zambia
	Golden Eagle 1,135oz nugget found, Coolgardie	Australia
1932	Magnitigorsk iron ore/steel complex starts production	Russia
1935	Fort Gouraud iron ore deposit discovered	Mauritania
	Gold production starts at Red Lake, Ontario	Canada
1939	Nickel/copper production starts at Norilsk	Russia
1942	1,549 miners die in coal dust explosion, Honkeiko	China
	CVRD (Vale) formed by central government	Brazil
1943	Williamson diamond pipe discovered	Tanganyika/Tanzania
1947	Nationalisation of British coal industry	UK
1948	Free State goldfield discovered at St Helena	South Africa
1950	Iron ore discovered in Western Australia	Australia
1952	Giant West Driefontein gold mine opens	South Africa
	Kedia d'Idjil iron ore mine established	Mauritania
	Nationalisation of Bolivian mining industry	Bolivia
1955	Discovery of Mir diamond pipe, Yakutia	Russia
1957	Major copper discovery near Legnica	Poland
1958	First sea diamonds mined, Orange River mouth	SW Africa
1960	Castro confiscates Freeport Sulphur's Moa Bay nickel mine	Cuba
1961	Gold discovered at Carlin, Nevada	USA
1964	Kidd Creek base metal deposit discovered, Ontario	Canada

1966	Western Mining discovers nickel at Kambalda	Australia
1967	Orapa diamond pipe discovered	Botswana
1968	ZIMCO acquires control of copper mines	Zambia
1969	Nickel discovered by Poseidon at Windarra, WA	Australia
	Impala platinum mine starts operation, the Bushveld	South Africa
1971	Allende nationalises US owned copper mines	Chile
1972	Bougainville copper/gold mine starts production	Papua New Guinea
	Gibraltar copper mine opened, Highland Valley BC	Canada
1975	Diamonds discovered at Argyle WA	Australia
	Olympic Dam discovered, Roxby Downs WA	Australia
	Indian coal mines merged into Coal Mines India (CMI)	India
1976	Richards Bay coal export terminal completed	South Africa
1977	Tara zinc mine opens, Navan	Irish Republic
1978	Carajas iron ore mines development decision	Brazil
1979	Gold reaches $500 for the first time	
1980	Sierra Pelada gold rush, Para	Brazil
	Hunt brothers try to corner silver market, price exceeds $40/oz	US/UK
1984	Amalgamation of four BC mines into Highland Valley Copper	Canada
1988	Construction of Escondida copper mine begins	Chile
1992	Diamonds discovered at Lac de Gras, Ontario	Canada
1993	Yanacocha gold mine opened	Peru
1994	Britain's coal mines de-nationalised	UK
1997	Bre X gold mining 'scam' collapses, Busang	Indonesia
1998	Tin mining finally finishes in Cornwall after 4000 years	England
2000	De Beers enacts supplier of choice diamond marketing strategy, ending 70 years of cartel control	UK/South Africa
2001	BHP merges with Billiton	UK/Australia

	Anglo American takes over De Beers	UK/South Africa
2002	Oyu Tolgoi gold/copper deposit discovered	Mongolia
2003	Economic black empowerment policy introduced	South Africa
2006	Vale acquires Inco	Brazil/Canada
	Freeport McMoran merges with Phelps Dodge	USA
2007	After 100 years South Africa loses position as world's largest gold producer to China	South Africa/China
	Australia ends ban on developing new uranium mines	Australia
	Rio Tinto acquires Alcan	UK/Canada
	Congolese mining licence review starts	DR Congo
2008	Maltby Colliery celebrates its centenary	England
	10% of FTSE 100 index now mining stocks with the admission of Ferrexpo	UK
2009	Gold reaches $1000 for the first time	
	Anglo American exits South African gold mining with sale of AngloGold stake	UK
	WR Grace successfully appeals against asbestos fines	USA
2010	Miners rescued from San Jose gold/copper mine	Chile
	BHPB's unsuccessful battle for Potash Corp	Canada
	First Quantum has its Kolwezi and Frontier mines expropriated	DR Congo
2011	De Beers continues retreat from London with Gabarone expansion	South Africa
	Glencore lists in London	UK
	ENRC corporate governance crisis	UK
	Oppenheimer family announces sale of its De Beers holding to Anglo American	South Africa
	Anglo American and Codelco clash over Los Bronces option deal	Chile
2012	Glencore bids for Xstrata which would create world's biggest miner	UK

GLOSSARY

Aeromagnetic Airborne survey of the Earth's magnetic field.

Alloying The melting and mixing of more than one metal to create a new element – copper and tin together make bronze.

Alluvial Loose soil or sediments moved by water but found in a dry, non-stream setting.

Amalgamation The combining of mercury with gold to liberate the gold from ore.

Anomaly Value or feature higher, lower or different to that expected, or to the norm.

Anthracite Hard, shiny coal with very high calorific value, difficult to light but clean burning and primarily used in domestic fires and boilers.

Assay The process of determining the amount of various elements/metals in a sample.

Auger drill Drill with screw bit which throws out the dirt/rock behind it as it drills down.

Basement Precambrian igneous and metamorphic rocks, the oldest rocks.

Bedrock Hard layer of igneous or metamorphic rocks beneath a near-surface layer of generally younger unconsolidated sediments. Also refers to the consolidated native rock underlying the broken surface rock above.

Beneficiation Crushing and separating ore into valuable fractions for further treatment and worthless fractions for disposal.

Bituminous coal Soft black coal, higher quality than lignite and can be upgraded into coke for steel making.

Block caving A mining method in which ore, usually friable, collapses under its own weight after having been undermined. The broken ore is then taken to the surface.

Bonanza gold Term used to describe extremely rich gold mineralisation.

Bronze An alloy of copper and tin.

By-product metal The residual (not prime) metals mined alongside the main metal in a multi-metal deposit.

Calorific value Amount of energy/heat released when fuel is completely combusted.

Cambrian Earliest period of the Palaeozoic era, 570 to 500 million years ago.

Carat (diamonds) The unit of weight for gemstones – one carat weighs 200 mgs.

Carat (gold) The measurement of purity in a gold item – 24 carat is pure gold, 9 carat is 36%.

Carbon capture	The process whereby carbon dioxide can be diverted from coal power station gas emissions and buried in underground storage areas such as caves.
Carburising	Diffusion of carbon into the surface layers of low carbon steel at high temperatures.
Cast iron	Re-melted iron and steel, wear and rust resistant, easy to shape but very strong in its finished state.
Cinnibar	The common ore of mercury, has a scarlet to brick red colour.
Claim	Area legally licensed to an entity to be explored for minerals.
Clipping coins	The practice of reducing the amount of gold in ancient coins whilst maintaining the face value of the coin.
Co-fusion	The ancient process of melting two or more metals in the same crucible to produce a metal variation – cast and wrought iron to produce steel.
Conquistadors	Spanish adventurers who conquered parts of central and south America in the 16th century.
Core	Cylindrical rock sample generally produced by diamond drilling.
Cover	Generally a near surface blanket of sediments that cover up basement.
Craton	A large portion of a continental shelf that has been undisturbed for over 500 million years.
Crawler	Undersea mining excavator.
Crosscut	Level in a mine driven to intersect mineralisation.
Crusher	Power driven machine which crushes large pieces of ore into smaller fractions.
Diamond drilling	Method of obtaining a cylindrical core of rock by drilling with a diamond bit.
Disseminated	Description of fine grained minerals widely dispersed through the enclosing rock.
Dragline	Huge mechanical excavators used to mine surface mineralisation particularly coal and iron ore.
Dredging	The excavation of mineralised sand/mud usually lying under shallow water.
Drift mining	Mechanical extraction of ore using a decline roadway excavated through an underground orebody.
Drive	Access to the mine workface.
Dyke	A sheet-like body of igneous rock cutting across other rocks at an angle.
Entrepot	A duty free trading area acting as a centre for the transport of both imports and exports through it to final customers at a profit to the entrepot's traders

Fault	Fracture in a rock sequence where one side has moved relative to the other.
Fire setting	Lighting a fire against a rock face in order to fracture the rock and liberate metals.
Footwall	The rock below the mineralised vein or block being mined.
Friable ore	Ore that is easily reduced to powder.
FTSE100/Footsie	The share index of the UK's 100 largest incorporated companies by market cap.
Galena	The natural mineral form of lead sulphide, may also contain silver and zinc.
Gallery	Widened out underground area where mining was undertaken.
Gem diamonds	Used in jewellery due to high quality (clarity and colour) and value.
Geochemical	Prospecting techniques which measure the content of certain metals in soils and rocks in order to define anomalies for further testing.
Geophysical	Prospecting techniques which measure the physical properties (magnetism, density etc) of rocks to define anomalies for further testing.
Geophysics	Study of physical properties of rocks, the Earth, and also exploration activities.
Grade	Measurement or estimate of the quantity of a metal in a sample.
Granite	Plutonic felsic igneous rock composed of quartz, feldspar and mica.
Greenstone	An igneous, metamorphic rock type often prospective for gold.
Ground electromagnetics	Ground based geophysical surveying using an induced electric current to measure variations in the local electromagnetic field of the Earth below.
Gulags	Soviet forced labour camps, often formed to exploit mineral deposits in Siberia, administered by the GULAG department during Stalin's rule.
Hanging wall	The mass of rock above the mineralised vein or block being mined.
Head frame	The structure above a mine shaft which houses the hoisting mechanism which lowers and raises miner and materials cages in the shaft.
Haematite	Iron oxide mineral.
Hydraulic mining	The use of high pressure hoses to liberate ore in its setting.
Hydrothermal	Process when hot water-rich solutions transfer materials or alter rocks.

Igneous	Rock formed by solidification from a molten rock or magma.
Induced polarisation	Ground geophysical surveying employing the passing of an electrical current into the ground and measuring the voltage decay.
Industrial diamonds	Low value stones used primarily in abrasives and cutting tools due to hardness.
Intrusion	A body of igneous rock that invades older rocks.
Ironstone	Fine grained, heavy, compact sedimentary rock containing iron oxide and clay.
Kimberlite pipe	An orebody often shaped like a flute or pipe which can contain diamonds.
Lamproite	Ultramafic alkaline volcanic rocks which occasionally, like kimberlites, contain diamonds which have travelled with the volcanic upthrust.
Laterite	An iron-rich rock formed at the surface due to intense weathering and leaching of underlying bedrock.
Lignite	A mineral halfway between coal and peat primarily used in power stations.
Lode	A vein of metalliferous ore that is embedded between layers of rock.
Mafic	A silicate mineral or rock rich in magnesium and iron.
Magnetic survey	Survey of the Earth's magnetic field from ground or air.
Marginal mine	A mine whose revenues and costs are uncomfortably close to each other.
Metallurgy	The physical and chemical behaviour of metallic elements.
Metamorphism	Process by which pre-existing rocks are changed by heat and pressure.
Mine pillars	Often un-mined ore left in pillars to provide roof support in mine galleries.
Mineralisation	Anomalous concentration of metals of potential economic interest.
Native metal	Metal that is uncombined/pure and can be worked from its natural state.
Native title	The claim by indigenous Australians that certain areas have been used by them historically.
Nitrates	White colourless salt used primarily in fertilizers.
Open cast/cut/pit	Mine where the orebody is shallow enough to be mined without sinking a shaft or excavating a drive to access the ore.
Ore	Rock that has economic quantities of metal embedded in it.
Outcrop	Where rock, possibly mineralised, breaks surface.
Overburden	The loose waste rock overlying an economic orebody.
Oxide mineralisation	Derived from alteration of primary sulphide minerals by oxidation in an exposed weathered zone.

Panning	The practice of taking possibly mineralised dirt and swishing it in a large shallow pan with water added so that any metal present settles on the bottom of the pan.
Pewter	A malleable metal alloy, largely of tin (95% plus) with hardeners such as copper and antimony.
Placer mining	The mining of an alluvial deposit often using an excavator.
Porphyry	A rock with separate dissolved and distinct layers of metals within the structure.
Pre Cambrian Shield	The oldest of the earth's rock strata formed by mountain erosion 600 million years ago.
Procurator	Government appointed financial agent in ancient Rome, usually well born.
Prospect	A mining property, target or anomaly which has potential for significant mineralisation.
Prospective	Describing a region thought to have potential for an ore discovery.
Proterozoic	Geological era from 2,400 to 570 million years: divided into Palaeo, Meso, Neo.
Province	Large area or region minerally unified.
Pyrite	Iron sulphide FeS_2, sometimes known as 'fools gold' because of its colour.
Pyrrhotite	An iron sulphide material which is slightly magnetic.
Quartz	Commonest mineral in the Earth's crust (silicon dioxide).
Radiocarbon dating	Radiometric dating using the natural radioisotope carbon ^{14}C to calculate the age of carbon materials up to 60,000 years old.
Refractory ore	Ore that is difficult to treat and does not allow metals to be recovered by standard concentration or leaching techniques.
Resources	Well-defined estimate of mineralisation requiring more drilling to confirm whether it is economic or not.
Reverse circulation	Drilling where cuttings are raised to surface by a stream of compressed air inside a metal tube.
Rock chip sample	Collection of a representative sample of small chips of rock.
Royalties	A financial impost on the value of mine turnover or profits collected by central/local government or by private royalty owners.
Salting	The (illegal) process by which 'foreign' metal is added to ore samples or drill core to enhance its value when it is assayed.
Sampling	Collection of a representative part of material for analysis.

Seam	Consolidated mineralised layer in rock structure as in coal seam.
Sediments	Solid organic or inorganic material, sand or mud, that has been deposited by the movement of wind, water or ice having previously been suspended in fluid.
Shaft	Dug from surface to provide access to underground mineralisation.
Slag	Waste rock from mining.
Smelting	The process whereby metal in ore is freed from the ore using heat from a reducing agent such as coal or charcoal in a hearth or furnace.
Staking	The process by which a prospector marks out mineral claim ground with pegs before registering the claim with the local mining authority.
Stamp mill/battery	A mill machine powered by water or steam that crushes ore by pounding rather than grinding in order for it to undergo further processing.
Stratigraphy	Classification of a series of layered rock or strata.
Stratum	A layer of sedimentary rock with internally consistent characteristics.
Stream sediments	Sediments within drainage channels – streams, rivers.
Strike	Direction of bearing of a bed or layer of rock in the horizontal plane.
Subcrop	Geological formation buried under shallow cover.
Sulphides	Minerals containing sulphur and metallic elements, often potential ore minerals.
Tailings	Material left over after processing when metal has been extracted from the ore.
Target	Location to be tested by drilling, often a survey anomaly or geological concept.
Tenement	Exploration/mining land title allowing various actions.
Thrust	Fault where rocks of lower stratigraphic position are pushed over higher strata.
Ton	Short ton - 2000 lbs, long ton - 2240 lbs
Tonne	Metric tonne – 2,205 lbs or 1,000 kgs.
Vaporisation	Metals being turned into steam due to furnace heat and then being captured for re-cycling when they cool back into waste dust.
Vein	Thin infill of a fissure or crack, often with quartz and sulphides.
Voortrekkers	Afrikaner farmers who left the Cape Colony, South Africa to establish communities in the north – Orange Free State, Transvaal and Natal.
Weathered	Decomposed by external natural forces such as wind and water.
Wrought iron	Malleable iron with a low carbon content, largely replaced by steel today.

BIBLIOGRAPHY

THE ANCIENT WORLD

Bowerstok, G.W. and others, *Late Antiquity: a Guide to the Postclassical World* (Harvard University Press, 1999)

Brewer-LaPorta, Margaret and others, *Ancient Mines and Quarries: A Trans-Atlantic Perspective* (Oxbow Books, 2010)

Craddock, Paul, *Ancient Mining* (Antiquity Publications, 1994)

Craddock, Paul and Lang, Janey, *Mining and Metal Production through the Ages* (British Museum, 2003)

Duncan, Lynne Cohen, *Roman Deep Vein Mining* (University of North Carolina, 1999)

Hauptmann, Andreas, *The Archaeometallurgy of Copper: evidence from Faynan, Jordan* (Springer, 2000)

Kennedy, B.A., *Surface Mining* (SMME, 1990)

Knauth, Percy, *The Metalsmiths* (Time-Life, 1974)

Lacy, Willard C. and Lacy, John C., 'History of Mining', *SME Mining Engineering Handbook* (1992)

Lewis, C. Andrew, *Prehistoric Mining at the Great Orme* (University of Wales, 1996)

Negev, Avraham and Gibson, Shimon, *Archaeological Encyclopedia of the Holy Land* (Continuum Publishing, 2001)

Ottens, Berthold, 'Mining in China: a 3000 year tradition', *The Mineralogical Record* (2005)

Schmidt, Robert G., Ager, Cathy M., Montes, Juan Gil, 'Roman Silver Mining', Metals in Antiquity Symposium (Harvard, 1997)

Stanczak, Marianne, *A Brief History of Copper* (CSA, 2005)

Sutherland, C.H.V., *Gold* (Thames and Hudson, 1959)

Tozer, Rev H.F., *Selections from Strabo* (Clarendon Press, Oxford, 1893)

Wagner, Donald B., *Iron and Steel in Ancient China* (EJ Brill, 1996)

THE MIDDLE AGES TO THE INDUSTRIAL REVOLUTION

'Early Coal Mining 1100-1500', *The Northern Echo* (2008)

'Jewelry and Gemstones', KHI Inc (2007)

'Spain: Mining Historical Heritage', Ministry of Culture, World Heritage UNESCO (2007)

Agricola, Georgius (Herbert and Lou Hoover translators), *De Re Metallica* (Dover Publications, 1950)

Bauer, Max, *Precious Stones* (Charles Griffin & Co., 1904)

Blanchard, Ian, *Mining, Metallurgy and Minting in the Middle Ages* (Franz Steiner Verlag, 2005)

Blanchard, Ian, *Mining, Metallurgy and Minting in the Middle Ages: continuing Afro European Supremacy Volume 2 1125-1225* and *Volume 3 1250-1450* (Franz Steiner Verlag, 2005)

Catelle, W.R., *The Diamond* (John Lane, 1911)

Claughton, Peter, *Silver Mining in England and Wales 1066-1500* (University of Exeter, 2008)

Cuadra, W.A., Dunkerley, P.M., 'A History of Gold in Chile', *Economic Geology* (1991)

Golas, Peter J., *Science and Civilisation in China: Volume 5 Mining* (Cambridge University Press, 1999)

Keen, Benjamin, Haynes, Keith, A History of Latin America (Houghton Mifflin Harcourt, 2009)

Lynch, Martin, *Mining in World History* (Reaktion Books, 2002)

Ofosu-Mensah, Emmanuel, Ababio, Historical overview of traditional and modern gold mining in Ghana (International Journal of Library, Information and Archival Studies, August 2011)

Pawloski, John, *Connecticut Mining* (Arcadia Publishing, 2006)

Postan, Michael Moissey, Postan, Cynthia and Edward Miller (ed.), *The Cambridge Economic History of Europe: Trade and Industry in the Middle Ages* (Cambridge University Press, 1987)

Ramberg, Ivar B., *The Making of a Land: Geology of Norway* (Norwegian Geological Association, 2008)

Streeter, Edwin, *Precious Stones and Gems* (George Bell & Sons, 1898)

Swieki, Rafal, 'Alluvial Placers', *Alluvial Exploration and Mining* (2008)

Thurston, Edgar, *The Madras Presidency: With Mysore, Coorg and the Associated States* (Cambridge, 1913)

Trueb, Lucien F., 'The Salsigne Gold Mine', *Gold Bulletin* (1996)

Wood, Michael, *Conquistadors* (University of California Press 2000)

THE INDUSTRIAL REVOLUTION

'Discussion on Mining Methods Past and Present in the Morro Velho Mine', Institute of Civil Engineers (1928)

'The California Gold Rush 1849' eywitnesstohistory.com (2003)

Bain, Harry Foster and Read, Thomas Thornton, *Ores and Industry in South America* (Harper & Bros, 1934)

Barton, Pierre, *Klondike* (McClelland & Stewart, 1972)

Blake, Robert, *Disraeli* (St. Martin's Press, 1967)

Brands, H.W., *The Age of Gold* (William Heinemann, 2005)

Bain, Harry and Read, *Thomas, Ores and Industry in South America* (Harper & Bros, 1934)

Calvert, Albert F., *Mineral Resources of Minas Geraes (Brazil)* (BiblioBazaar, 2008)

Caron, Francois, *An Economic History of Modern France* (Methuen, 1979)

Chanter, Laura, *Coal Face* (Languedoc News, 2009)

Colebatch, Hal, *Claude de Bernales, The Magnificent Miner* (Hesperian Press, 1996)

Collier, Simon and Sater, William F., *A History of Chile: 1808-2002* (Cambridge University Press, 2004)

Crisp, Robert, *The Outlanders* (Peter Davies, 1964)

Culter, Suzanne, *Managing Decline: Japan's coal industry restructuring and community response* (University of Hawaii Press, 1999)

Curle, James Herbert, *The Gold Mines of the World* (BiblioBazaar, 2008)

Disraeli, Benjamin, *An Enquiry into the Plans, Progress and Policy of the American Mining Companies* (John Murray, 1825)

Dunn, Matthias, *Belgium Coal Field* (Pattison and Ross, 1844)

Dunn, Matthias, *An Historical, Geological and Descriptive View of the Coal Trade of the North of England* (Pattison & Ross, 1844)

Eakin, Marshall C., *A British Enterprise in Brazil* (Duke University Press, 1989)

English, Henry (stockbrokers), *A General Guide to the Companies formed for working Foreign Mines* (Boosey & Sons, 1825)

Erman, Erwiza, 'A Case Study of Tin Mining on Bangka Island', Green Governance Conference

Fletcher, Steve, 'Lead Mining in Spain in the 19th century: Spanish industry or British adventure', *Bulletin of the Peak District Mines Historical Society* (1991)

Gardner, A. Dudley and Flores, Verla R., *Forgotten Frontier: A History of Wyoming Coal Mining* (Westview Press, 1989)

Harpending, Asbury, *The Great Diamond Hoax* (James H Barry Press, 1913)

Harrison, Joseph, *An Economic History of Modern Spain* (Manchester University Press, 1978)

Harvey, Charles E., *The Rio Tinto Company* (Alison Hodge, 1981)

Hocking, Geoff, *Gold* (Five Mile Press, 2006)

Holborn, Hajo, *A History of Modern Germany 1840-1945* (Eyre & Spottiswoode, 1965)

Lang, John, *Power Base* (Jonathan Ball, 1995)

McMahon, Gary and Remy, Felix, *Large Mines and the Community* (IDRC/World Bank, 2001)

Martin, C.H., *The History of Coal Mining in Australia* (Australian IMM, 1993)

Nimura, Kazoa and Gordon, Andrew, *The Ashio Riot of 1907: a social history of mining in Japan* (Duke University Press, 1997)

Porter, G.R., *Statistical View of Mining Industry in France* (Literary Gazette, 1838)

Rector, John L., *The History of Chile* (Greenwood Press, 2003)

Rickard, T.A., *A History of American Mining* (McGraw-Hill, 1932)

Roberts, Brian, *The Diamond Magnates* (Hamish Hamilton, 1972)

Rosenthal, Eric, *Gold! Gold! Gold!* (Macmillan, 1970)

Schwartz, Sharron P., *Creating the cult of Cousin Jack: Cornish Miners in Latin America 1812-1848* (Institute of Cornish Studies, 1999)

Sherwood, Alan and Phillips, Jock, *Coal and Coal Mining in the 19th century* (Encyclopedia of New Zealand, 2009)

Sippel, Patricia, *International and Technological Change in Japan's Economy* (Routledge, 2005)

Thomson, Tessa, *Paddy Hannan A Claim to Fame* (Thomson Reward, 1993)

Ui, Jun, *Industrial Pollution in Japan* (United Nations University Press, 1992)

Uren, Malcolm, *Glint of Gold* (Robertson & Mullens, 1948)

Walters, Rhodri Havard, *The economic and business history of the south Wales steam coal industry* (Arno Press, 1977)

Webb, Martyn and Webb, Audrey, *Golden Destiny* (City of Kalgoorlie-Boulder, 1993)

Wheatcroft, Geoffrey, *The Randlords* (Weidenfeld & Nicholson, 1985)

Wilkening, Kenneth E., *Acid Rain Science and Politics in Japan* (MIT Press, 2004)

Wilkins, Mira and Schroter, Harm G., *The Free Standing Company in the World Economy, 1830-1996* (Oxford University Press, 1998)

Williams, Gardner, *The Diamond Mines of South Africa Vol 2* (BF Buck & Co., 1905)

Wills, Walter H, Barrett, RJ, *The Anglo African Who's Who and Biographical Sketch Book* (George Routledge & Sons, 1905)

THE MODERN AGE

'An Integrated Mining and Exploration House', Hancock Prospecting Pty Ltd. (2008)

'European Mineral Statistics 2004-2008' (British Geological Survey, 2010)

'Iron Ore in Brazil: Restructuring with Growth', Brazilian Development Bank (BNDES) (2000)

Bancroft, J.A., *Mining in Northern Rhodesia* (British South Africa Co.,1961)

Barnes, Michael, *Fortunes in the Ground* (Stoddart, 1993)

Barnes, Michael, *More Than Free Gold* (General Store, 2008)

Blainey, Geoffrey, *The Rush that never Ended* (Melbourne University Press, 2003)

Bruce, J. Todd, 'Rustenburg and Johnson Matthey: an enduring relationship', *Platinum Metals Review* (1996)

Denning, Dan, 'Iron Ore in Australia: A History of Red Gold in the Pilbara', *The Daily Reckoning* (2007)

Epstein, Edward, *The Diamond Invention* (Hutchinson, 1982)

Fecteau, Jean-Marie and Harvey, Janice, *Agency and Institutions in Social Regulation* (University of Quebec Press, 2005)

Gregory, Theodore, *Ernest Oppenheimer and the Economic Development of Southern Africa* (Oxford University Press, 1962)

Hoffman, Arnold, *Free Gold* (Associated Books, 1947)

Hoffman, Carl, 'How a Rogue Geologist Discovered a Diamond Trove in the Canadian Arctic', *Wired Magazine* (2008)

Holland, John, *The History and Description of Fossil Fuel, the Collieries and Coal Trade of Great Britain* (Whittaker & Co, 1835)

Hunter, Janet and Storz, Cornelia, *Institutional and Technological Change in Japan's Economy: past and present* (Routledge, 2006)

Kettell, Brian and Timmins, Howard, *Gold* (Oxford University Press, 1982)

Kirshenbaum, Noel W., *TA Rickard and his California Connections (Society for Industrial Archaeology*, Samuel Knight Chapter, 1999)

Kowalski, Stanislaw J., 'Kolyma: The Land of Gold and Death', *Aerobiological Engineering* (1997)

Kusnir, Imrich, *Gold in Mali, Acta Montanistica Slovaca* (1999)

Leskov, Mikhail, Schetinsky, Roman, Kryuchkova, Anna, 'Mining Investment Climate in the CIS', (LMBA Alchemist, April 2009)

Levinson, Olga, *Diamonds in the Desert* (Tafelberg, 1983)

McDonald, Iain and Tradoux, Marian, *The History of the Waterberg Deposit* (Applied Earth Science Maney Publishing, 2005)

MacDonald, Bertrum H. and Connor, Jennifer J., *The History of the Book in Canada 1840-1918* (University of Toronto Press, 2005)

MacDonald, Catherine, Roe, Alan, *Tanzania (Country Case Study) The challenge of mineral wealth* (International Council Mining Metals, 2007)

Machens, Eberhard W., *Platinum, Gold and Diamonds: the Adventures of Hans Merensky's Discoveries* (Schweizerbart, 2009)

Mukonoweshuro, Eliphas G., *Colonialism, Class Formation and Underdevelopment in Sierra Leone* (University Press of America, 1993)

Mupimpila, Christopher and van der Grijp, Nicolien, *Global product chains: northern consumers, southern producers, and sustainability: the case of copper in Zambia* (IISD, 1997)

Newbury, Colin, *The Diamond Ring: Business, Politics and Precious Stones in South Africa 1867-1947* (The Clarendon Press, 1989)

Prochaska, David, *Making Algeria French: Colonialism in Bone 1870 to 1920* (Cambridge University Press, 1990)

Przeworski, Joanne Fox, *The Decline of the Copper Industry in Chile and the Entrance of North American Capital* (Arno Press, 1980)

Rothchild, Donald and Curry Jr., Robert L., *Scarcity, Choice and Public Policy in Middle Africa* (University of California Press, 1978)

Sherwood, Alan and Phillips, Jock, 'Coal and Coal Mining', *Encyclopedia of New Zealand* (2009)

Stoneman, William H., *The Life and Death of Ivar Kreuger* (Bobbs-Merrill Co., 1932)

Sykes, Trevor, *The Money Miners* (Wildcat Press, 1978)

Testa, Angie and Decarli, Bill, *A Dead Man's Dream* (Hesperian Press, 2005)

Thapa, Bishal and Kumar, Sandeep, 'The Indian Coal Sector: a tale of promise and problems', *World Coal* (2006)

Thomson, Elspeth, *The Chinese Coal Industry* (Taylor and Francis, 2002)

Thomson, Elspeth, *The Chinese Coal Industry: an economic history* (RoutledgeCurzon, 2003)

Tweto, Ogden, 'Ore Deposits in the United States 1933-67', American Institute of Mining Engineers (1968)

Vergara, Angela, *Copper Workers, International Business, and Domestic Politics in Cold War Chile* (Pennsylvania State University Press, 2008)

Wagner, Donald, *The Traditional Chinese Iron Industry and its Modern Fate* (Curzon Press, 1997)

Waszkis, Helmut, *Dr Moritz (Don Mauricio) Hochschild 1881-1965. The Man and his Companies. A German Jewish Mining Entrepreneur in South America* (Iberoamericana and Vervuert Verlag, 2001)

Waszkis, Helmut, *Mining in the Americas* (Woodhead Publishing, 1993)

Wells, Jennifer, *Bre-X* (Orion Business Books, 1998)

Williams, Roger, *King of Sea Diamonds* (WJ Flesch, 1996)

Wilson, A.J., 'The Pick and the Pen', *Mining Journal Books* (1979)

INDEX

www.ingramcontent.com/pod-product-compliance
Ingram Content Group UK Ltd.
Pitfield, Milton Keynes, MK11 3LW, UK
UKHW022311120125
453379UK00003B/17/J